nonlinear circuits handbook

designing with analog function modules and IC's

by
The Engineering Staff of
Analog Devices, Inc.

Edited by
Daniel H. Sheingold

Published by
Analog Devices, Inc.
Norwood, Massachusetts 02062 U.S.A.

Library of Congress Catalog Card No. 75–42559

ISBN: 0–916550–01–X

Additional copies may be ordered from Analog Devices, Inc., P. O. Box 796, Norwood, Mass 02062.

PREFACE

This book's origins lie in the solid ground of Analog Computing theory and practice. Until recently, the design and use of circuit configurations based on the conscious employment of nonlinear analog computing elements and formulations has been the recondite province of a few diehard experts in specialized areas of application.

Now this area of endeavor, so lately in the shade of digital technology, is becoming increasingly illuminated as the sparks thrown off by the commercial availability of modular devices with improved performance and decreased cost, complexity, and size, ignite the tinder of latent applications. It is our present objective, stemming from motives that range from a sense of technological mission to some that might be considered far more crass (yet eminently human), to provide an agency by which these small flames might be fanned into a bright and cheerful blaze, by the light of which all users of electronic circuits might gain useful knowledge, and benefit by a rediscovery of analog technology.

There are books available on analog computing and operational-amplifier applications, and there exists a smattering of publications on nonlinear circuitry and its applications; yet it is all but impossible to find a single source that combines information on principles, circuitry, performance, specifications, testing, and application of the class of devices specifically designed to be purchased for use in nonlinear applications. It is to this task that we have addressed ourselves in preparing this volume. Beyond the simple provision of information, however, we have felt it necessary to seek to communicate some of our excitement over the possibilities of analog nonlinearity and its often-cooperative role in systems involving both digital and analog data.

To make it all fit, we have had, in some cases, to sacrifice: rigor for vigor, specific details for general principles (and hints of possibilities), and evenhanded treatment for a particular orientation — that of a major manufacturer of the devices in question. In each case, though, the sacrifice is offset by a pragmatic gain, for the reader as well as for ourselves. For example, the citing of specific instances of commercially-available entities (manufactured by Analog De-

i

vices) tends to impart a sense of reality and practicality that a more-austere orientation might lack.

Furthermore, to avoid imposing today's technology (in this rapidly-changing field) on tomorrow's reader, we have repeatedly urged, and here reiterate, that the interested reader arm himself with Product Guides, data sheets, application notes, and other propaganda from all responsible and innovative manufacturers of nonlinear circuits, and seek to be on their mailing lists.

It is our hope that this volume will meet whatever need the reader brings to it: education for the tyro, ideas for the experienced practitioner, gap-filling for the engineer or scientist whose competence lies in other fields, practical advice for the theorist, and a source of ready reference for all.

HOW TO USE THIS BOOK

The prospective users of this book, whether students or experienced design engineers, naturally have a wide variety of backgrounds, interests, and needs. Although it is not expected that any reader will be totally satisfied, all who seek enlightenment, ideas, or guidance on matters having to do with nonlinear analog circuits should find something of value.

Whatever his interest, the reader will find the self-explanatory structure of the book laid bare in the Table of Contents, which every reader should explore thoroughly before proceeding further.

One can read through this book sequentially, but it is not necessary to do so; browsing is encouraged. Each unit is essentially self-contained, with occasional references to importantly-related units. Though this involves some redundancy, it also enables a topic to be approached from several points of view. The encyclopedic Index should be useful when exploring any topic in depth.

The Bibliography is a brief and eclectic assortment of sources of information on various topics covered within the book. Each item is chosen, either because of its specific practical value or timely interest, or because it in turn has a reference section that will "fan out" and give the reader large coverage from a small base. Design engineers should use this Handbook in conjunction with data sheets on specific products of interest, and with the most recent edition of the comprehensive *Analog Devices Product Guide.* In addition to its up-to-date contents and much data (with prices) on specific products, it also contains a wealth of technical information on related subjects and products, not all of which is duplicated in these pages.

Readers are urged to communicate to us their comments and suggestions for future editions of this Handbook, as to content, *errata,* omissions believed significant, and new applications ideas.

ACKNOWLEDGEMENTS

The writing of this book was a cooperative effort by the engineering staff at Analog Devices. Contributions have come from a wide distribution of Divisions and Departments, and reflect the efforts of design engineers, applications engineers, marketing engineers, test engineers, and others.

Major contributors include Lewis Counts (Chapters 3-2, 3-3, 3-6, and 3-7), Fred Pouliot (Chapters 4-1, 4-2, and 4-3), John Cadogan (Chapter 3-1), Barry Hilton (Chapter 3-5), and Walter Borlase (Chapter 4-4). Chapter 3-4 was written by Dr. Modesto A. Maidique, Stanley Harris, Richard Wagner, and Fred Mapplebeck. Besides the writing of a portion of Chapter 3-4, many good ideas and useful comments, plus advice and encouragement, were supplied by Barrie Gilbert. Important contributions were also provided by Richard Burwen, Fernando Romero, and Donn Blomerth.

Advice, stimulation, and encouragement, plus the vital means of getting the book published, were provided by Ray Stata (President of Analog Devices), Lawrence T. Sullivan, Robert Peterson, John Corsi, and Joseph Codispoti.

The book was prepared by our Publications department, under Marie Etchells, profusely illustrated by Ernest Lehtonen, Penny Brian, Eileen Solari, Shu Ngon Chau, and Joseph Furbush; much of the typesetting was done by Camilla O'Brien. A goodly portion of the manuscript was typed by Edna Godfrey.

Prior to publication, portions of the book have appeared in *Instruments and Control Systems, Electronic Engineering Times, EDN,* and *Electronic Design,* with yeoman assistance from Richard Goldberg.

Organization, editing, many ideas, and much of the writing were furnished by the undersigned, who also accepts the responsibility for any shortcomings.

January 1, 1974   *D. H. Sheingold*
Norwood, Mass.

TABLE OF CONTENTS

tance Multiplier, Nonlinear Errors, Dynamics of the Transconductance Multiplier, Linearizing he Transconductance Multiplier; Log-Antilog Multipliers, Offsetting a 1-Quadrant Multiplier for Operation in 4 Quadrants, Pulse-Modulation Multipliers, Multiplier Specifications, Checklist of Multiplier Parameters, Testing: Test Equipment, Test Circuits

PART FOUR: AIDS FOR THE DESIGNER

NONLINEAR CIRCUITS HANDBOOK

Designing With Analog Function Modules and IC's

Basic Operations

This chapter offers a brief historical and philosophical perspective of the nonlinearity scene and summarizes the principal features of useful nonlinear phenomena.

LINEARITY VS. NONLINEARITY

An ideal linear device is one for which cause and effect are proportional* for all values of inputs and output. For the great majority of analog circuits, *linear* relationships are sought. Natural devices, though, are in general nonlinear. But they are often linear enough over limited ranges to be quite useful. Much design effort is expended in finding and using devices having exceptional linearity, and in conditioning and normalizing signals to match the linear range of a device with that of the signal.

Because of the often exquisite difficulty of finding or designing devices with sufficient linearity for many tasks, the word *nonlinear* has come to have a pejorative connotation.

But devices and circuits can be designed to have nonlinear relationships that are well-defined, controllable, stable, available at low cost, and, what's more, *useful*. Examples of such relationships include multiplication, square-law, log ratios, and controlled discontinuities. Just a few applications include modulation, power mea-

---

*The *IEEE Standard Dictionary of Electrical and Electronic Terms* (Wiley-Interscience, 1972) defines *linearity* as "a property describing a constant ratio of incremental cause and effect" and a *linear system or element* as one for which "if y_1 is the response to x_1 and y_2 is the response to x_2, then $(y_1 + y_2)$ is the response to $(x_1 + x_2)$, and ky_1 is the response to kx_1."

surement, signal shaping, and simulating or correcting for the non-linearity of measuring devices. Many more applications are described in these pages.

As nonlinear devices with improved characteristics (i.e., stable, repeatable fidelity to the ideal nonlinear relationship) become available, better understood, and easier to manufacture and use, their already low cost will decrease further (especially through the use of monolithic integrated circuits), and they will be more widely and readily used by electronic circuit designers. The objective of this book is to help accelerate the trend.

NONLINEAR DEVICES AND ANALOG COMPUTING

The search for devices that embody many kinds of nonlinear relationships faithfully surfaced in the heyday of the analog computer, when it was necessary, at times, to simulate multiplication, division, limits, hysteresis, calibration curves, and a host of other nonlinearities. Servomechanisms, curve tracers, and diode function generators were the most popular ways of approaching these problems. (Of these, diode function generators have shown the best chances for widespread usage and future survival, but better approaches are now available.)

As analog-computer techniques became absorbed into instrumentation and general analog-circuit design, as transistors were perfected and integrated circuits appeared, functions that had been performed by expensive and unwieldy rack-mounted packages began to become available as modular (and eventually integrated-circuit) components, starting with the operational amplifier. They soon came to include multiplier-dividers, log circuits, diode function generators, and increasingly-complex operations. Multiplication and logarithmic circuit designs could be based on inherent natural properties of transistors, operational amplifiers could be used freely to sharpen diode thresholds, and good, fast, simple comparators and electronic switches became available.

This revolution in cost, simplicity, improved performance, and the ready availability of useful nonlinear devices is quite recent. The time is ripe to stand back and take inventory of some of these riches, viewed in the perspective of their family relationships.

A LIST OF USEFUL NONLINEAR OPERATIONS

Here are a number of nonlinear operators that are useful as building blocks in circuits, apparatus, instruments, and systems. They are based on practical devices that owe their functional efficacy to one or more of a few basic properties of transistor and operational amplifier circuits.

1. *Transconductance* as a linear function of collector current (multipliers)
2. *Base-emitter forward voltage* as a logarithmic function of collector current (logarithmic devices)
3. *Presence or absence of current* as a function of polarity of the applied voltage (switches and comparators).
4. *Near-perfect temperature compensation* as a consequence of monolithic matching of devices
5. *Near-ideal transconductance and transresistance* inherent in op amp circuitry (voltage-to-current and current-to-voltage conversion)
6. *Crisp switching* as a result of high gain in op amp and comparator circuits

Nonlinear devices may be classified according to their smoothness. If the function is smooth and differentiable (except perhaps at its extremities), it may be classed as a *continuous function*. If it has one or more discontinuities or "jumps" (e.g., comparators), or if its first derivative has discontinuities (e.g., piecewise-linear functions), it is classed as a *discontinuous function*.

Basic Continuous Functional Operations
    Multiplication
    Division (ratios)
    Squaring
    Square-Rooting
    Logarithms
    Exponentials (antilogarithms)

Basic Discontinuous Functional Operations
Ideal Diode
Controlled Switches
Comparators

Derived Continuous Functional Operations
Arbitrary Exponents
True Root-Mean-Square
Log Ratio (two variables)
Sinh^{-1} ("AC Logarithm")
Vector Sum
Trigonometric Functions

Derived Discontinuous Functional Operations
Absolute Value
Bounds
Dead Zone
Jump and Window Functions
Hysteresis

For the most part, the devices described in this chapter, in the order listed, are discussed in terms of their ideal "black box" response. That is, their inputs and outputs are in terms of voltage, and they are free from loading errors. Practical device characteristics are discussed in succeeding chapters. It is important to bear in mind that since most of these useful functions either involve operational amplifiers or are generated by transconductances, *current* may be a basic (and probably accessible) input or output.

MULTIPLICATION

A two-input *multiplier,* in response to two input voltages, supplies their product, multiplied by a dimensional (V^{-1}) constant

$$E_0 = \frac{V_1 \cdot V_2}{V_r} \qquad (1)$$

A commonly-used range of voltages is ±10V for both inputs and the output. In this case, $V_r = 10V$. (Note that $10 \times 10/10 = 10$.)

If the output and both inputs can have either positive or negative polarity, and if the polarity relationships are consistent, the multiplier is called a "4-quadrant" multiplier. If response to only one of the inputs is bipolar, it is a 2-quadrant multiplier; if all signals are of a single polarity, it is a 1-quadrant device. The "quadrants" are those that would be found in the V_1-V_2 plane if one pictures the output axis as perpendicular to the plane of the paper.

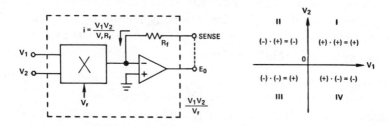

Multipliers are usually furnished with an extra terminal that allows the feedback path around the output amplifier to be completed externally. In addition to facilitating gain adjustment, this terminal permits the multiplier to be used as a divider or square-rooter, as will be shown. Multipliers often have one or more differential inputs, to deal with off-ground signals.

Besides multiplication in analog computing, multipliers can be used for squaring, modulation, dynamic gain setting, and power measurement. They are available at low cost in I.C. form, and in a wide variety of performance specifications (and prices) in the form of compact discrete modules.

The scale factor, $1/V_r$, though usually fixed (with allowance for trim), is often manipulable (in some versions) by an externally-applied voltage or current, which is, in fact, a third input.

Complete multipliers involve many techniques of design. The major weight of discussion in this book is given to transconductance, logarithmic, and pulse-width-and-height-modulated types. These cover a wide range of performance capability; they are compact, low-cost, and reliable. Such other all-analog types as quarter-square, magnetic, modulated triangular-wave, servo, Hall-effect, and "slaved" multiplier designs, though historically feasible for design and widely used in specialized applications, are outside the scope of this book. Multiplying D/A converters are discussed in the *Analog-Digital Conversion Handbook**, and elsewhere in the literature. They also are outside the scope of this book, which is almost entirely devoted to purely-analog technology.

There are available on the market integrated-circuit quasi-multipliers that require a large amount of external circuitry to operate as ideal "black-box" multipliers. They, and such other specialized devices as "balanced modulator" chips, are considered only as circuit elements that embody (incompletely) the fundamental principle of transconductance multiplication.

Though it might appear at first glance that much of the technology has been thus foreclosed arbitarily, the reader will find that the practical aspects of understanding and applying multipliers covered in this book will stand him in good stead if he wishes to consider other approaches.

DIVISION AND RATIOS

A *divider* usually has two inputs. The output is the ratio of the two inputs, multiplied by a dimensional (V) constant.

$$E_0 = V_r \frac{V_2}{V_1} \tag{2}$$

The commonly-used ranges for division are $\pm 10V$ for V_2, 0^+ to $+10V$ (or 0^- to $-10V$) for V_1, and $\pm 10V$ for E_0. For such applications, V_r is 10V.

Practical dividers are either two-quadrant or 1-quadrant, depending on whether or not the numerator may be bipolar.

Three techniques are widely employed for division: the use of multipliers in feedback loops, multiplier designs in which the scale factor is variable, and open-loop division using logarithmic elements.

---

*Analog Devices, Inc., 1972, 402pp. illustrated, $3.95.

Fast, accurate multipliers, when used in feedback loops, provide fast, accurate 2-quadrant division over small dynamic ranges of denominator. This approach is useful in ratiometric measurements, where it is necessary to correct for minor variations in a measurement that are caused by variations of the reference, e.g., in strain-gage and other bridge-type measurements. The weakness of the feedback approach is that errors tend to be inversely proportional to the magnitude of the denominator; dynamic ranges greater than about 30:1 are untenable, even if high-accuracy multipliers are used.

The variable scale constant and logarithmic approaches, on the other hand, permit wide excursions of the denominator (as long as the output can be expected to remain within bounds), and a considerably closer approach to zero. Accuracy is moderate, but response tends to be slow at low levels. Variable scale-constant multipliers are two-quadrant devices; however, with log devices alone, best results are obtained in single-quadrant operation.

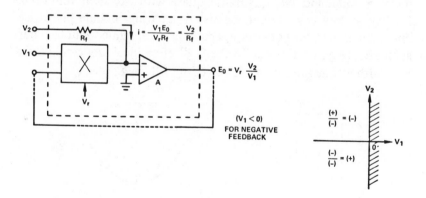

One should not expect an analog divider to be capable of division by zero, or by bipolar numbers. (With switching, 4-quadrant division is feasible — away from zero.) Generally, the problem can be restated to eliminate such anomalous operations. A divider can, of course, compute reciprocals if the numerator is held fixed at an appropriate constant value.

Division and multiplication are often combined in a single device (i.e., V_r is a variable input signal).

SQUARING

A *squarer* provides at its output a voltage proportional to the square of the input, multiplied by a dimensional (V^{-1}) constant.

$$E_0 = \frac{V_{in}^2}{V_r} \qquad (3)$$

Typical ranges for a 2-quadrant squarer are ±10V for V_{in}, 0 to 10V for E_0, and 10V for V_r. In 1-quadrant squaring, V_{in} is of only one polarity. A 1-quadrant squarer, preceded by an absolute-value circuit ("full-wave rectifier"), will provide 2-quadrant squaring.

A squarer is useful in such operations as power measurement, wave-shaping, frequency doubling, and —used as a feedback element— square-rooting. Until simpler means of multiplication became available, one of the most important applications of squarers used to be in "quarter-square" multiplication.*

A transconductance multiplier with identical input voltages produces a wideband two-quadrant square with excellent functional fidelity at low cost. Other means of squaring include piecewise-linear diode-resistor-network approximations, semiconductor characteristics (e.g., FET's) and "dithering" of an ideal diode characteristic with a triangular wave, followed by filtering.

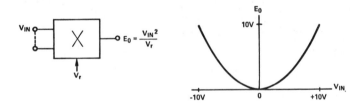

The "odd-function" square $x|x|$ has an output that is proportional to the square of the input but takes on the polarity of the input. It can be generated with two 1-quadrant squarers and a few op amps, or by interposing an absolute-value device between the two inputs of the multiplier used as a squarer.

*$x \cdot y \doteq \frac{1}{4}(x+y)^2 - \frac{1}{4}(x-y)^2$

SQUARE-ROOTING

A *square-rooter* is a 1-quadrant device that computes either the positive or the negative square root of an input voltage multiplied by a dimensional (V) constant of appropriate polarity.

$$\pm E_0 = \sqrt{V_r \cdot V_{in}} \quad \text{or} \quad \pm E_0 = \sqrt{-V_r \cdot V_{in}} \qquad (4)$$

For 10V full-scale input and output, V_r is 10V. Since the slope of the square-root is theoretically infinite at zero, one might expect the largest errors to occur near zero, with slow response, and perhaps even hysteresis. Furthermore, since real square roots occur only for positive arguments, if the input signal can change sign, it may be necessary to constrain either the input or the output (or both) to prevent "lockup." Such constraints are mandatory if square-rooting is achieved by feedback around a divider that in turn involves feedback around a multiplier.

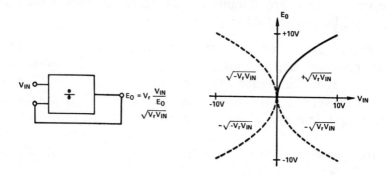

Square-roots are used in root-mean-square and vector computations, physical measurements, and simulation of fluid-flow parameters. The most popular means of square rooting are log-antilog ($\sqrt{x} = \epsilon^{\frac{1}{2}\log \epsilon^x}$) and feedback around a divider ($z = ky/z$). The log-antilog approach produces good accuracies, wide dynamic range, and benign behavior through zero. The divider approach is capable of higher speed and better accuracy near full scale and over modest dynamic ranges. Square roots computed by feeding back around $V_A V_B / V_C$ devices can use the second multiplicative input for obtaining the geometric mean ($\sqrt{V_A V_B}$).

The "odd function" square root $x/\sqrt{x}$ has the polarity of the input and an output proportional to the square-root of the input. It can be generated by an odd-function squarer in a feedback loop.

LOGARITHMIC CIRCUITS

The ideal inverting voltage-to-voltage logarithmic circuit generates the function

$$E_0 = -K \log_B \left(\frac{V_{in}}{V_r} \right) \tag{5}$$

where V_r is the normalized unity reference, the value of V_{in} for which $E_0 = 0$. V_r is usually set arbitrarily, for either full-scale input, mid-scale output, or elsewhere. It is of appropriate polarity to make the argument positive: i.e., if V_{in} is positive, V_r is also positive; if V_{in} is negative, V_r is also negative. (The logarithm of a negative argument is not defined in terms of real values.) If the logarithmic device accepts a current input, I_{in}, the dimensional constant, V_r, is replaced by a current reference, I_r.

K (also a dimensional constant) is the scale factor, the number of volts corresponding to a ratio equal to the base B. For example, if B is 10, K is the number of volts corresponding to $V_{in}/V_r = 10$ (i.e., 1 decade). If B is ϵ, K is the number of volts corresponding to the ratio ϵ. If B = 2, K is the number of volts per *octave*. Popular values of K are 1 or 2 volts per *decade*. To compute the equivalent K for any other base (B'), multiply K by $\log_{10} B'$ to obtain K'; e.g., 1V/decade = 0.3010V/octave.

Logarithms are useful in signal compression, measurement of quantities having wide dynamic range, displaying information in "decibel" form, linearization of logarithmic data, computation of powers and roots, and wide-range division.

Today's logarithmic circuits are almost universally based on the relationship between the collector current and V_{BE} in a diode-connected transistor in the feedback circuit of an op amp. The transistor may be connected either as a two-terminal diode*

*A high-β transistor should be used.

(collector tied to base) or as a three-terminal element with one terminal (usually the collector, for β-independence) fixed at the summing-point potential of the op amp.

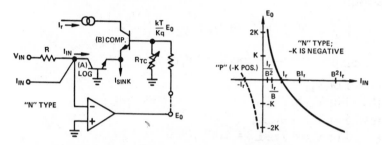

Since a diode's I_r and K both vary substantially with temperature (about 8%/°C and 0.8%/°C, referred to the current), it is important to compensate for temperature variations in practical applications. I_r variations are substantially cancelled by driving a fixed current through a matched diode (usually a twin on a monolithic chip) and taking the difference of the V_{BE}'s. K increases linearly with temperature at 1/3%/°C of its value at +27°C. In log operation, it is usually compensated for by a resistive divider having an equal temperature coefficient.

Logarithmic elements are commercially available in several forms having varying degrees of flexibility (and cost):

(1) As complete compensated current- or voltage-to voltage log modules (e.g., Model 755)

(2) As internally-compensated voltage-to-current antilogarithmic feedback elements, requiring an external operational amplifier (Model 752), and

(3) As matched transistor pairs with compensating resistive divider (Model 751), and as monolithic transistor pairs (AD818).

To ensure proper polarity relationships without excessive use of external operational amplifiers, logarithmic devices are available with a choice of polarities: "P" devices (-K positive, V_r and V_{in} negative) utilize PNP transistors to supply positive current to a summing point in response to negative inputs; "N" devices (-K negative, V_r and V_{in} positive) utilize NPN transistors to sink current at a summing point in response to positive inputs.

ANTILOG CIRCUITS

The inverse of the logarithm, the *exponential* relationship, is of the form

$$E_0 = {}^{\prime}V_r B^{-V_{in}/K} = V_r \log_B{}^{-1} \left(\frac{V_{in}}{-K}\right) \tag{6}$$

For example, if B = 10, and K = 1V/decade,

   (a) For $V_{in} = 0$, $E_0 = V_r$
   (b) For $V_{in} = K$, $E_0 = 10V_r$
   (c) For $V_{in} = -K$, $E_0 = V_r/10$
   (d) For $V_{in} = 2K$, $E_0 = 100V_r$

Since real exponentials are always positive, E_0 is of the same polarity as V_r. In practical circuits that use inverting operational amplifiers, K and V_r are of the same polarity; -K is positive for "P" devices, and negative for "N" devices (see Logarithmic Circuits).

Antilog devices are usually employed in connection with operations on logarithmic variables. For example, if a number of input variables are to be raised to powers, multiplied and divided, they might be individually converted to logarithmic form, then summed or differenced, with weights corresponding to the individual exponents, and anti-logged.

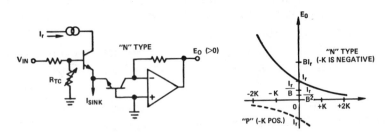

Antilog devices use the same circuitry as log devices; the distinction lies in the way they are connected. The log transistor and its temperature-compensating circuitry develop a current that is proportional to the antilogarithm of the applied voltage. If the applied voltage is the input signal, the op amp, with a feedback resistor, develops an exponential output voltage; if the applied voltage is the amplifier's *output*, it will be constrained at the value required to balance the input current (that is, the output will be proportional to the log of the ratio of I_{in} to I_r).

THE "IDEAL DIODE" OPERATOR

For switching purposes, the "ideal diode" is a one-way switch that is open when the imposed voltage is of one polarity and closed when the polarity is opposite. The *ideal diode operator* is a voltage-to-voltage circuit that would have the same response as a circuit that used an ideal diode as the switching element: the output voltage is zero for one polarity; it increases linearly with input when the polarity changes. The ideal diode operator can also be considered as a "zero-bound" circuit.

Ideal-diode operators are useful in precision dead-zone, bounds, and absolute-value circuits, and in function fitting with piecewise-linear approximations. Previously unthinkable in terms of economy, such circuits now benefit by the linearity, stability, and very low cost of IC operational amplifiers.

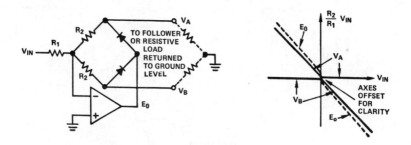

CONTROLLED SWITCHES

Though not strictly nonlinear devices (one considers switches as elements of linear time-dependent functions, controllable by outside influences), and although their characteristics and applications would require a monograph many times the size of this volume, we mention them here briefly, because they are pertinent to useful nonlinear functions.

Switches, often operated by comparators, are used locally in non-linear circuitry to establish new conditions when thresholds have been crossed by an input or an output voltage or current. For example, they may change a gain, reverse a polarity, or initiate a new mode of operation, thus producing an overall nonlinear response. Analog switches are often an integral part of nonlinear devices, e.g., pulse-height-width multipliers, V-to-f converters. They are of course essential to multiplexers and most types of D/A and A/D converters.

Switches come in a wide variety of forms, ranging from electro-mechanical, optoelectronic, and Hall-effect relays, for switching-with-isolation, to straight logic-operated electronic devices. These last are of two basic kinds: *voltage* switches and *current* switches. Examples of voltage switches, which open or close a circuit, include MOS and other FET types, as well as saturated bipolar transistors. Current switches, which switch by diverting a current, usually in-volve diodes or bipolar transistors, operating in the linear region. They are capable of high speeds.

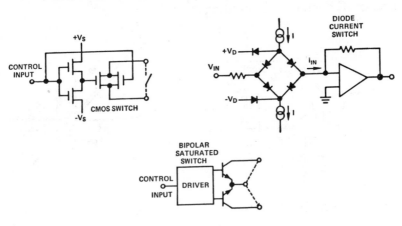

COMPARATORS

Comparators are devices that have two stable output states. They signal whether an input current or voltage has crossed a threshold imposed by one or more other currents or voltages, either fixed or variable.

They are used as polarity sensors, as digital inputs to analog-controlled logic systems; they operate switches, sharpen transitions, quantize analog voltages, and —followed by switching or precision bounding— they can serve as elements of waveform generators.

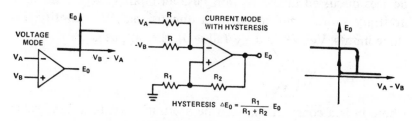

A comparator is similar to an operational amplifier, in that it usually has high gain and a sensitive low-drift differential input circuit. Comparators are essentially open-loop devices; therefore internal frequency compensation against external loop closure (an important feature of op amps) is not needed. The result is higher switching speed than might be obtained with an op amp used as a comparator. Since only small amounts of negative feedback are needed for a comparator to oscillate, great pains must be taken, both in design and use, to separate the input and output circuits, electrically, physically, thermally, and at the power terminals.

Though comparators are differential devices, the common-mode range permitted with some types may be quite small. Depending on the design, the two stable comparator output levels may be compatible with standard digital logic levels (e.g., TTL), with full-range op-amp output levels (e.g., $>\pm10V$, for $\pm15V$ supplies), or with high-voltage, high-current swing capacity, for relay drive. To prevent ambiguity of the switching level due to noise, hysteresis can be effected by feeding back a small fraction of the output to the positive input terminal. In addition, some comparators can be latched in response to a digital signal.

Comparators are used either in a *voltage* mode, with two voltages to be compared applied to the two inputs, or in a *current* mode, where the voltage developed by passive summation of two or more currents is compared with a reference level, usually near "ground".

ARBITRARY EXPONENTS

The antilog device is an exponential function. That is, it raises a constant base to a power determined by the input signal. The devices discussed in *this* section raise an input voltage ratio to an arbitrary *power,* multiplied by a dimensional (V) quantity. For three inputs, V_Y, V_Z, and V_X (all $\geqslant 0$), the output is

$$E_o = V_Y \left(\frac{V_Z}{V_X}\right)^m \qquad (7)$$

where *m* is a constant that can be arbitrarily set to a fixed value, typically in a range specified as $m_{MAX} \geqslant m \geqslant 1/m_{MAX}$. For m > 1, the ratio is raised to the power *m*. For m < 1, the $1/m$th root is obtained. To obtain negative powers, (-*m*), the roles of V_Z and V_X are simply interchanged.

Arbitrary exponents are useful in analog computing that involves exponents, in linearizing nonlinear data, and in developing power-series approximations, either with conventional integral powers or roots, or with non-integral powers (e.g., 1.211, 0.735).

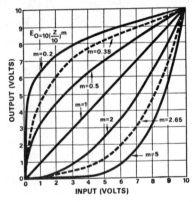

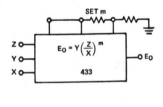

The design of an arbitrary-exponent device is straightforward: The log of the ratio (V_Z/V_X) is obtained, it is multiplied by the factor *m*, then the antilog is obtained. The function can be built up with log/antilog building blocks, or it can be purchased as a single, low-cost device specifically designed to do the job (e.g., the Analog Devices' Model 433). For convenience, such a device may make available a fixed low-TC reference voltage to furnish any constant inputs that may be required. The overall device gain may be keyed to this constant, e.g., for 10-volt ranges and a predetermined* reference, V_r,

$$E_0 = \frac{10}{V_r} \cdot V_y \cdot \left(\frac{V_z}{V_x}\right)^m \tag{8}$$

"TRUE" ROOT-MEAN-SQUARE

An ideal root-mean-square device computes the average of the squared input over a given interval, then takes the square-root of the average, viz.,

$$E_{RMS} = \sqrt{\frac{1}{T} \int_0^T (V_{in})^2 \, dt} \tag{9}$$

True RMS is useful in evaluating AC signals (including noise), in control loops (e.g., automatic gain control), and as front-end signal conditioning for AC inputs to analog and digital panel meters and data-acquisition systems.

In practice, instead of the definite integral, a "running average" is taken, usually approximated by a first-order RC lag circuit. This approximation is valid for stationary waveforms if the filter time constant is sufficiently large, and if enough time is allowed for the output to settle.

In conventional AC-instrumentation practice, a further approximation has been employed:

$$E_{RMS} \cong 1.111 \cdot \frac{1}{T} \int_0^T |V_{in}| \, dt \tag{10}$$

This latter approximation (the mean absolute value) is valid primarily for sine waves. It leads to gross errors if applied to noise, square waves, pulses of arbitrary duty cycle, and —of course— all waveshapes of unpredictable nature, including fluctuating DC. Though widely used, it can be greatly misleading.

*by the manufacturer

The straightforward way of computing true RMS is to perform the operations in the order indicated: Square, Average, Root. It has the disadvantages of complexity, cost, and loss of resolution because of the doubled order of dynamic range $[(100:1)^2 = 10,000:1]$.

A Better Way is to use a 2-quadrant $V_1 V_2/V_3$ device, together with an op amp connected as a simple low-pass filter, to solve the implicit equation

$$E_{out} = Ave\ (V_{in} \cdot V_{in}/E_{out}) \qquad (11)$$

Since E_{out} can be assumed to be constant for stationary waveforms,

$$(E_{out})^2 = Ave(V_{in})^2 \qquad (12)$$

whence

$$E_{out} = \sqrt{Ave(V_{in})^2} \qquad (13)$$

If mean absolute value can adequately measure a known waveform (equation 10, with an appropriate multiplying factor), one can either use a full-wave rectifier with filtered output, or compute the average of $\sqrt{V_{in}^2}$ by using the *input* of the filter (i.e., the output of the multiplier) as the denominator in equation 11. Though this latter approach might seem a bit roundabout, it may be useful for apparatus requiring a choice of RMS or MAV.

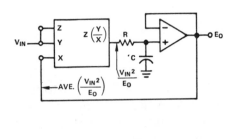

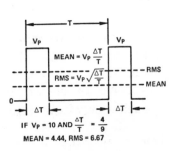

LOG RATIO

Log ratio devices can measure ratios of either voltages or currents

$$E_0 = K \log_B \frac{V_1}{V_2} \text{ or } K \log_B \frac{I_1}{I_2} \text{ or } K \log_B \frac{V_1/R}{I_2} \qquad (14)$$

They are useful where measurements involve exponential data (e.g., light measurements), where the inputs or the ratio itself can have a wide dynamic range, for gain measurements displayed in log form (e.g., "dB"), for generation of powers and roots, and for signal compression.

Log ratio devices are available as complete entities, e.g., Analog Devices' Model 756. They can also be assembled from log amplifiers, or with the more-elementary log devices mentioned in the section on Logarithmic Circuits.

Since real values of the logarithm do not exist for negative arguments, logarithmic functions* of bipolar signals (e.g., AC signals) must be obtained in terms of properties of the signal rather than the signal itself. Examples of such properties include mean-absolute, RMS, and peak measurements.

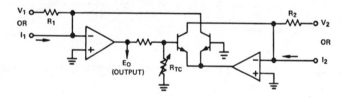

SINH⁻¹ OR "AC LOG"

The inverse hyperbolic sine characterizes the output of an op amp that has two complementary antilog transconductors (e.g., 752) paralleled in its feedback path.

---

*Log ratio is computed by subtracting logarithms. However, even if the ratio were obtained first, the rules regarding zero and bipolar denominators must be observed.

$$-2\,\frac{V_{in}}{R} = I_r \exp(E_0/\text{-}K) - I_r \exp(E_0/K) = 2I_r \sinh(E_0/\text{-}K) \qquad (15)$$

whence

$$E_0 = K \sinh^{-1}\left(\frac{V_{in}}{RI_r}\right) \qquad (16)$$

This useful device has logarithmic behavior over wide ranges of $+V_{in}$ and $-V_{in}$, and well-behaved linear behavior through zero. It will compress bipolar signals logarithmically in a symmetrical and predictable manner.

Graded null meter and wide-range bipolar analog panel meter are two DC applications; non-saturating signal compression is a typical AC application. With AC signals, the choice of I_r is compromised by the conflicting demands of bandwidth and dynamic range.

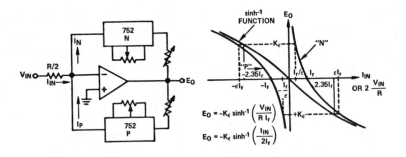

If the antilog elements are connected in the input path, the inverse function, i.e., the hyperbolic sine, is generated.

In the simplest (but least stable) form of this device, the antilog elements may be a pair of diodes connected in parallel, back-to-back.

VECTOR SUM (MAGNITUDE)

A vector-sum device computes the square-root of the sum of the squares of the inputs

$$E_0 = \sqrt{V_1{}^2 + V_2{}^2 + \ldots + V_n{}^2} \qquad (17)$$

Typical applications include summation of orthogonal measurements, such as length, force, or voltage vectors, and measures of random statistical quantities. Geometrical quantities, such as sums of areas, and diagonals of rectangular n-dimensional figures are also obtainable.

There are two popular approaches to computing vector sums: direct and implicit. The *direct* approach requires that each input be individually squared, the sum be taken, and the result square-rooted. While the method is straightforward, the expansion of dynamic range inherent in squaring imposes serious limitations on accuracy if E_0 is to vary over a wide dynamic range. For n variables, n squarers, a square-rooter, and a summing amplifier are required.

The *implicit* approach, which is capable of yielding more-accurate results, calls for implementation of the equation

$$E_0 = \frac{V_1{}^2}{E_0 + V_n} + \frac{V_2{}^2}{E_0 + V_n} + \dots + \frac{V_{n-1}{}^2}{E_0 + V_n} + V_n \quad (18)$$

It requires n-1 devices that compute $V_a \cdot V_b / V_c$, and two summing amplifiers.

The special (and most-frequently encountered) case, in which n = 2, requires a single $V_a \cdot V_b / V_c$ device and 2 modest-performance op amps. This compares favorably in cost, complexity, and performance with the direct approach, using 2 squarers, one rooter, and an op amp.

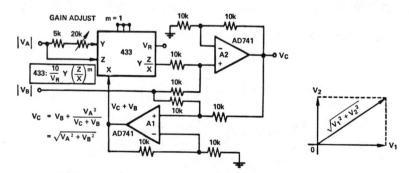

TRIGONOMETRIC FUNCTIONS

Useful trigonometric functions include A $\sin\theta$, A $\cos\theta$, r $\sin\theta$, r $\cos\theta$, their combinations in sums and products, and $\tan^{-1}(V_y/V_x)$.

They are applied in vector resolution and composition, coordinate transformations, waveshaping, and function generation.

Trigonometric relationships among analog variables can be simulated roughly by simple circuits involving FET or transistor characteristics, and to greater accuracy by piecewise-linear diode function fitting or by power-series approximations.

A recent significant development in power-series approximations is the use of non-integral powers, computed by adjustable-exponent devices, such as the Model 433. This approach can significantly decrease the number of power terms required for a given level of accuracy. For example, $\sin\theta$ can be approximated by $(x - x^3/6.79)$ to within 1.35% over the range 0 to $\pi/2$; but the approximation $(x - x^{2.827}/6.28)$ has less than 0.25% error over the same range of angle.

In analog-digital function fitting, trigonometric relationships may be stored in read-only memories (ROM's) and returned to analog form by D/A conversion.[1]

The most important consideration (other than the basic fit) affecting cost and complexity of designs involving trigonometric functions is the range of angle. Since both the input and the output of a "$\sin\theta$" device are usually voltages, the function that is really being fit is (for example) $V_{FS} \sin \frac{\pi}{2} (V_{in}/V_r)$. V_{FS} is the voltage cor-

TYPICAL APPROXIMATIONS:

$$\sin\theta \cong \theta - \theta^{2.827}/6.28$$

$$\cos\theta \cong \left[\frac{\pi}{2} - \theta\right] - \frac{1}{6.28}\left[\frac{\pi}{2} - \theta\right]^{2.827}$$

$$\tan^{-1}\frac{V_B}{V_A} = \theta \cong \frac{\pi}{2}\frac{(V_B/V_A)^{1.2125}}{1 + (V_B/V_A)^{1.2125}}$$

[1] See *Analog-Digital Conversion Handbook*, Analog Devices, 1972, page I-65 *et seq.*

responding to the sine of 90°, and V_r is the voltage corresponding to 90°.

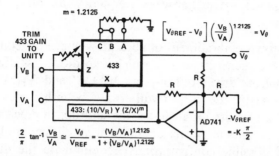

The simplest designs are those involving a single quadrant, $0° \leqslant V_{in} \leqslant V_r$. Those involving two quadrants, e.g., $-V_r \leqslant V_{in} \leqslant V_r$, are not too much more difficult (the linear term alone is within 0.25% to -14°). But if many quadrants are required, and especially if the angle can increase without limit, some form of switching and polarity sensing is necessary to continually translate the function to the first (two) quadrant(s), maintaining appropriate polarity relationships.

ABSOLUTE VALUE

An absolute-value device, otherwise known as a "full-wave rectifier," measures the instantaneous magnitude of the departure of a voltage from zero. The output may be assigned an arbitrary positive or negative polarity, depending on the circuit application.

$$E_0 = \pm |V_{in}| \qquad (19)$$

Applications in precision instrumentation include AC measurements, function fitting, inputs to single-quadrant devices (squarers or vector sums), triangular-wave frequency doubling, and error measurements.

The best-known embodiment of absolute value is the conventional "full-wave rectifier" circuit, in which two diodes, sharing a common terminal, are driven out-of-phase by $+V_{in}$ and $-V_{in}$. Depending on diode polarity, the output will always be either plus or minus the magnitude of the input, less 1 diode drop. In precision instrumentation, circuits involving operational amplifiers are used (in a number of possible configurations) to eliminate the diode drop (and its variations with current and temperature).

BOUNDS

A bounding circuit has an output that is linear for inputs up to a preset value, and unchanging beyond it. *Upper* and *lower* bounds are often used together; either may be fixed or variable, depending on the application. For an input, V_{in}, and upper and lower bounds V_U and V_L,

$$
\begin{array}{lll}
E_O = -V_{in} & (V_L \leqslant V_{in} \leqslant V_U) & \\
E_O = -V_U & (V_{in} \geqslant V_U) & (20) \\
E_O = -V_L & (V_{in} \leqslant V_L) &
\end{array}
$$

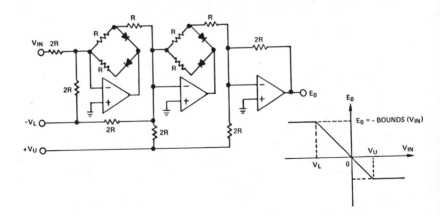

Precision bounds are useful in setting, or simulating, limits to voltage or current (or its rate-of-change), in establishing thresholds of ranges of operation in piecewise-linear function fitting, and (preceded by comparators) in establishing precise voltage staircase functions (i.e., quanta).

Bounds can be implemented with diodes and/or transistors, employing matched pairs to provide first-order threshold cancellation. However, greater accuracy can be obtained by using diodes with operational amplifiers to form "ideal diode" circuits, in which the diodes serve only as switches, with their thresholds corrected-for by the inherent properties of high-gain operational-amplifier loops.

DEAD ZONE

In a dead-zone operation, the output is typically a linear function of the input, except for a band that is insensitive to the input. That is, for an input V_{in},

$$\begin{aligned}
E_0 &= 0 & V_L &\leqslant V_{in} \leqslant V_U \\
E_0 &= V_{in} - V_U & V_{in} &\geqslant V_U \quad (21) \\
E_0 &= V_{in} - V_L & V_{in} &\leqslant V_L
\end{aligned}$$

Dead-zone is related to bounds

$$E_0 (Z) = V_{in} - E_0 (B) \qquad (22)$$

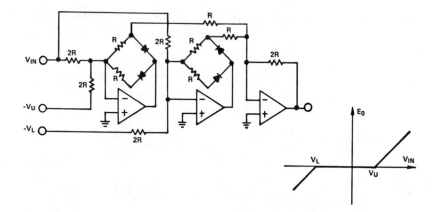

Like bounds, dead-zone is useful in setting thresholds for piecewise-linear function generation. It is also useful in suppressing noise in the vicinity of a null, in generating "linear" hysteresis functions, and in stabilizing the amplitude of "limit cycle" oscillations. It has also been used for velocity modulation of oscilloscope intensity, and —with dither— for fitting parabolic functions.

Dead zone, like bounds, can be established simply but not very accurately with diodes and transistors, or, —with great accuracy, at low frequencies— by ideal-diode circuits.

JUMP AND WINDOW FUNCTIONS

A "jump" function is simply the output of a comparator

$$E_0 = V_A \quad (V^+ - V^-) > 0$$
$$E_0 = V_B \quad (V^+ - V^-) < 0 \tag{23}$$

The outputs of two TTL-compatible comparators "hard-wired" together can produce a "window" in the response to a common input, a band for which the output is at one level, with the response everywhere else at the second level.

$$E_0 = V_A \quad V_1 < V_{in} < V_2$$
$$E_0 = V_B \quad V_2 < V_{in} < V_1 \tag{24}$$

Although the comparator output levels are dependent on loading and temperature, they can be shifted and rendered quite accurate and stable by the use of precision bounds.

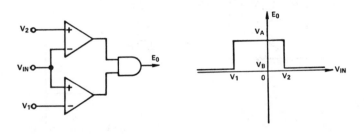

Jump functions are useful in sensing polarity, in precisely locating thresholds, and for all of the applications mentioned under Comparators. The window function can be used for precise quantization, for grading and sorting, and for pulse-width modulation. A comparator and a window can be used in conjunction to initiate a function that depends on the sign and quantized magnitude of a voltage.

HYSTERESIS

A device that exhibits hysteresis has an output that is a two-valued function of the input, over a portion of the input range. That is, the response is not uniquely dependent on the value of the input; it depends also on its history. A plot of output vs. input shows the characteristic "loop" behavior.

A familiar form of hysteresis is seen in a two-level comparator with fractional positive feedback. Once the output has switched, the input must be backed off to a second threshold beyond the switching point to cause the device to switch back to the original state. Flip flops and "latchup" phenomena rely on large amounts of hysteresis.

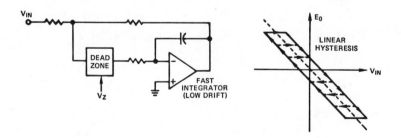

Besides two-level hysteresis, there is also "linear" hysteresis, best-known to electrical engineers in the form of magnetic "remanence," and to mechanical engineers as "backlash" in gear trains. Devices with linear hysteresis follow the input in one direction; when the direction of input is reversed, a dead band must be traversed before the output follows.

This form of hysteresis can be simulated by closing a feedback loop around an integrator, preceded by a dead zone. As the input increases, the integrator will follow (first-order lag), with the dead-zone output at the forward threshold. When the input is reversed, it must traverse the dead band (the integrator *holding*) before the output can be made to reverse direction.

CONCLUSION

We have seen here a wide variety of useful nonlinear phenomena, based on fundamental building blocks. In Part Two, we shall see ways in which these phenomena are, can be, and perhaps *should* be employed in solving an even greater variety of real analog design problems.

Function Fitting

Chapter 1

Function fitting, in general, is the translation of a mathematical or empirical relationship (between a *dependent* variable —output— and one or more *independent* variables —inputs) from one medium, such as a table, a mathematical formula, or a set of curves, to another medium, usually a physically-realizable device or system having an output and one or more inputs. A function may be fit by an "exact" relationship, or it may be approximated.

There are two basic steps in function fitting (Figure 1). The first is the establishment of a close-enough approximation in terms of ideal building blocks, that is, a conceptual model. The second is the successful employment of actual devices to embody the function within an acceptable set of constraints, such as range, scale factor, accuracy, drift, response time, complexity, cost, etc.

In this chapter, we shall consider functions that can be realized by circuits which embody "instantaneous dc" relationships between sets of voltages having a limited (not infinite) dynamic range of variation. For the most part, we consider single-valued functions, i.e., each set of input values creates a unique value of output, independent of history. (Through switching, or hysteresis, though, single-valued functions can become multi-valued.) We omit, to a great extent, the fitting of dynamic response relationships, such as linear transfer functions, or filter characteristics, whether in the frequency domain or the time domain, since the circuit theorists and filter designers have explored that area with great enthusiasm and have produced a profuse output of published material.

We also assume that the functions to be fit by analog techniques are statistically satisfactory. That is, we do not consider the class of problem in which the data points characterizing a relationship are scattered. Any necessary statistical smoothing has already been accomplished. However, it is possible to use random noise multiplicatively to simulate operators having appreciable stochastic fluctuation.

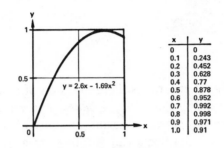

x	y
0	0
0.1	0.243
0.2	0.452
0.3	0.628
0.4	0.77
0.5	0.878
0.6	0.952
0.7	0.992
0.8	0.998
0.9	0.971
1.0	0.91

a. Curve table, and equation

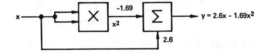

b. Block diagram

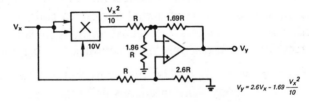

c. Circuit, scaled to 10V full scale, $V_y = 10y$, $V_x = 10x$

Figure 1. Analog function fitting

The limitations of space and time do not permit as thorough a coverage of the subject as it deserves, but we do hope to leave the reader with the outlines of analog function-fitting techniques (both as a bag of tricks and a guide to further thinking), some pointers for successful application, and a few examples.

WHY ANALOG FUNCTION FITTING?

We indicated in the introductory chapter that measurements of phenomena that one might wish to obtain linearly often come out nonlinearly. For example, a thermocouple is a cheap and simple (but low-level) means of measuring temperature differences. It can consist of as little as three segments of wire, with two thermal junctions at different temperatures. But its output voltage departs significantly from a linear function of temperature, depending on the materials it is made of, and the temperature range under consideration. Other examples might include the Wheatstone bridge, a simple way of measuring resistance deviation, but nonlinear for large deviations, and the deflection of an oscilloscope's spot, a nonlinear function that depends on the electron gun, the shape of the tube face, and the voltages applied to the X and Y axes.

It is possible to linearize these measurements (i.e., obtain an output reading that linearly represents the desired indication) by using nonlinearity to either compensate for or obtain the inverse of a nonlinear transducer function. The use of function-fitting techniques for such applications is discussed in Chapters 2-3 and 2-5.

Other applications of analog function fitting include calibration, simulation of nonlinear relationships in analog computers and computer-based instruments, translating indirect measurements into useful form economically, and generating time functions of arbitrary shape.

Nonlinear function fitting can also be performed digitally by read-only memories (ROM's) —often in conjunction with A/D and D/A converters— and by combinations of hardware and software. Decreasing costs and wide availability of digital hardware, plus an ever-increasing library of algorithms would appear to make this approach seem increasingly tempting, despite its inflexibility and complexity. But the cost of analog IC's and modules (both op amps and functional operators) has also decreased dramatically; analog approaches are still a "best buy" for simple relationships, and lower cost (for improved accuracy and increased circuit sophistication) makes them more competitive for increasingly-complex relationships.

RATIONAL FUNCTIONS

The simplest functions to fit, in concept, are those that can be expressed "exactly" by an equation involving basic operations: squaring, rooting, multiplication, addition, logarithms, and arbitrary power and roots. The basic operators mentioned in the introductory chapter can of course be used to fit functional relationships that are identical to the operations they perform.

They can also be combined to fit a wide range of explicit functions of 1 or more variables, such as

$$u = 1 + 0.3w^2 \tag{1}$$

$$u = v \, (\mathrm{r} - w) \tag{2}$$

$$u = v^{\mathrm{m}} + w^{\mathrm{n}} \tag{3}$$

$$u = \frac{v \cdot w}{1 + w^{\mathrm{k}}} \tag{4}$$

$$u = (v + \mathrm{m}) \, (w + \mathrm{n})^{\mathrm{r}} \tag{5}$$

$$u = \frac{v}{1 + \dfrac{v}{2}} \tag{6}$$

where u, v, and w may be variables, and the other terms fixed or adjustable constants. The summation operations are often performed with op amps.

In some cases, equations can be rewritten for embodiment in several ways. For example, equation (2) can also be written $u = v \cdot \mathrm{r} - v \cdot w$ (Figure 2). If r is a constant, either equation can be embodied with a single multiplier and a single subtractor. In general, one tends to choose the equation that will give the best compromise of error vs. cost. In this instance, if w is always small in magnitude compared to r, the configuration of Figure 2b is probably a better choice, because the more-important term can be handled with linear circuit elements, and errors of the $v \cdot w$ term will be of lower order. On the other hand, if w can be comparable to r in magnitude, it is probably better to take the difference of two large numbers *before* performing a nonlinear operation; therefore one would use the configuration of Figure 2a. The underlying assumption here (nearly always justified) is that accurate linear operations are cheaper than nonlinear operations of comparable accuracy.

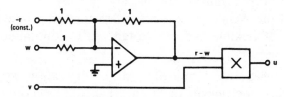

a. u = v (r - w), r = constant, preferred when vr and vw are of comparable magnitude

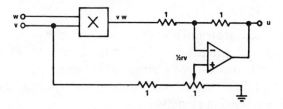

b. u = vr - vw, r = constant, preferred when vr >> vw, as when fitting a function that is essentially linear, with a small deviation.

Figure 2. The way an equation is written affects both circuit configuration and performance

For more-complicated functions, range and error analyses of all the available alternatives should always be performed, either analytically or empirically (if one happens to have "worst-case" components on hand).

SCALE FACTORS

Having optimized the circuit and chosen a set of devices likely to implement it economically, one is confronted by the *scaling* problem, i.e., determining the exact relationship between the electrical circuit and the function it fits, including constants (gains and biases) and the ranges of all voltages.

Every accessible voltage or current in the analog circuit corresponds to a variable in the original functional operation. If it is an important term, requiring good accuracy, its range should be close to full scale of the device producing (or accepting) it; but it should not appreciably exceed the full-scale range, for any combination of inputs or outputs (unless it happens to be a Bounds function). Input and output ranges are usually predetermined by other elements of the overall system, but ranges at intermediate locations are somewhat flexible and can be tailored for optimum dynamic range.

Since the ranges are determined by the configuration, scale factors (i.e., electrical coefficients or "gains.") should be chosen after the configuration has been adopted. It should be noted, though, that availability of appropriate scale factors may be a factor in the choice of configuration.

The experienced instrumenter can usually derive scale factors directly from the equation to be implemented, based partly on intuition, partly on common sense, and partly on a set of well-learned but perhaps unverbalized rules. Others may benefit by observing the following principles:

1. The original equation should be dimensionally correct.

2. If not already in dimensionless form, it can be normalized by multiplying and dividing each variable by a multiple of its range, usually 100%. Consider this example: for

$$y = Ax \sin\theta + K \quad \text{(Figure 3a)} \tag{7}$$

$$Y_m \frac{y}{Y_m} = A X_m \frac{x}{X_m} \sin\theta + K \tag{8}$$

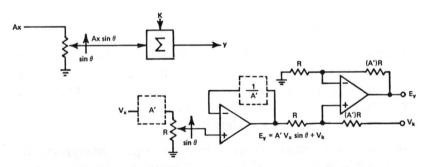

Figure 3a. Block diagram and electrical schematic for linear scaling example

where Y_m, X_m, are the full-range values of x and y. Defining the ratio-to-its-range as the dimensionless variable, and dividing both sides of the equation by Y_m, the equation becomes, in terms of dimensionless variables, y' and x'

$$y' = A\frac{X_m}{Y_m} x' \sin\theta + \frac{K}{Y_m} \tag{9}$$

3. Write the equation of the analogous electrical circuit (Figure 3a), using (unknown) coefficients A' and K' to relate the various voltages, and including any known electrical scale factors (such as those inherent in multipliers and log devices).

$$E_y = A'V_x \sin\theta + K'V_{Km} \tag{10}$$

4. To determine the unknown constants, multiply and divide by the expected maximum values of voltage, to normalize the electrical equation:

$$E_{ym}\frac{E_y}{E_{ym}} = A'V_{xm}\frac{V_x}{V_{xm}}\sin\theta + K'\,V_{Km}$$

$$\frac{E_y}{E_{ym}} = A'\frac{V_{xm}}{E_{ym}}\frac{V_x}{V_{xm}}\sin\theta + K'\frac{V_{Km}}{E_{ym}} \tag{11}$$

The normalized equations, 9 and 11, must be identical, therefore

$$A'\frac{V_{xm}}{E_{ym}} = A\frac{X_m}{Y_m}, \text{ therefore } A' = A\frac{X_m}{Y_m}\frac{V_{xm}}{E_{ym}}$$

and

$$K'\frac{V_{Km}}{E_{ym}} = \frac{K}{Y_m}, \text{ therefore } K' = \frac{K}{Y_m}\frac{E_{ym}}{V_{Km}}$$

5. The electrical constants are now substituted in the electrical system equation (10). The process is by no means as formidable

as it may appear at first glance, because usually, $V_{xm} = E_{ym} = V_{Km} = 10V$; so

$$A' = A\frac{X_m}{Y_m} \text{ and } K' = \frac{K}{Y_m}$$

The scaling, as described in this example, has (so far) applied to an essentially linear circuit, corresponding to the case in which $\sin\theta$ is a dimensionless gain (e.g., a potentiometer setting). The process differs somewhat if the sine function is the output voltage of a sine operator (Figure 3b), and must be multiplied by V_x

$$V_s = V_{sm}\,\sin\frac{\theta_m}{V_{\theta m}}V_\theta \tag{12}$$

The electrical equation becomes

$$E_y = A'\frac{V_x}{V_r}\,V_{sm}\,\sin\frac{\theta_m}{V_{\theta m}}V_\theta + K'V_{Km} \tag{13}$$

where V_r is the multiplier's scale constant. The equation normalizes to

$$\frac{E_y}{E_{ym}} = A'\frac{V_{xm}}{E_{ym}}\frac{V_x}{V_{xm}}\frac{V_{sm}}{V_r}\sin\frac{\theta_m}{V_{\theta m}}V_\theta + K'\frac{V_{Km}}{E_{ym}} \tag{14}$$

Again, recognizing that equations 9 and 14 are identical,

$$A'\frac{V_{xm}}{E_{ym}}\frac{V_{sm}}{V_r} = A\frac{X_m}{Y_m}, \text{ and } A' = A\frac{X_m}{Y_m}\frac{E_{ym}}{V_{xm}}\frac{V_r}{V_{sm}}$$

If $V_{xm} = E_{ym} = V_r = V_{sm} = 10V$,

$$A' = A\frac{X_m}{Y_m} \text{ and } K' = \frac{K}{Y_m}$$

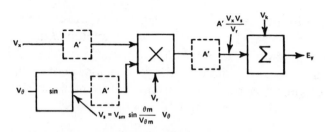

Figure 3b. Electrical block diagram of nonlinear version, showing possible locations for scale factor

It is important to note that the scale factor A', when associated with a multiplication, can be applied at any one of the three terminals, or distributed among two or more of them, if necessary to optimize the dynamic range for both inputs and the output. For example, if $V_{sm} = 5V$ and $V_r = V_{xm} = E_{ym} = 10V$, A' will be doubled. Most likely, a factor of 2 should be applied between the output of the sine operator and its input to the multiplier, if it is desired to make full use of the multiplier's input range.

After the scale factors have been computed, they should be checked, by considering various extremes of input and output signals; any indicated modifications should be made. While the approach suggested here works, it is no better than the assumptions. Awkward assumptions will lead to awkward dynamic ranges.

5. Note that the assignment of a "maximum" value E_{ym} to E_y does not automatically *guarantee* that V_y will not exceed full scale, unless the set of normalizing voltages is fully consistent. With practice, one will develop a near-intuitive feeling for proper scale factors and will find much of the above procedure unnecessary to plow through in detail. Incidentally, time-dependent devices may also be scaled in this manner. Where time appears in an equation, it is multiplied and divided by a nominal "unit value," usually 1 second, but often the characteristic time of the slowest integration, in high-speed analog computing devices or systems. $(T\frac{t}{T} = T\ t')$

In chapter 2–3, it is shown how one might develop and scale a thermocouple-compensation circuit.

INVERSE FUNCTIONS

If $u = f(v)$, the inverse function, $v = f^{-1}(u)$, may be obtained (in concept) by the use of $f(v)$ in a high-gain negative feedback circuit (Figures 4 and 5a). This is already widely exploited in:

(a) the generation of logarithmic operations. The exponential I-V relationship of a diode in the feedback path of an operational amplifier matches the input current, enforcing a logarithmic output voltage (see Figure 4c and 4d),

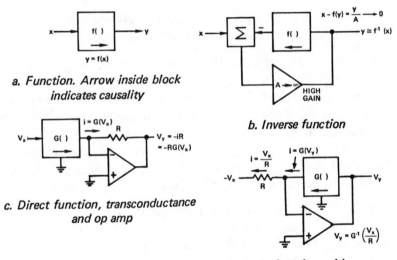

a. Function. Arrow inside block indicates causality

b. Inverse function

c. Direct function, transconductance and op amp

d. Inverse function, with op amp

Figure 4. Direct and inverse functional operations

(b) the use of multipliers for division. The product of one input and the output is made to equal the second input, therefore the output is proportional to their ratio,

(c) the use of squarer-connected multipliers for square-rooting. The product of the output multiplied by itself is made to equal the input, hence the output is the square-root of the input (Figure 6a).

Such schemes can be applied to combined functional operations for generating operators that are more easily obtainable in the inverse form. For example, if $y = x + \log x$, there is no closed-form solution to this transcendental equation if one desires x. One configuration for obtaining x, given y, is a high-gain feedback loop around $x + \log x$, as shown in Figure 5a.

There are a number of evident restrictions to the use of this technique:

1. The net incremental feedback must be negative over the range of interest.

2. Instabilities resulting in oscillation or "latchup" should be ruled out. Stabilizing and range-limiting circuitry may be necessary, with possible restriction of range, bandwidth, or accuracy. Loop gain and phase shift must be examined under all conditions. Adjustably-offset random noise may be employed as an input to detect sensitive frequency and amplitude bands.

3. In general the functions should be single-valued and monotonic in the range of interest. For example, $\sin^{-1}(x)$ should be limited to within a range of $\pm 90°$.

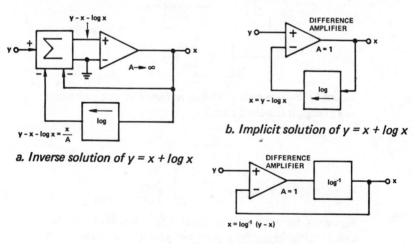

a. Inverse solution of $y = x + \log x$

b. Implicit solution of $y = x + \log x$

c. Another implicit solution of $y = x + \log x$

Figure 5. Inverse and implicit solutions. Note that (a) can be implemented identically to (b), in this case, but (b) is conceptually simpler.

IMPLICIT SOLUTIONS ($x = f\{x, y,...\}$)

A powerful feedback technique for solving for any variable that can be made to appear twice in an equation (by non-redundant summation, factoring, or other trickery) is the *implicit* use of the variable in solving for itself, without necessitating the explicit use of high gain to enforce the feedback constraints. Figure 5b shows an implicit solution as an alternative to the inverse for obtaining x in $y = x + \log x$. Here are a few additional examples of this use of algebraic analog computing.

1. Square rooting (Figure 6b) Any divider may be used as a square-rooter; non-feedback types tend to be the most successful. If the input, x, is divided by the output, y, to obtain y,

$$y = \frac{x}{y}, \text{ or } y^2 = x \tag{15}$$

and

$$y = \sqrt{x} \tag{16}$$

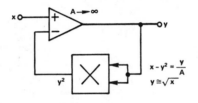

Figure 6a. Square root as an inverse function, using a squarer in a high-gain feedback loop

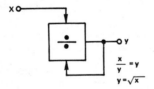

Figure 6b. Square root as an implicit function. If the divider uses a multiplier fed back with high gain, the configuration is identical to 6a. But a device specifically designed for division will retain low error over a much wider dynamic range.

2. Root mean-square (see page 17) A multiplier-divider (uv/w) may be used, followed by an averaging filter, to compute the average:

$$y = \text{ave. } (x^2/y) \tag{17}$$

For stationary waveforms, and using a filter having a sufficiently long time constant, y will be constant, and

$$y = \sqrt{\text{ave. } (x^2)} \tag{18}$$

3. Vector sum and difference (see page 21) If $w = \sqrt{u^2 + v^2}$, one can compute w as follows, using a multiplier-divider

$$w^2 - u^2 = v^2 = (w + u)(w - u) \tag{19}$$

Dividing by $(w + u)$

$$\frac{v^2}{w + u} = w - u \tag{20}$$

and

$$w = u + \frac{v^2}{w + u} \tag{21}$$

Additional variables may be embraced simply by adding terms; for example, to compute $w = \sqrt{u^2 + v^2 + x^2 + y^2}$,

$$w = u + \frac{v^2}{w + u} + \frac{x^2}{w + u} + \frac{y^2}{w + u} \tag{22}$$

Given w and u, the vector *difference* $v = \sqrt{w^2 - u^2}$ may be computed by dividing equation 19 by v, whence (Figure 7a)

$$v = \frac{(w + u)(w - u)}{v} \tag{23}$$

4. Bridge linearization The output of a Wheatstone bridge configuration with one leg variable is of the form

$$y = \frac{x}{1 + x} \tag{24}$$

This response is linear only for small values of the deviation, x. It can be linearized by solving implicitly for x (Figure 7b)

$$x = y + x \cdot y \tag{25}$$

Since the deviations are usually small, a multiplier with very modest specifications may be used, provided that the signals are scaled to use near-full-scale capability, and that drift of the second (i.e., correction) term is low.

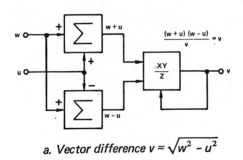

a. Vector difference $v = \sqrt{w^2 - u^2}$

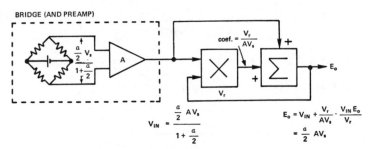

b. Linearizing a Wheatstone-bridge output

Figure 7. Applications of implicit feedback

There are three principal reasons for using implicit solutions, when they are appropriate:

a. To simplify the block diagram. In the case of vector summation, a $u \cdot v / w$ multiplier and two operational amplifiers replace two squarers, a square-rooter, and a summing amplifier.

b. To avoid expansion of dynamic range. Squaring a signal with 100:1 dynamic range results in a signal with 10,000:1 dynamic range. Noise, drifts, and reduced bandwidth can impair overall accuracy. The $u \cdot v / w$ operation, on the other hand, remains net first order.

c. To provide an improved fit with few additional components.

Figure 14 and the appendix to this chapter show the great improvement in fitting sin x due to using modified equations involving an additional feedback term.

Like inverse functions, implicit functions must be single-valued. Unlike inverse functions, they need not always be monotonic. For example, sin x can be approximated from $-\pi$ to $+\pi$ with greatly improved accuracy, using feedback.

FITTING ARBITRARY FUNCTIONS

For the purpose of this section, "arbitrary functions" are defined to include all functions that cannot be fit "exactly" by a conceptual closed-form equation, whether explicit or implicit. In other words, there is almost always a residual theoretical error, which must be considered along with the device errors to determine the overall closeness of fit.

Besides such obviously arbitrary functions as empirically-determined circuit and system nonlinearities requiring calibration curves, "arbitrary functions," by the above definition, include such analytic but non-rational functions as sin θ, $\tan^{-1} x$, and a whole host of functions characterizable by infinite series.

In general, to be fittable by more-or-less simple analog circuits, a functional operation must be bounded (i.e., defined within a finite range of all variables involved), single-valued in terms of inputs, and free from singularities (except where they can be satisfactorily fit by diode breakpoints, switching, or comparator "jump" functions). To be practical for analog circuit elements, there is the additional constraint that circuit complexity (consequently the cost) must be competitive with digital function generation (ROM's alone, or ROM's plus digital processing, plus at least one step of conversion).

Relationships are smoothly fit using logarithmic, exponential, or power-law elements, or they may be fit more-or-less directly with a set of straight-line segments produced by diode breakpoints (Figure 8). The former technique requires more-sophisticated mathematics and error analysis, but the output is differentiable, and the error function is satisfyingly smooth. The latter technique is well-suited to quick, empirical fitting of functions of one variable, but the error consists of a series of cusps that can be troublesome if differentiation is used or if the function is employed within a feedback loop. Also, arbitrary functions of 2 or more variables (which are difficult to fit, in any event) run into structural limitations, due to the sheer number of "piecewise-planar" elements; the influence of each faceted element also poses tricky visualization problems. Smooth functions, in linear, or nonlinear combinations —on the other hand— pose no interpolation problems. Any point is readily calculable, though not necessarily an accurate fit.

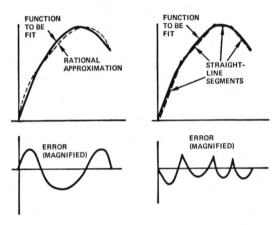

Figure 8. Smooth fit vs. piecewise-linear fit

SMOOTH APPROXIMATIONS

Although these approaches require more mathematics than do piecewise-linear approximations, the manipulations are of a kind that is not difficult if one is armed with a mechanized calculating device, such as an HP-35 pocket calculator. For mechanized optimization of the fit, a programmable engineering calculator, or access to computers, is helpful (but not essential unless one has the job of fitting many similar-but-different functions, or it is necessary to solve a large number of simultaneous equations to obtain a high-accuracy, high-order fit).

The ready availability of large amounts of calculating power, combined with the ready availability (at low cost) of today's multipliers, dividers, power-and-root devices, and logarithmic elements, makes smooth analog approximations (with errors typically varying from 0.1% to 1%) far more practical now than they have ever been.

The approach to fitting a function $y = f(x, A, B, C,...)$, where x and y are variables, and $A, B, C,...$ are constants, to the desired prototype shape, involves the following steps.

1. It is helpful to start with the data in normalized form.

2. Postulate a function that is likely to have the "right shape."

3. Write the equation for as many specific points as there are constants to be solved-for in the approximating function. The fit will be exact at those points.

4. Solve the set of simultaneous equations for the constants. Plug them into the equation, and check, by substituting the specific values chosen for "exact fit" into the equation.

5. Try out the equation at other intermediate values of x. Solve each for y and subtract the expected value of y to obtain the error. It may be helpful to plot an error curve.

6. If the errors are of reasonable magnitude, but are greater between one pair of calculated points than another, new intermediate points may be selected, and the equations written, solved, and tried-out for the new points. This process may be repeated as often as necessary to give (for example) equal maximum errors in all ranges. (Interpolation formulas may be used to shorten the process.)

If the errors are obviously too large, a different function may be tried. The reader will recognize that both experience and creativity will be of great help in proposing a function that has small inherent errors for a given shape. Here are some suggestions that those unfamiliar with the process may try as a starter:

- Try to find a "natural law" (e.g., logarithmic response)
- Try to fit deviations from linearity or from simple functional relationships, having a somewhat similar shape to the curve in question, such as $\log x$, $1/x$, e^x, x^m, etc.

- Try truncated power series $(A + Bx + Cx^2 + Dx^3$, etc.)
- Try series involving non-integral exponents, e.g., $A + Bx + Cx^m$
- Try implicit functions, e.g., $y = (A - y) x^m = Ax^m /(1 + Ax^m)$. While the two expressions are identical, the first uses fewer elements.
- Try the "Hoerl equation" $y = Ax^B \epsilon^{Cx} = A \ln^{-1} (B \ln x + Cx)$
- Try a more-easily-fit complementary function, such as $\cos x = \sin(\pi/2 - x)$

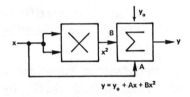

$$y = y_o + Ax + Bx^2$$

Figure 9. 2nd-degree polynomial using single multiplier. Op-amp configuration depends on polarity of constants in this figure and those that follow.

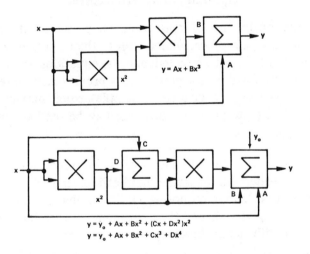

$$y = Ax + Bx^3$$

$$y = y_o + Ax + Bx^2 + (Cx + Dx^2)x^2$$
$$y = y_o + Ax + Bx^2 + Cx^3 + Dx^4$$

Figure 10. Odd-function 3rd-degree polynomial and generalized 4th-degree polynomial, using 2 multipliers. For complete generality, the origin may be offset along the X-axis by an amount h by adding a bias to the input. The x's then become $x' = x - h$.

POLYNOMIALS AND POWER SERIES

Polynomials can be modeled with multipliers and operational amplifiers. The minimum number of multipliers required to fit truncated power series of various degrees are:

2nd degree (involves x^2) ... 1 (Figure 9)

4th degree (involves x^4 and lesser powers) .... 2 (Figure 10)

8th degree (involves x^8 and lesser powers) ... 3 (Figure 11)

However, if implicit feedback is used, any of these truncated series may be converted into an *infinite* series, convergent over a limited (but adequate) range (Figures 12 and 13). The resulting enrichment can greatly improve the theoretical fit.

For example, a cubic ($y = Ax + Cx^3$) can fit sin x to within ±0.6% of full scale, from $\pi/2$ to $-\pi/2$, or within ±13.2%, from π to $-\pi$. But with the simple addition of a feedback term ($y = Ax + Cx^3 + Ex^2y$), the theoretical error becomes less than ±0.01% ($\pi/2 > x > -\pi/2$); and over the wider range of angle (π to $-\pi$), the error is still less than ±1.2%. The following example shows how dimensionless coefficients are derived, and the appendix to this chapter provides comparative details of a variety of sine-function-fitting schemes.

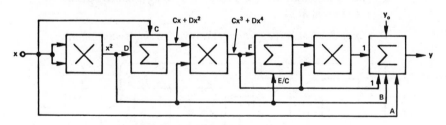

$$y = y_0 + Ax + Bx^2 + Cx^3 + Dx^4 + (\frac{E}{C}x^2 + F Cx^3 + F Dx^4)(Cx^3 + Dx^4)$$

$$= y_0 + Ax + Bx^2 + Cx^3 + Dx^4 + Ex^5 + (F C^2 + \frac{E}{C}D)x^6 + 2F C Dx^7 + F D^2 x^8$$

$$= y_0 + a_1 x + a_2 x^2 + a_3 x^3 + a_4 x^4 + a_5 x^5 + a_6 x^6 + a_7 x^7 + a_8 x^8$$

Figure 11. Generalized 8th-degree polynomial, using 3 multipliers. This configuration obtains its relative simplicity at the cost of 2 degrees of freedom (a_7 and a_8 are functions of a_3, a_4, a_5, a_6, and a_6 is not independent of a_3, a_4, a_5).

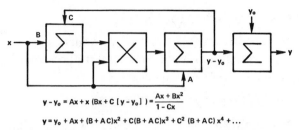

$$y - y_o = Ax + x\,(Bx + C\,[\,y - y_o\,]\,) = \frac{Ax + Bx^2}{1 - Cx}$$

$$y = y_o + Ax + (B + AC)x^2 + C(B + AC)x^3 + C^2\,(B + AC)\,x^4 + \ldots$$

a. Second-degree polynomial with implicit feedback produces infinite series, convergent for $Cx < 1$, has three degrees of freedom.

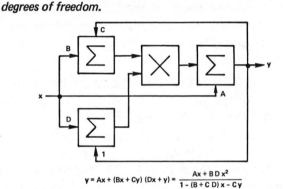

$$y = Ax + (Bx + Cy)\,(Dx + y) = \frac{Ax + BDx^2}{1 - (B + CD)x - Cy}$$

b. Second-degree polynomial in both x and y has four degrees of freedom, but coefficients are derived with greatly-increased difficulty. If $y(0) \neq 0$, y_0 is added outside the loop, as in 12a.

Figure 12. Implicit approximations with a single multiplier.

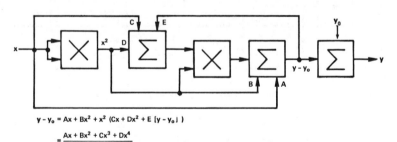

$$y - y_o = Ax + Bx^2 + x^2\,(Cx + Dx^2 + E\,[\,y - y_o\,]\,)$$

$$= \frac{Ax + Bx^2 + Cx^3 + Dx^4}{1 - Ex^2}$$

$$y = y_o + Ax + Bx^2 + (C + AE)x^3 + (D + BE)x^4 + E(C + AE)x^5 + E(D + BE)x^6 + \ldots$$

Figure 13. Fourth-degree polynomial using 2 multipliers with implicit feedback produces infinite series, convergent for $Ex^2 < 1$, has up to five degrees of freedom. For odd function, $B = D = 0$; for even function, $A = C = 0$.

Earlier in this chapter, we have mentioned the practical limitations to the degree of fit; device cost and performance, and circuit complexity. To these must be added the difficulty (even in simple, low-cost configurations) of coping with the many degrees of freedom as the number of coefficients is increased. Three coefficients is a reasonable maximum for an engineer with a hand calculator, unless he is of a mathematical bent and enjoys solving this kind of problem. If mechanized stored-program calculators and computers are at hand, the device cost and performance limitations become more significant.

AN EXAMPLE: $y = f(x) \simeq \sin x$ $\left(0 \leq x \leq \frac{\pi}{2}\right)$

The appendix to this chapter, as mentioned, shows a number of equations and configurations that provide theoretical fits (of $\sin x$) to varying degrees of accuracy and suitability. We will derive here, as an example of the function-fitting process, the simplest of the approximations, a quadratic polynomial, using a single multiplier,

$$y = Ax + Bx^2 \quad [y(0) = 0] \tag{26}$$

and compare it with a more-accurate version, still using a single multiplier, but adding an implicit feedback

$$y = \frac{Ax + Bx^2}{1 - Cx} = Ax + x(Bx + Cy) \tag{27}$$

To obtain a trial set of coefficients in (26), substitute y and x (in radians) at two points. Let us use the end point, $x = \pi/2$, and an experimental intermediate point, $x = 1$ rad $= 57.296°$:

$$\sin \pi/2 = 1 \quad = A\,(\pi/2) + B\,(\pi/2)^2 \tag{28}$$

$$\sin 1 = 0.8415 = A + B \tag{29}$$

Solving simultaneously for A and B, we find that B = -0.3589 and A = 1.2004; hence, $y = 1.2004x - 0.3589x^2$.

In testing this approximation over the range of angles 0 to $\pi/2$,* maximum error (y - sin x) appears at 21.6° (error < 3.4% F.S.) and at 74.5° (error = - 0.96% F.S.) The error is zero at 0, 1 radian, and $\pi/2$ radians.

By choosing a different value of angle for intermediate zero error in (29) and solving for new coefficients, testing them, repeating, etc., it is possible to arrive at a "best" fit, with symmetrical maximum errors of about ±2.1%. This approximation is

$$y = 1.155x - 0.33x^2 \qquad (30)$$

Intermediate zero-error occurs at about 42.2°. The maximum errors occur at 17.4° and 68.6°. Error plots appear in Figure 14.

The block diagram of a configuration that would produce this approximation is shown in Figure 9. By adding an implicit feed-

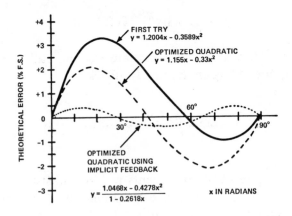

Figure 14. Errors of quadratic approximation to y = sin x $(0 < x < \frac{\pi}{2})$. Percent error = 100 (f(x) - sin x).

---

*Maximum error can be determined by an error plot, or by differentiating the error equation ($f(x)$ − sin x) and solving for the values of x at which the derivative is zero.

back, the configuration of Figure 12a, characterized by equation 27, is achieved, resulting in a reduction of the errors to less than ±0.5% (Figure 14).

The coefficients of equation (27) can be calculated by a similar process to that outlined above. However, the additional degree of freedom provided by C requires that two intermediate zero-error angles be determined, and we can expect *three* error maxima. The solution of three simultaneous equations makes the process somewhat more arduous, but still manageable, and the results are rewarding, since a fourfold reduction of error is obtained, at the cost of one additional op amp. The new equation, with optimized coefficients, is

$$y = \frac{1.0468x - 0.4278x^2}{1 - 0.2618x} \quad 0 < x < \pi/2$$

$$= 1.0468x - x(0.4278x + 0.2618y) \tag{31}$$

Maximum theoretical errors occur at 11.5° (0.42% F.S.), 47.1° (-0.44% F.S.), and 80.4° (0.44% F.S.), with zero error at 0°, 28°, 65.5°, and 90°.

It should be noted here that some reduction of maximum error could be obtained by allowing non-zero error at the end points.

In theory, this is tenable, though it necessitates an additional constant (and hence an additional simultaneous equation), but in practice it can be disastrous, since it makes calibration more difficult, and magnifies sensitivity to variations in device tolerances.

It should also be noted that errors discussed here are expressed in percentage of full-scale, rather than percent of the ideal value of $\sin x$. To convert the plotted errors to the latter form, they should be divided by $\sin x$. The ratio errors will be found to be larger, the error maxima will occur at different values of angle, and they will no longer be equal in magnitude. To test the approximations with the aim of minimizing ratio-to-ideal-value errors, the error function is $f(x)/\sin x - 1$.

A much better theoretical fit can be obtained, using two multipliers, with the additional bonus that it can be made to work in two quadrants (i.e., $-\pi/2 < x < \pi/2$). The simple cubic (Figure 10) gives $\pm0.6\%$ maximum error, but with implicit feedback, the theoretical error can be reduced to less than $\pm0.01\%$, which probably represents greater accuracy than might be expected from any of the devices currently available to implement the approximation at reasonable cost. That is, the error is limited by the devices, rather than the approximation. For single-quadrant fitting, a single $U \cdot V^m$ device (such as Model 433), with m set at 2.0, can replace the two multipliers.

Finally, using a $u \cdot v^m$ device in the same configuration, but with m set at a non-integral value, the maximum theoretical error can be reduced to $\pm0.15\%$ F.S. open-loop, and $\pm0.004\%$ with implicit feedback. The above approximations are all included in the appendix to this chapter.

PIECEWISE-LINEAR FUNCTION FITTING
(A Brief Introduction)

As Figure 15 shows, a nonlinear relationship is fit by summing gain segments (S_1, ΔS_2, ΔS_3, etc.) that have zero contribution until a threshold is crossed. Beyond the threshold, the output of a given segment contributes linearly. The nature of the ideal contribution (and the errors, too) is determined by both the location of the thresholds and the incremental gains attributed to the segments, as well as the means of implementation.

The simplest "diode function generators," or DFG's, as such devices are commonly termed, use segments that provide either zero or linear response (from the threshold to full-range input), as shown. Since the contributions accumulate, sharp reversals require a large amount of gain to overcome the accumulated gain of earlier segments. Circuits have been built using truncated segments to avoid the accumulation of gain, but they tend to lead to an unwieldy amount of circuitry; in addition, they require careful matching of break points to avoid "glitches" where one segment leaves off and another starts.

The conceptually-simplest segment is obtained with a biased diode and a precision resistor, but its temperature sensitivity leaves much to be desired. Chapter 3-5 discusses several more-practical approaches to individual segments. "Ideal-diode" op-amp circuits are an obvious possibility, because of their stable thresholds, sharp corners, and precise gains, as well as the low cost of operational amplifiers.

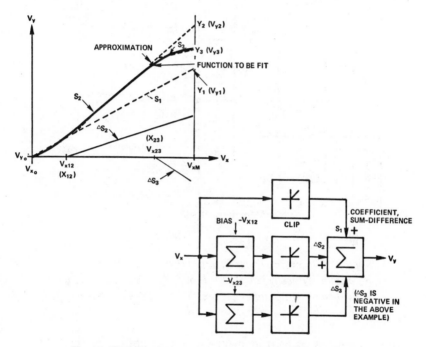

Figure 15. Basic 3-segment piecewise-linear function fitter.

Positive or negative contributions from individual segments are obtained by the use of a subtractive output circuit, usually consisting of an inverting output amplifier and an intermediate current inverter. For special-purpose function fitting (which comprises the great majority of applications), gains and thresholds may be computed, and fixed resistance values —with minor "tweaks"— are used. For general-purpose function fitting, potentiometers typically are used for setting each threshold (bias) and gain. To obtain the gamut of positive-to-negative gains, potentiometers straddle the positive and negative summing buses (Figure 16).

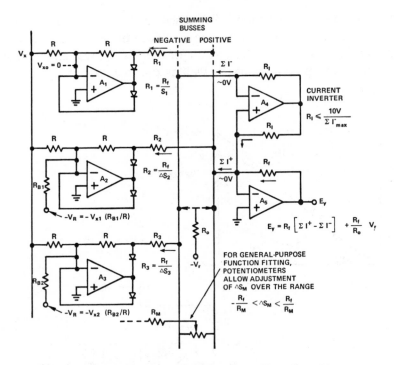

Figure 16. Piecewise-linear function fitter. For adjustable thresholds, typically the R_B's = R and $V_R = V_{xj}$, adjusted by individual potentiometers connected to a stable reference source.

Fitting functions using the piecewise-linear approach is, to begin with, a paper exercise. It can be done graphically or numerically. The graphic approach simply involves a large normalized plot of the function to be simulated. Draw the minimum number of connected straight lines to fit the curve to the required accuracy. (Figure 15). This fixes the break points, X_{12}, X_{23}, X_{34},... The straight lines are all extended to the ordinate corresponding to the maximum value of x, X_m. This permits the incremental gains to be computed accurately, even for short segments. The incremental gains, S_1, ΔS_2, ΔS_3, for the example shown are

$$\frac{Y_1}{X_m}, \frac{Y_2 - Y_1}{X_m - X_{12}}, \frac{Y_3 - Y_2}{X_m - X_{23}} \tag{32}$$

When the circuit has been assembled, final setting of the coefficients of the function*

$$V_y = V_{Y0} + S_1(V_x - V_{x0}) + \Delta S_2(V_x - V_{x_{12}}) + \Delta S_3(V_x - V_{x_{23}}) \quad (33)$$

can be simply done as follows:

1. Set the thresholds, $V_{x_{12}}$, $V_{x_{23}}$, etc.
2. With $V_x = V_{x0}$, and all gains at zero, set the output bias $V_y = V_{y0}$.
3. With $V_x = V_{xm}$, adjust S_1 for $V_y = V_{y_1}$, ΔS_2 for $V_y = V_{y_2}$, etc., in that order, keeping all gains at zero until the previous gains have been set. Use overall output attenuation (temporarily-reduced R_f), if necessary to keep V_y within reasonable limits.

Because all the adjustments are made with $V_x = V_{xm}$, there is a tendency for cumulative gain errors to be reduced. The function can now be checked at the intermediate points. If the breakpoints are not sharp, this factor should be taken into account on the paper plot before establishing the values of V_{y_1}, V_{y_2}, etc. The fit can be refined, if necessary, by minor adjustments to the thresholds, and repeating step 3.

A WORD ABOUT SUMMING-AMPLIFIER CONFIGURATIONS

Sum-and-difference amplifiers are well-known, having been discussed in just about every textbook and tutorial article on the basic applications of differential op amps.

In function-fitting applications, there is usually an amplifier that bears the brunt of summing a number of arbitrary inputs with a variety of gains of either polarity. The choice is usually between a differential amplifier and two inverting op amps in a subtracting configuration.

FOR SMOOTH APPROXIMATIONS, the inputs are usually taken from either op amps or nonlinear modules (or IC's), which have low-impedance operational-amplifier outputs. For these applica-

---

*$\Delta S_j = 0$ for $V_x - V_{xij} = 0$

tions, either the differential subtractor or the inverting subtractor of Figure 16 may be used. Because the inverting subtractor operates at ground level, it is more suitable for applications where gains must be adjustable, and the impedance changes associated with a specific gain adjustment must not disturb the other gains. However, if the gains are fixed, the differential subtractor is somewhat less costly; resistance ratios are easy to compute if the basic rule associated with Figure 17b is observed. See also *Electronics*, June 12, 1975, pp 125–126.

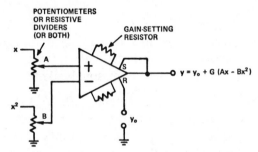

Figure 17a. Use of fixed-gain differential amplifier for 2-variable system with coefficients of opposing polarities (see Figure 9).

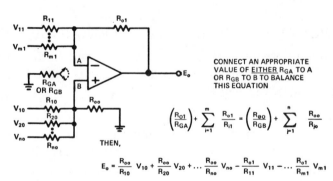

Figure 17b. Use of differential op amp for summing and differencing an arbitrary number of inputs with arbitrary fixed gains.

FOR PIECEWISE-LINEAR APPROXIMATIONS, the source impedance of the additive terms is usually nonlinear, being low in the conducting state and high in the open state. Thus, the isolation afforded by the summing-point of an inverting amplifier is not only desirable, but necessary, to avoid interaction.

A close look at Figure 16 will disclose the interesting fact that the positive input of A4, instead of being grounded, is connected to the summing-point of A5. In this connection, A4 serves as a *current inverter or reflector*, rather than as a voltage inverter. The purpose becomes clear if one considers that the summing point of A5 is loaded by the high output impedance of a current source rather than the usually-low resistance R_I, thus minimizing the closed-loop gain of A5, increasing bandwidth, and reducing the amplification of drift and noise.

PRACTICAL MATTERS

We have dealt with an "ideal building-block" approach to function-fitting, while appearing to ignore the practical characteristics of the building blocks that are to be used. The purpose was to avoid interjecting issues that, while highly appropriate, would tend to serve as digressions and dilute the main course of the argument. Also, each case must be analyzed in terms of the specific functional operations, their configuration, and the allowable ranges of input and output. Since the variety of permutations and combinations is broad, it is virtually impossible even to begin to cover them all in the detail they deserve in the available space. Nevertheless, this chapter would be incomplete if it didn't provide some guidance toward practical implementation of the ideas.

Practical considerations include scaling, component choice, errors (and their sensitivity to parameter variation and drift), response speed, and (for feedback configurations) stability.

If the reader is mathematically gifted, he will have little difficulty determining the scaling, the sensitivity to parameter tolerance (within the limits specified for the real devices and passive elements), or computing the approximate speed of response. While stability may be investigated theoretically, it is perhaps better to explore it experimentally.

More typically, the reader will have sufficient mathematical facility to compute the constants and perform the scaling (an example is given in the appendix), but may have difficulty with the mathematical formulations involved in error and stability analysis. In that case, as regards error analysis, he should perform a series of "brute force" calculations involving changes in the constants to find out which are the most tolerant and those that are the most sensitive. One approach is to make (say) a 0.1% change in a given constant, and determine the effect (magnitude and direction) on the maximum error. Another is to make (say) a 0.1% change in an input variable, and determine its effect on the output error.

It may be useful, in this day of calculators, to perform the computations of theoretical constants to many significant digits, then to round off, one digit at a time, until a significant effect on the error is seen. The theoretical examples given in this chapter and its appendix have all been worked out to an excessive number of places for the accuracy involved.

In any case, the reader should study the chapters in Part 3 that pertain to the *devices* to be used, and those in Part 4 that pertain to their application in the specific *operations* to be used (e.g., multiplication, division, logs, etc.). Naturally, familiarity with the data sheets for the devices actually to-be-chosen is essential, to be sure that they are physically and electrically compatible with the rest of the system and that there are no unpleasant surprises in the list of specifications.

Performance should always be checked on a "breadboard" that includes a facility for investigating response, stability, and the effects of parameter variations in those portions of the circuit that analysis (or intuition) suggests are most sensitive.

The resistors can be chosen at the next-lower (for example) standard values, with appropriate tolerances, temperature-sensitivity, and cost; the effects of parameter variations can be studied experimentally by "tweaking" incremental resistances connected in series.

Dynamic responses and stability are best studied experimentally, using large and small sine and square waves (and perhaps noise) biased at various levels. Response can often be improved, especially where subtraction is involved, by seeking to match the approximate responses of branches being summed by delaying the faster using an R-C lag circuit.

X-Y plots on the oscilloscope screen (input horizontal, output vertical), using sine- or triangular waves, can be quite helpful in observing the shape of the curve(s), determining that there have been no gross errors of fit, finding amplitude-sensitive instabilities, and (by frequency adjustment) determining simultaneously both amplitude and "phase" response. Not only the output behavior can be observed; one can also observe behavior at intermediate stages.

If the function involves a deviation from linearity, errors can be more-sensitively explored by subtracting the output from a signal proportional to the input, and observing just the deviation, plotted against the input.

Errors can also be determined point-by-point, using voltage sources and precise digital voltmeters, or by comparing X-Y plots on a chart recorder with hand-plotted curves. Where large numbers of identical functions are to be monitored, or trimmed, computer-test techniques can be brought into play in various ways, for example, by programming the input, and comparing the output with the stored "correct" values, either digitally (go-no), or with an analog readout established by computer graphics.

CONCLUSION

This chapter has sought to introduce the reader to the basic ideas and techniques relating to analog function fitting, and to encourage the increased application of low-cost nonlinear analog devices in calibration, compensation, and measurement. Some of these ideas will reappear, perhaps in amplified form, in subsequent chapters. The concentration in this chapter has been on the development of conceptual models. The following chapters will utilize some of these ideas in the context of their applications.

APPENDIX TO CHAPTER 2-1

Analog Approximations for sin x with Ideal Devices

1. Quadratic, one-quadrant, single-multiplier
   A. Explicit function: $y = 1.155\ x - 0.33\ x^2$

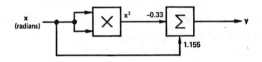

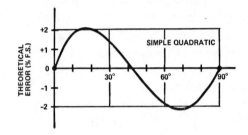

B. Implicit function: $y = 1.0468\ x - x\ (0.4278\ x - 0.2618\ y)$

$$= \frac{1.0468\ x - 0.4278\ x^2}{1 - 0.2618\ x}$$

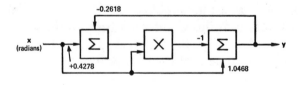

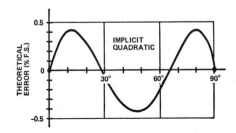

2. Cubic, two-quadrant, 2-multiplier, or one-quadrant UV^2

A. Explicit function: $y = 0.98252\ x - 0.14019\ x^3$

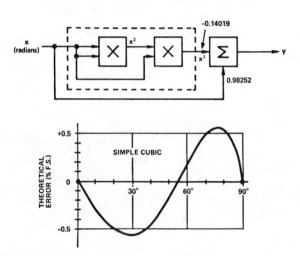

B. Implicit function:

$$y = 1.00042\ x - x^2\ (0.111382\ x + 0.056646\ y)$$

$$= \frac{1.00042\ x - 0.111382\ x^3}{1 + 0.056646\ x^2}$$

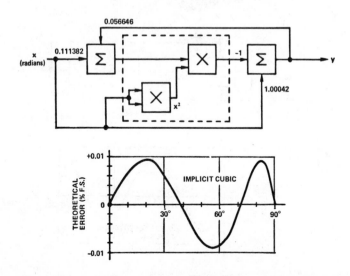

3. Non-integral exponent, one-quadrant, single UV^m

A. Explicit function: $y = 1.0095\,x - 0.169\,x^{2.7525}$

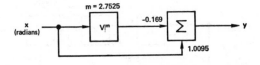

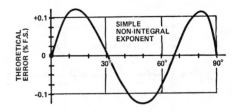

B. Implicit function:

$$y = 0.999642\,x - x^{2.02}\,(0.1073254\,x + 0.0604426\,y)$$

$$= \frac{0.999642\,x - 0.1073254\,x^{3.02}}{1 + 0.0604426\,x^{2.02}}$$

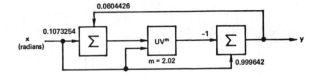

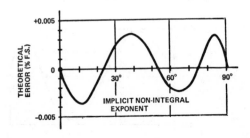

4. Extended angular range

A. Implicit cubic, 4-quadrant $-\pi \leqslant x \leqslant +\pi$, 2 multiplications

$$y = \frac{1.0287\,x - 0.10423\,x^3}{1 + 0.0904\,x^2} \text{, Circuit similar to 2(B)}$$

$$= 1.0287\,x - x^2\,(0.10423\,x + 0.0904\,y)$$

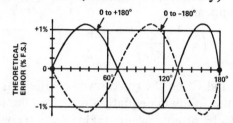

B. Implicit non-integral exponent, 2 quadrant, $0 \leqslant x \leqslant \pi$, single UV^m

$$y = \frac{0.9790\,x - 0.0657\,x^{3.36}}{1 + 0.0814\,x^{2.36}} \text{, Circuit similar to 3(B)}$$

$$= 0.9790\,x - x^{2.36}\,(0.0657\,x + 0.0814\,y)$$

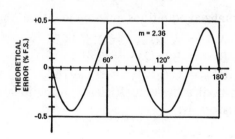

SCALING EXAMPLE

To demonstrate the application of the scaling principles discussed in the chapter, electrical coefficients for example 2B will be derived, using the following assumptions:

1. 10V full-scale input corresponds to $\pi/2$ radians.
2. 10V full-scale output corresponds to $\sin \pi/2$.
3. Multiplier transfer functions are $V_1 V_2/10 = E_{out}$

$$y = \frac{A\,x - B\,x^3}{1 + C\,x^2} = A\,x - x^2\,(B\,x + C\,y) \cong \sin x$$

where

$$A = 1.00042$$
$$B = 0.111382$$
$$C = 0.056646$$

Since the maximum value of y is 1, y is already normalized. However, though x is dimensionless (radians), it is not normalized. To normalize x to its maximum value, $\pi/2$, multiply and divide by $\pi/2$, wherever x appears.

$$y = A \frac{\pi}{2} \frac{x}{\pi/2} - \left(\frac{\pi}{2}\right)^2 \left(\frac{x}{\pi/2}\right)^2 \left[B \frac{\pi}{2} \frac{x}{\pi/2} + C y\right]$$

If we let A′, B′, and C′ be the (unknown) coefficients of the electrical equation, the following equation describes the ideal performance of the electrical equivalent, taking into account the multiplier transfer functions.

$$E_y = A'V_x - \frac{V_x^2}{10} \cdot \frac{B'V_x + C'E_y}{10}$$

Normalizing,

$$\frac{E_y}{10} = A' \frac{V_x}{10} - \left(\frac{V_x}{10}\right)^2 \left[B' \frac{V_x}{10} + C' \frac{E_y}{10}\right]$$

Because the normalized equations must be identical,

$$A' = A(\pi/2) = 1.571456$$
$$B' = B(\pi/2)^3 = 0.431693$$
$$C' = C(\pi/2)^2 = 0.139768$$

A circuit that embodies these coefficients, using ideal multipliers, op amps, and resistors, is:

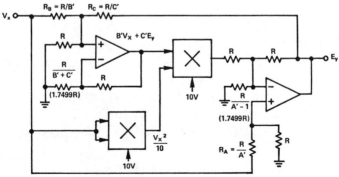

$R_B = 2.31646R, \; R_C = 7.15471R, \; R_A = 0.636353R$

Time—Function Generation

Chapter 2

Two products that revolutionized electronic instrumentation in the '30's and '40's were the oscilloscope and the sine-wave generator. The latter applied stimuli to systems or devices under test; the former permitted observation and time-domain measurement of the response. Since then, time-function generators as instruments have become greatly sophisticated; today, digitally-programmed sine- and square-wave, pulse, triangular, and even ROM-determined arbitrary function generators are available, in speeds from mHz to MHz.

As the uses of general-purpose function generators spread, the possibilities for low-cost, compact, in-house-designed *special-purpose* function generators for use in specialized equipment became apparent. The availability of operational amplifiers at low cost enabled some of these possibilities to become realities, and now the collateral availability of low-cost circuit elements with controlled, predictable nonlinearity should stimulate the greatly-increased use of function generators in OEM equipment.

A few examples of applications for function generation include establishing "profiles" (temperature, flow, velocity) in control systems, adjusting programmed parameters in test and instrumentation systems, and providing time bases of special form (e.g., logarithmic) in chart-recorder and oscilloscopic readout devices. Other applications for nonlinearity include variable-frequency polyphase oscillators, voltage-controlled filters, and low-cost signal generation with precise control of amplitude, frequency, and/or phase. These last include, of course, the classical sine-, square-, and triangular-wave generators, variable duty-cycle pulse generators, and one-shots, as well as the famous phase-locked loop. Also, one should not forget random-noise generators.

In this chapter, we discuss some of the principles of function generation and suggest ways of accomplishing a few basic functions with available standard building blocks. There exists, of course, a voluminous body of publications describing a gamut from simple circuits—involving transistors and passive elements—to the catalogues of manufacturers devoted to test instrumentation. Our aim is to neither replace nor surpass these efforts. Rather, it is to provide the designer with a modest indication of the range of rôles that controlled nonlinearity can play in function generation, with the thought that it will form a respectable complement to his bag of design tools, tricks, and ideas.

FUNCTION GENERATORS ARE MULTI-FACETED

It is possible to conceive of an extremely-wide range of function generators, classified in many different ways. What is common to them all is the use of nonlinearity: it is quite difficult to imagine a means of independently generating time functions, starting with a dc power source, without in some purposeful way involving non-linear devices.

While this chapter deals with only a few specific examples of function generation, the following inclusive inventory of function-generator properties may be helpful to the reader who is seeking insight into remote (as well as better-known) aspects of this all-embracing field.

1. PERIODICITY: Aperiodic (single-shot), Stationary, Modulated, Random. Although the familiar connotation of "single-shot" is a pulse generator that delivers a single pulse in response to a stimulus, the term also should suggest such possibilities as a single half-sine, or a damped exponential train, or an arbitrary velocity or torque profile for a dynamometer test. *Stationary* waveforms are those having statistical properties that do not change with time. In practice, if a determinate waveform's amplitude, frequency, phase, or shape, or a random waveform's mean, variance, amplitude, distribution, and frequency spectrum are constant over a lengthy period

before, during, and after a measurement it is involved in, it may be considered stationary. *Modulated* refers to the variation of some property of the waveform during any observation interval (or from interval to interval) in response to a signal, for example, amplitude, phase, frequency, pulse-width, pulse position, presence or absence. It specifically includes voltage-to-frequency conversion, which can be viewed either as generation of a signal having a voltage-determined frequency, or modulating a signal about a fixed frequency.

2. SPEED: Very Low (fractions of 1Hz), Low, Audio, High, Video (> 1MHz). These distinctions are largely qualitative, but they are important insofar as they affect the choice of components or approach, the criticality of design, the limitations of accurate behavior, and the difficulty of use and measurement. The easiest portions of the spectrum to design for are in the middle, from about 1Hz to 30kHz. Suitable passive elements are small and cheap, and active devices have low drift and noise, as well as reasonable bandwidths. Thermal effects, that plague the low end, and stray capacitance and inductance, that complicate life at the high end, are rather manageable in the middle. Since this is not a complete text on the design of function generators, most of the specific circuits suggested have their best performance in the low-to-audio range of speeds and frequencies.

3. SHAPING: Simple vs. Complex. Simple functions are those that Nature allows to be achieved (in concept) with a minimum of basic hardware. They include sine-waves, as produced by resonant elements, square waves (produced by switching), triangular waves (often a by-product of square waves, or vice versa), exponential waves (also a by-product of square waves), and pulse trains. *Complex* functions involve operations on simpler functions, including modulation, filtering, and nonlinear function fitting (analog or digital) applied to simple waveforms. Analog function fitting involves ramps and function fitters; digital involves pulse trains, read-only memories (ROM's), and D/A converters with appropriately-filtered output; both can be combined to advantage (see Figure 11, this chapter). Most of today's commercial sine-square-triangle generators obtain the sine in a *complex* fashion: a triangular wave is applied to a fitted sine operator (Figure 1).

Random noise can be generated simply (for example, by amplifying resistor or junction noise) or in complex fashion (by generating a pseudo-random waveform having sufficiently-low autocorrelation, using a pulse train, tapped shift-register, exclusive-or'd feedback, D/A conversion, and filtering).

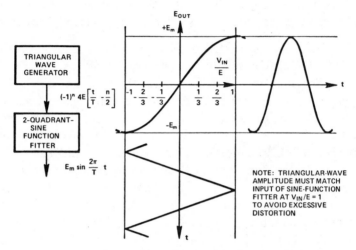

Figure 1. Generating a sine wave by function fitting

4. CONTROLLED PARAMETERS: Amplitude, Frequency, Phase, Mark/Space Ratio, Ranges, Shape, Measure. These parameters are affected by the manner of function generation; their fidelity to the desired behavior comprises the basic performance specifications of the function generator. Mark-space ratio takes on a broader meaning than just pulse on-off time: as a measure of symmetry, it also refers to a ratio between up-going and down-going intervals of ramps and sweeps. Departure from specified *shape* may be specified as "distortion." *Measure* indicates a form of average measurement that may be used instead of, or in addition to, *amplitude,* to characterize the waveform; for example, RMS or mean absolute-value. Crest factor is the ratio of peak amplitude to RMS. (With random noise, it is easier to measure RMS repeatably, than to observe peak amplitudes; the probabilities of various crest factors are a function of the noise distribution.)

5. FORM OF PARAMETER VARIATION: Fixed, Manually Adjustable (Continuously, or continuously, in ranges), Discretely, Digitally-Programmed, Auto-Ranged, Continuously Variable (modulated). This is entirely determined by the application, but it has profound effect on the design. Depending on this choice, for example, a parameter may be set by a resistor, by a pot (and switched fixed-resistors), by resistance-decade switches, by a D/A converter, or by an analog multiplier.

6. PARAMETRIC ACCURACY CLASS: 0.01%, 0.1%, 1%, Externally-Calibrated. This category is a catchall that includes such terms as "absolute accuracy," relative accuracy, precision, repeatability, stability. The above numbers represent orders of magnitude of *error* and each of these desirable characteristics is generally specified by a small number that represents the *deviation* from perfection. A function generator designed for a given application may have parameters that differ widely in error magnitudes; for example, frequency may be held to within parts-per-million, but amplitude variations and shape distortion may be of the order of 10%. The widespread availability of low-cost multipliers and D/A converters has made it possible for test-systems to be built that depend, not on costly fixed calibration of every generator used, but rather on a single programmed reference against which all generators are automatically calibrated and computer-adjusted to the desired settings before each measurement for which each is used.

7. INDEPENDENCE: Free-Running, Synchronized, Slaved. A free-running function generator depends for its accuracy and timing entirely upon its own internal reference sources, and to some extent (usually minimized) on the supply voltage. A synchronized device is allowed to free-run most of the time, but is from time-to-time brought "up-to-speed." A slaved device follows its speed reference, cycle by cycle.

8. FREQUENCY-DETERMINING ELEMENT: Resonant, Level-Controlled, External. Function generators that use internal crystal oscillators, Wien bridge, phase-shift or integrator-loop oscillators are *resonant*. Those that switch phase when a threshold has been

crossed are *level-controlled*. Though level-controlled types (multivibrators, one-shots, etc.) can be low in cost, their timing usually depends on an RC time constant, a reference supply, and a comparator; resonant types depend only on linear parameters, such as RC time constant. All types must of course take into account amplifier phase shifts and parasitic reactances. The amplitude-control arrangements for resonant types affect the damping, and may thereby marginally affect the frequency.

BASIC TRIANGULAR/SQUARE-WAVE GENERATOR

Figure 2 shows the configuration common to many varieties of level-controlled oscillators. It consists of a hysteretic comparator and an integrator. The output of the hysteresis element has two stable states, E_{o+} and E_{o-}; it switches to E_{o+} when the input exceeds V_{1+}, and it remains in that state until the input is less than V_{1-}, whereupon it switches to E_{o-}. It remains in that state until the input once again exceeds V_{1+}.

Suppose that the output has just switched to E_{o+}; it is applied to the integrator input. The integrator's output, starting from V_{1+}, decreases linearly with time at a rate E_{o+}/RC. At the end of the interval

$$\Delta t_1 = RC \frac{V_{1+} - V_{1-}}{E_{o+}} \qquad E_{o+} > 0 \qquad (1)$$

the output of the integrator is V_{1-}, and the output of the hysteretic comparator switches to E_{o-}. The integrator's output now *increases* linearly with time at the rate $-E_{o-}/RC$, until the output of the integrator is once again V_{1+}, which occurs at the end of the interval

$$\Delta t_2 = RC \frac{V_{1+} - V_{1-}}{-E_{o-}} \qquad -E_{o-} > 0 \qquad (2)$$

The period is

$$T = \Delta t_1 + \Delta t_2 = RC \left(\frac{V_{1+} - V_{1-}}{E_{o+}} \right) \left(1 - \frac{E_{o+}}{E_{o-}} \right) \qquad (3)$$

The frequency is

$$f = \frac{1}{T} = \frac{E_{o+}}{\left(1 - \frac{E_{o+}}{E_{o-}}\right)\left(V_{1+} - V_{1-}\right) RC} \tag{4}$$

The mark-space ratio of the square-wave is

$$M/S = \Delta t_1 / \Delta t_2 = - E_{o-} / E_{o+} \tag{5}$$

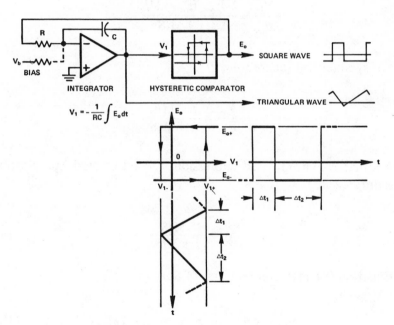

Figure 2. Basic triangular/square-wave generator

The peak-to-peak amplitudes of the triangular wave and the square wave are $(V_{1+} - V_{1-})$ and $(E_{o+} - E_{o-})$, respectively. For amplitude symmetry of the triangular wave, $V_{1+} = - V_{1-}$; For amplitude symmetry of the square wave, $E_{o+} = - E_{o-}$.

If a symmetrical square wave is desired with a mark/space ratio other than unity, a suitable bias V_b may be added to the integrator input. This bias is added to both E_{o+} and E_{o-} for computing the periods and mark/space ratios. However, the output levels of the hysteretic comparator are unaffected. For example, if $E_{o+} = - E_{o-} = +10V$, and a mark/space ratio of 2:1 is desired,

$$2 = \frac{-(E_{o-} + V_b)}{(E_{o+} + V_b)} = \frac{10 - V_b}{10 + V_b} \tag{6}$$

Solving, $V_b = -10/3$ volts, or $-E_{o+}/3$. In general,

$$\frac{V_b}{E_{o+}} = \frac{-E_{o-}/E_{o+} - M/S}{1 + M/S} \tag{7}$$

For symmetrical square waves,

$$\frac{V_b}{E_o} = \frac{1 - M/S}{1 + M/S} \tag{8}$$

Another commonly-used expression, related to mark/space ratio, is *duty cycle*, η,

$$\eta = \frac{M/S}{1 + M/S} \tag{9}$$

Equation (8), rewritten in terms of duty cycle, is

$$\frac{V_b}{E_o} = 1 - 2\eta, \text{ or } V_b = E_o - 2\eta E_o \tag{10}$$

If the bias voltage added at the integrator input is a constant, E_o, less a variable, $V_m = 2\eta E_o$, the duty-cycle will be a linear function of V_m (linear pulse-width modulation). Unfortunately, the frequency will not remain constant; it will be a function of V_m.

An additive bias at the integrator input may also be used to obtain time symmetry (M/S = 1) if the comparator has asymmetrical output levels. An additive bias is essential to meet the constraints of (1) and (2) if the comparator has unipolar output, e.g., if its outputs are in the TTL logic range (say, 5V & 0.5V).

If the output is symmetrical, the frequency can be linearly controlled by introducing a multiplication between the comparator output and the integrator input (Figure 3). For manual control, the "multiplier" can be a potentiometer; for voltage-control, it can be a multiplier; and for digital control, it can be a multiplying D/A converter. If a multiplier with a 10V scale constant is used, the frequency is

$$f = \frac{V_f\,E_o}{20\,RC} \cdot \frac{1}{V_{1+} - V_{1-}} \qquad (11)$$

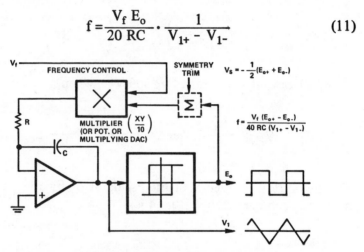

Figure 3. Controlling the oscillator frequency

OPERATIONAL AMPLIFIER AS HYSTERETIC COMPARATOR

Figure 4 shows a simple operational amplifier circuit, using positive feedback to develop hysteresis. Amplifiers that limit "hard", within a volt or so of the power supply, are especially useful for these circuits. For greater stability, a temperature-compensated zener diode regulator circuit could be used. This stabilizes, not only the amplitude of E_o, but also the frequency and mark/space ratio, and the triangular-wave amplitude, all of which depend on E_o.

To illustrate how it works, consider that the output has just switched to E_{o+}, as V_1 reached the threshold V_{1+}, V_1 then decreases linearly and will continue to do so until the voltage at the amplifier's positive input terminal goes negative. That occurs when V_1 reaches V_{1-}

$$-V_{1-}\left(\frac{R_2}{R_1+R_2}\right) = E_{o+}\left(\frac{R_1}{R_1+R_2}\right) \quad (12)$$

The output switches to E_{o-}, the integrator's output starts back up, and continues to climb until the amplifier's input terminal goes positive (when V_1 reaches V_{1+})

$$V_{1+}\left(\frac{R_2}{R_1+R_2}\right) = -E_{o-}\left(\frac{R_1}{R_1+R_2}\right) \quad (13)$$

Thus, the output switches at V_{1+} and V_{1-}, when

$$V_1 \geqq -\frac{R_1}{R_2}E_{o-} \text{ and when } V_1 \leqq -\frac{R_1}{R_2}E_{o+} \quad (14)$$

The theoretical frequency of the oscillator of Figure 2, using the hysteretic comparator of Figure 4 is

$$f = \frac{R_2}{R_1} \cdot \frac{1}{RC} \cdot \frac{-\dfrac{E_{o+}}{E_{o-}}}{\left(1-\dfrac{E_{o+}}{E_{o-}}\right)^2} \quad (15)$$

The triangular-wave amplitude is

$$\left(V_{1+} - V_{1-}\right) = \frac{R_1}{R_2}\left(E_{o+} - E_{o-}\right) \quad (16)$$

The frequency may be controlled independently of the triangular of square-wave amplitudes, by adjusting RC, or by placing a gain adjustment in the feedback path to the integrator input. Symmetry of the triangular-wave amplitude may be controlled by introducing a bias current at the hysteresis summing point, via resistor R_0, connected to a voltage source of appropriate polarity. For fine trim, if the comparator output is nearly symmetrical, the adjustment may be connected between the supplies, with R_0 fairly large. If, on the other hand, a large offset must be dealt with (as when the comparator output swings between 0.5 and 5V), the adjustment may be a variable resistance in series with a fixed resistance.

For that case, if V_s is a negative voltage, at symmetry,

$$-\frac{V_s}{R_0} = \frac{1}{2}\left(\frac{E_{o+} + E_{o-}}{R_2}\right) \tag{17}$$

The biasing of the triangular wave doesn't affect its amplitude, frequency, or mark/space ratio.

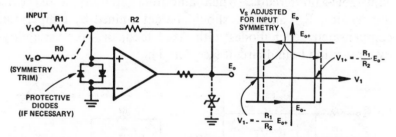

Figure 4. Operational amplifier as hysteretic comparator

A PRACTICAL OSCILLATOR CIRCUIT[1] (Figure 5)

This circuit, using low-cost components, provides square waves of about ±14V, with near-unity mark/space ratio, and triangular waves of about ±10V, with reasonable symmetry, at about 100Hz, for the values given. Frequency, triangular-wave amplitude, symmetry, and mark/space ratio may all be adjusted, by the means discussed above. Because A1 is a FET-input amplifier, frequencies as low as 0.1Hz and less are feasible, using large values for C and R

[1]"Triangular and square-wave generator has wide range," by R.S. Burwen, *EDN* Magazine, December 1, 1972.

(10MΩ for 0.1Hz) and/or an attenuator ahead of R. Bias current and offset voltage in A1 act in the same way as an external bias of $V_{os} + I_b R$, producing a slight modification of the mark/space ratio. Square-wave rise time is about 1.5μs, and fall time is about 0.5μs.

The frequency is affected by the saturation voltages of A2, and by the power-supply voltages. However, as equation (15) can show, sensitivity to symmetrical power-supply variations is quite small, and even individual variations as large as 20% cause no more than a couple-of-percent change. If stable passive components are used, a frequency stability of ±0.02%/°C is attainable. Capacitor C is preferably a polycarbonate type for stability, and also to ensure linearity of the triangular wave.

Although frequency stability is excellent, amplitude stability depends on the power supplies, the output-transistor saturation voltages, and the load (and their variations with temperature). For most applications, however, the outputs would be followed by adjustable-gain circuits. When amplitude stability is of critical importance, E_{o+} and E_{o-} should be determined by temperature-compensated zener diodes with fixed load, or —for variability— by a precision bound circuit (see Part 1).

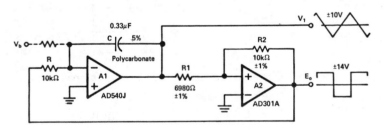

Figure 5. Practical oscillator circuit

Besides the inherent square-wave and triangular wave, and the variety of pulse widths, other functions, including sine waves, may be generated by feeding the output of the triangular-wave generator into one of the many varieties of function fitter described in Chapter 2-1. Trapezoidal waves may be generated by feeding the triangular-wave output into a set of bounds. Triangular pulses may be produced by feeding the triangular wave into a dead zone (Figure 6). Exponential responses are obtained with simple high- or low-pass filters.

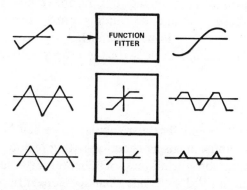

Figure 6. Shaped triangular waves: arbitrary, trapezoid, tri-angular pulses

ONE-SHOT (Figure 7)

The circuit of Figure 5 may be slightly modified to obtain a one-shot. A diode across the feedback capacitor prevents the output of A1 from ever becoming sufficiently negative to switch the output of A2 to E_{o-}. A capacitively-coupled positive-going logic pulse, applied to the negative amplifier input, starts the cycle by switching the output to E_{o-}. The integrator output ramps upward until it reaches V_{1+}, switching the output of A2 to E_{o+}. The integrator then ramps back through zero and stops at one negative diode-drop. The circuit is then ready for the next *start* pulse. The driving rate should be slow enough to allow each cycle to be completed, and the pulses should be narrow, compared to the cycle time.

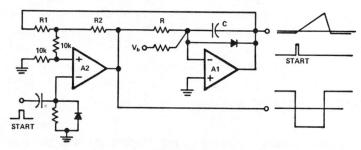

Figure 7. One-shot function generator

The ramp output may be used as an input to a functional operator to generate arbitrary functions of time that occur but once. The diode drop can be biased out in the input stages of the associated circuitry. If the descending ramp is not desired, the change-of-state of A2 can operate a switch to disconnect the output of A1. (The change back, with the *start* pulse, can reconnect it.)

SINE-WAVE OSCILLATOR

We have already noted that a triangular wave can be shaped to a low-distortion sinusoid, using function-fitting techniques. Some of the smooth-function techniques of Chapter 2-1 may yield considerably more-faithful sinusoids than the conventional piecewise-linear diode shaping networks. It has also been noted that the frequency (and symmetry) of such oscillators is dependent on voltage thresholds, as well as RC time constants.

For some purposes (for example, if the waveform is to be differentiable, with low distortion), an oscillator that relies solely on passive components for frequency control may be more desirable. The class of oscillator that uses RC networks includes the Wien bridge, phase-shift oscillators, twin-T oscillators, and *state-variable* oscillators.

This last type is an analog-computer equivalent to an L-C circuit. It consists of two integrators in a negative-feed-back loop, with damping appropriate to maintain amplitude control (Figure 8). It has some interesting features: first, since integrators have a fixed 90° phase shift, with unity gain at the frequency of oscillation, it is inherently a two-phase oscillator, producing both sin ωt and cos ωt; second, two analog multipliers or dividers will allow a voltage to set (or modulate) the coefficients that determine frequency or period (respectively); third, the system can either free-run or be started at any arbitrary point in the cycle, determined by preset initial conditions; fourth, the damping can be set to produce exponentially-decreasing or increasing waveforms.

For the free-running case (stationary amplitude), a slight amount of regenerative damping ensures that the oscillation will build up.

When one of the outputs reaches a level established by comparison
with an amplitude reference, degenerative damping is applied at
the peaks, reducing the last increment of buildup, and maintaining
successive peaks at the same amplitude. While this introduces some
distortion, it is integrated (smoothed) before appearing at one of
the outputs and is integrated again before appearing at the other
output.

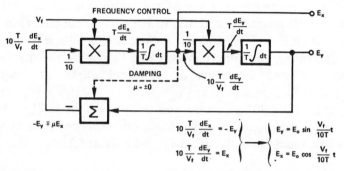

Figure 8. Block diagram of variable-frequency 2-phase sinusoidal
oscillator. For fixed frequency, replace multipliers by coefficients. If not free running, apply initial conditions to integrators
in SET. If driven from summing-point as 2nd-order filter, $\mu > 0$,
E_y is low-pass output, E_x is band-pass, and output of Σ is
high-pass.

A PRACTICAL 2-PHASE SINE-WAVE OSCILLATOR

Low-cost, high-performance complete-on-a-single-chip IC multipliers, such as the AD533, make it feasible to build oscillators
having two-phase sine-wave output, with frequency controllable
by a voltage. The frequency may be varied over a wide range,
depending on the dynamic range of the multiplier, for frequency-
sweep applications, or it may be centred about a fixed frequency
for highly-linear frequency modulation. While an IC multiplier is
used for the example of Figure 9 because of its low cost, there is
no inherent barrier to using a wideband multiplier, such as the
429, for increased bandwidth, or a high-accuracy multiplier for
increased low-frequency accuracy and resolution, or even multiplying D/A converters, for digital control of frequency.

The oscillator shown in Figure 9[2] delivers a 2-phase sine-wave output tuneable over a 10:1 frequency range by means of the DC control voltage. The output amplitude is stabilized by zener reference diodes at about 7Vrms and maintained constant within 1dB over the range of frequencies.

The oscillator system consists of two integrators, A1 and A2, and a unity-gain inverter, A3, forming a negative feedback loop. The effective time constants (T = a RC) of the integrators are varied by a pair of multipliers, M1 and M2, which serve to (in effect) increase the conductance of R1 and R2 as the control voltage is increased, thus decreasing the time constant and increasing the natural frequency. Viewed in terms of gain and phase, at frequency $f_n = 1/(2\pi aRC)$, with $a = 1$, ($V_f = 10V$, $a = V_f/10V = 10/10$) both integrators have 90° phase lag and unity gain, the multipliers also have unity gain, and there are three sign inversions, all of which looks like a loop gain of $1\underline{/0°}$ at f_n (and only at f_n).

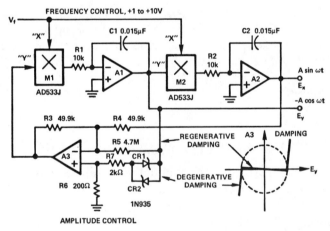

Figure 9. Practical version of configuration shown in Figure 8

To ensure sufficient regeneration to start and maintain the oscillation, a small amount of positive feedback is fed from the output of A1 through R5 to the input of A3. This causes the oscillation to build up until one or both of the zener diodes CR1, CR2, begin to

[2]"Frequency Modulator" by R.S. Burwen, *Analog Dialogue*, Volume 5, No. 5.

conduct at the tips of the waveform and produce increased negative feedback via the positive input of A3. The positive feedback must be kept small enough to provide buildup at a reasonable rate without requiring a large amount of negative feedback to keep the amplitude under control, since the zener diodes introduce some distortion. (Fortunately, this small distortion is integrated once in A1 and again in A2, so that the output of A2 is quite clean, and that of A1 is "oscilloscope-clean.")

With the values shown, the oscillator can be tuned from 100Hz to 1kHz. Distortion at the cosine output was measured at 0.74% at 100Hz and 0.46% at 1kHz. At the sine output, distortion was 0.64% at 100Hz and 0.18% at 1kHz. Distortion, especially at the lower end of the tuning range, is somewhat affected by the nonlinear feedthrough in the multipliers.* Multiplier nonlinearity and drift (using low-cost IC's) placed a limit on the useful tuning range.

It is easy to modify this design to operate with frequency modulation about a fixed frequency. For example, to operate at 1kHz, with ±10% frequency variation linearly controlled by V_f (±10V range), change R1 and R2 to 100kΩ, and add 10kΩ resistors between the output of A3 and the input of A1, and between the output of A1 and the input of A2.

SWEEP CIRCUITS

Linear sweeps, like the output of a triangular-wave generator, are usually produced by an integrator within a feedback loop; but instead of a linear retrace, a fast return is obtained by "dumping" the capacitor charge through a switch. The retrace is blanked (oscilloscope) or the pen lifted (recorder) during the retrace interval.

Nonlinear sweeps are desirable for some purposes. For example, in swept-frequency measurements, either the sweep may be logarithmic, or the frequency may be varied exponentially by applying an exponential input to control a variable-frequency oscillator.

---

*This distortion can be reduced by use of the "cross-feeding" technique for improving multiplier linearity, as discussed in Chapter 3-2.

(An ordinary linear display sweep may be used, since equal increments of time will represent equal ratios of frequency.) Starting at the high-frequency end, such a sweep can be obtained by passing a step through a simple RC coupling element (Figure 10). The output is $V_{in}e^{-t/RC}$. If it is used to control a frequency, the frequency will decrease by equal ratios in equal intervals of time.

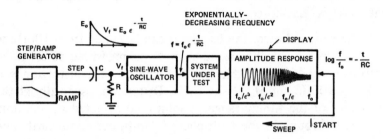

Figure 10. Use of logarithmic sweep for frequency-response measurements.

In this example, and throughout the chapter, it is tacitly considered that the rate of variation of "frequency" is so slow compared to variations *at* the frequency being controlled that there is little difficulty with the assumption that the waveform is stationary. Since this chapter deals with techniques rather than analysis, it must be assumed that for clearly interactive situations, in which frequency must be defined incrementally, the reader has an understanding of the mathematical implications and can deal with their consequences. The circuits, little caring about the complexity of the mathematics that describes their behavior, will perform nevertheless.

MARRYING ANALOG AND DIGITAL CIRCUITS

It is possible to generate linear sweeps of precisely-maintained amplitude and frequency, with arbitrary resolution, independent of the properties of capacitors and analog comparators, by driving a D/A converter with a counter that is itself driven by a train of pulses from a clock generator. The clock may be crystal-controlled, with frequency adjusted by counting down or a binary-rate multiplier, or it may be a simpler circuit.

Frequency depends only on the ratio of the clock rate to the total number of counts used, and amplitude can be scaled at the output of the converter. If the converter is a multiplying type, the sweep amplitude can be scaled by a voltage. The upper limit on speed is determined by the maximum clock rate, resolution, and settling-time of the converter. The converter circuitry should be "glitch-free," that is, there should be no large spikes at major-carry transition points (e.g., from 0 1 1 1 1 to 1 0 0 0 0).

A digitally-generated sweep, of appropriate resolution, with (or without) filtering may be applied to an analog function-fitter circuit (Chapter 2-1) to generate waveforms of any shape, in the same way that a purely-analog sweep might be applied (Figure 11). This is often a good deal less costly and more versatile than using a read-only memory (ROM) for shaping. Yet, like a ROM, it has the added possible benefit of being completely under the time control of the system. Not only is it slaved to the clock frequency– it can be started, stopped, *held* indefinitely, and reset, with simple logic circuitry. This would appear to be a happy combination of the best of analog and digital technology, characterized by simplicity, low cost, and versatility.

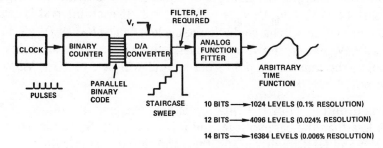

Figure 11. Arbitrary analog waveforms synchronized to digital clock

VOLTAGE-TO-FREQUENCY CONVERSION

The circuits of Figure 3 and Figure 5 are, in a sense, voltage-to-frequency converter circuits, but they have several limitations. Perhaps the most serious is that the range of continuous variation is limited, at best, to about 100:1. Also, they cannot be easily synchronized without some means of "dumping" capacitor charge.

Figure 12 shows a more-sophisticated circuit that is capable of 1:10,000 resolution and nonlinearity, gain stability (with external reference) to within 10ppm/°C, and practically negligible sensitivity to the dc power supplies. It is operated by a 100kHz clock, to which the output is synchronized. For an input variation of 0 to -10V, the output frequency varies proportionally from 0 to 50kHz. A synchronized 50kHz pulse train is also available, as a frequency reference.

$$f = 5 \times 10^4 \left(\frac{V_{in}}{-V_r} \right) \tag{18}$$

The AD301A amplifier operates in the linear mode; that is, the negative input terminal tracks the voltage at the positive input. Therefore, the current through R, equal to V_{in}/R, flows toward the capacitor. Q1A is a switching transistor that either has zero collector current, or a current equal to $V_r/\frac{1}{2}R$, flowing *away* from the capacitor. When Q1A turns on, the capacitor is charged by the net current $(2V_r + V_{in})/R$; When Q2 turns off, the capacitor discharges at $^-V_{in}/R$. Thus, to maintain equilibrium, for each time Q1 charges, the number of equal intervals spent discharging must be $(2V_r + V_{in})/-V_{in} = -2V_r/V_{in} - 1$. If each interval is 10$\mu$s, the total time per charge-discharge cycle is 10μs $(1 - 2V_r/V_{in} - 1) = -20V_r/V_{in}\mu$s. If, now, each charge-discharge cycle produces a pulse, the number of pulses per second will be $5 \times 10^4(-V_{in}/V_r)$, as noted in (18).

When output Q of the flip-flop is low, the emitter voltage of Q2 is less than the base voltage, and it is turned off. Since the bases of Q1A and Q1B are driven together, and the emitter circuitry is identical, their collector currents should track rather precisely. Thus, the collector current of Q1A should be equal to $2V_r/R$. When output Q of the flip-flop is high, Q2 is able to conduct; it furnishes enough current through the emitter resistor to raise the emitter voltage of Q1 above the base line, turning off the collector current.

Whenever the output of A1 is slightly below the threshold of the D input of the flip-flop, the next pulse causes $\overline{Q}$ (the output of the circuit) to go high. It also causes current to flow through Q1A, and a large increment of charge to raise the output of A1 by $\Delta V_1 = I_1 \Delta t / C$ (where $I_1 = (2V_r + V_{in})/R$). The next clock pulse finds the output of A1 high, $\overline{Q}$ goes low, and Q goes high, cutting off the flow of current through Q1A. The decrease of charge during this interval is $\Delta V_2 = I_2 \Delta t / C$, (where $I_2 = -V_{in}/R$). At the next clock pulse, unless $V_{in} = -10V$, the output of A1 is still high, $\overline{Q}$ remains low, and Q remains high, allowing a further decrease of charge. This process is repeated until the output of A1 is again slightly below the threshold of the D input, a cycle has been completed, and a new cycle begins.

When $V_{in} = -10V$, the charge and discharge periods are equal in number, and the output is at a 50kHz rate. The second half of the flip-flop counts down by 2, so that the reference pulse train is also at 50kHz.

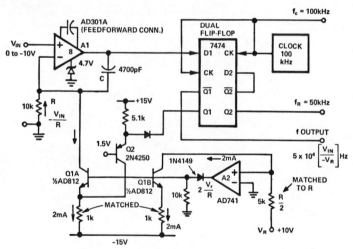

Figure 12. High-accuracy synchronized voltage-to-frequency converter

CONCLUSION

This chapter has sought to give an overview of function generators in general, and to provide details of a few useful circuits in particular. The objective is to arouse interest in special-purpose function generation, with particular emphasis on the cooperative rôles of linear and nonlinear analog devices and the possibilities of their fruitful collaboration with digital circuits.

Instruments & Data Acquisition

Chapter 3

The design of instruments and "front-end" circuitry for data-acquisition systems is perhaps the area of greatest prospective payoff to users of nonlinear computational devices.

The circuits discussed in this chapter produce analog information that may be either directly read out by a human operator, or digitized and transmitted from a remote location to a control center, without requiring further interpretation. The analog data-reduction circuits covered here are simple and more-or-less universally applicable. The closely-related treatment of measurement and control circuits in Chapter 2-5 complements (and to some extent overlaps) the material presented here, and is somewhat similar in basic form; but it tends to include more-ramified analog computation, applied to situations that are more specialized.

ANALOG DATA REDUCTION

The primary goal of the configurations discussed here is to *reduce* data by analog techniques. To *reduce* data, as used here, means to extract significant information from one or more analog inputs, and transmit it —either to the human eye or to an interface— as meaningful, compact, well-paced data.

For a single variable, data-reduction can consist of extracting the peak, average, RMS, mean-square, or some other measure that is consistent in the presence of large numbers of individual data points. If the process is *stationary*, it may involve an average; if one-shot, it may call for a peak, integral, or final-value. If the measurement is nonlinear, it may call for *linearization*; if wide-ranging, it may call for *compression*.

If the measured data comprises many variables, further combination may be in order: summing and differencing (linear, vector, or root-square), ratios or products (linear, log, or otherwise), multiplexing.

The reduced data may be read out via analog or digital panel meters. Of it may be digitized (perhaps by a digital panel meter, that also provides a readout) and transmitted in digital form, to a remote control station (for further processing or remote printout on a teletypewriter or CRT terminal) via some compatible system, such as SERDEX*.

Figure 1 shows a single-channel data-acquisition subsystem, typical of those encountered in the *Analog-Digital Conversion Handbook*.[1] Whereas much space is given, in that volume, to pre-amplification, grounding, conversion, sample-hold, and analog multiplexing, this chapter (and related chapters) will be concerned with the blocks in which analog data is transformed into more-useful (but still analog) forms to meet specific needs.

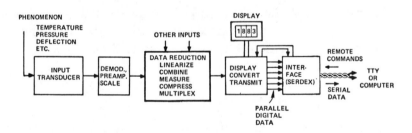

Figure 1. Typical data-acquisition channel

*SERDEX: SERial Data EXchange (Analog Devices trade-name), a means of simply controlling conversion processes, and transmitting data and commands in serial ASCII format under control of a teletype keyboard (or a computer programmed in a high-level language, such as BASIC) via an isolated current loop employing a simple twisted-pair of wires. While not strictly within the scope of this volume, it is nevertheless of great potential usefulness to the hardware-oriented analog-digital system designer. Complete data and applications information is available from Analog Devices, Inc.

[1]*Analog-Digital Conversion Handbook*, edited by D.H. Sheingold, Analog Devices, Inc., 1972, 402pp.

LINEARIZING

The system designer must strike an economic balance between convenience of measurement and convenience of dealing with the measured information. The simplest and most convenient transducers frequently have a nonlinear relationship between the variable being measured and the electrical output. Linear transducers, if available, often turn out to be less sensitive, more costly, or difficult to implement. Linearization can make it possible to obtain greater sensitivity by using nonlinear regions that are usually shunned.

For example, the simple Wheatstone bridge, a 4-terminal device used in a wide variety of pressure, force, strain, and electrical measurements, has an inherent nonlinearity (Figure 2a), which increases with sensitivity (e.g., it is 50% at $K = 1$, $\alpha = -1$). By opening one leg, and using a readout operational amplifier to drive a portion of the bridge, one can obtain linear response (Figure 2b).

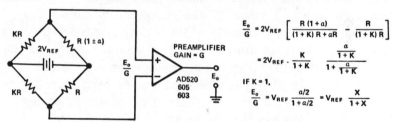

$$\frac{E_o}{G} = 2V_{REF} \left[\frac{R(1+a)}{(1+K)R + aR} - \frac{R}{(1+K)R} \right]$$

$$= 2V_{REF} \cdot \frac{K}{1+K} \cdot \frac{\frac{a}{1+K}}{1 + \frac{a}{1+K}}$$

IF K = 1,

$$\frac{E_o}{G} = V_{REF} \frac{a/2}{1 + a/2} = V_{REF} \frac{X}{1+X}$$

a. Nonlinear response of Wheatstone bridge

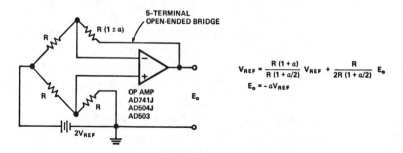

$$V_{REF} = \frac{R(1+a)}{R(1+a/2)} V_{REF} + \frac{R}{2R(1+a/2)} E_o$$

$$E_o = -aV_{REF}$$

b. Linear version of bridge using operational amplifier

Figure 2. Nonlinear and linear bridge circuits

However, there are a number of significant costs: First, an amplifier *must* be used (whereas a Wheatstone bridge can be read out with a passive analog meter); also, five terminals are necessary, and the cable connecting the amplifier with the transducer affects loop stability; in addition, if gain is needed, an extra amplifier is needed; finally, 4-terminal bridges are cheap, widely available, and standard in many transducers —which leaves the designer with no alternative at the transducer level.

Fortunately, bridge nonlinearity is described by a simple mathematical relationship, and it can be compensated for completely by the use of a multiplier and an operational amplifier, as we have indicated in Chapter 2-1. The simplest approach is to use the configuration of Figure 3a, where implicit feedback is used to obtain the inverse of the bridge nonlinearity function. It has the benefit of summing a purely-linear term with a correction term. It is also possible to compute the inverse directly, using division (Figure 3b). Although this approach makes good use of a divider (the maximum dynamic range of the denominator is only 3:1), it relies on the inherent linearity of the divider over the whole range of variation. Since, at full scale ($\alpha = 1$, $K = 1$), the correction term is 50% of the output (Figure 3c), multiplier nonlinearities in the circuit of Figure 3a are in effect attenuated by 50%, while the divider nonlinearities are not attenuated. On the other hand, if V_{REF} (in the denominator) is the actual bridge-reference voltage, the divider circuit will also compensate for reference-voltage variations.

The correction terms should be scaled to represent the portion of the range of resistance variation represented by α. Usually, a transducer is chosen to operate over the most linear portion of the bridge's range (small α and large K or large α and small K) to avoid the need for linearization. But this means throwing away sensitivity and signal-to-noise ratio for the sake of linearity, since the output is in either event a small fraction of the supply voltage. *A major advantage of linearization* is the prospect of using a more sensitive (albeit grossly nonlinear) bridge, in which the variable arm can conceivably go from zero to more than 200% of the fixed resistance, to deliver outputs comparable in magnitude to the

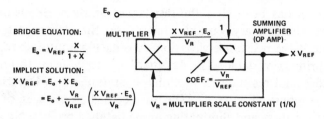

a. Bridge linearizer using implicit solution

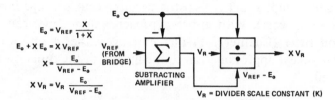

b. Bridge Linearizer using divider. Note that gain of this ratiometric circuit can be made independent of V_{REF}

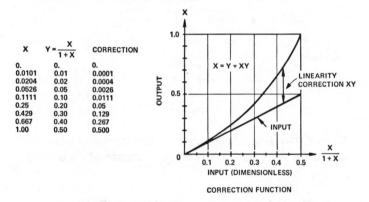

c. Tabulation and plot of bridge linearization function

Figure 3. Bridge linearization circuits

bridge-supply voltage.

It is evident that, for a high-level signal, a preamplifier is unneeded. If the bridge supply is floating, the multiplier and summing amplifier can be single-ended. If the bridge supply is returned to system ground, one can use amplifiers and multipliers that have differential inputs.

But bridge linearization alone may not be enough. The tacit

assumption has been that the resistance variation, αR, is proportional to the primary variable that causes the resistance to vary. But what if the resistance variation, α, is itself a nonlinear function of the primary variable? The designer has two choices: to linearize the bridge and resistance functions separately, or to linearize the overall response (Figure 4). The former has advantages of using standard circuitry and eliminating immediately a predictable source of nonlinearity (usually the major one); the latter has the advantage of possibly simpler and less-costly circuitry (but perhaps involves greater setup cost). With either approach, the designer can use a function fitter (Chapter 2-1) that employs either a smooth or a piecewise-linear approximation to the inverse of the function to be linearized.

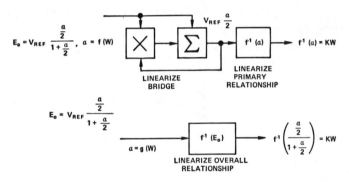

Figure 4. Two ways of linearizing a bridge-transducer measurement when the deviation is a nonlinear function of the primary variable (W)

LINEARIZATION EXAMPLE: THERMOCOUPLE

Figure 5 includes a tabulation of the relationship[2] between temperature and output voltage of a nickel-chromium X copper-nickel (Chromel-Constantan) thermocouple, with $0°C$ reference junction, over the range from $0°$ to $661.1°C$ (0 to 50mV). From the plot, it can be seen that the output is linear within $\pm1°C$ from about $340°C$ to beyond $650°C$. The deviation from linearity increases at lower temperatures to about $40°C$ at zero.

[2]The figures in the table are based on a tabulation in *The Omega Temperature Measurement Handbook* (1973), page A-9, published by Omega Engineering, Inc., Stamford, Connecticut 06907, based on 1971 figures from the National Bureau of Standards.

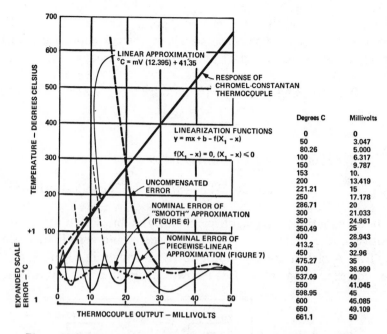

Figure 5. Nonlinear thermocouple response and theoretical residual errors, using two different linearizing functions

If, for any reason, it is necessary to obtain linear temperature measurement within ±1°C over the indicated temperature range, using this device, a linearizing circuit must be used to obtain an output voltage proportional to temperature, given the millivolt output (mV) of the thermocouple. Following preamplification, two approaches may be used to compensate for the nonlinearity at the lower end: a smooth fit, or a piecewise-linear fit. Examples of the errors experienced with a smooth cubic fit, and with a 5-segment piecewise-linear approximation, are shown in Figure 5.

In both cases, the desired response is fit by an operation of the form

$$°C = \underbrace{(\text{slope})(mV) + (\text{intercept})}_{\text{linear portion}} - \underbrace{f(mV_o - mV)}_{\text{correction}} \qquad (1)$$

For values of mV greater than the threshold, mV_o, the correction term is zero. In both cases, a "breakpoint" enforces this condition. For the specific case considered here, the correction functions providing the theoretical error plots in Figure 5 are:

$$f = 0.2391(28.943 - mV) + \left[0.09464(28.943 - mV)\right]^{3.512} \quad (2)$$

and

$$f = 0.6356(23 - mV) + 1.021(13.419 - mV)$$
$$+ 1.473(6.317 - mV) + 1.17(1.495 - mV) \quad (3)$$

In both equations, the coefficients of the bracketed terms are positive if the bracketed terms are positive, and zero if they are negative. A circuit corresponding to (2) is shown in Figure 6, and one corresponding to (3) is shown in Figure 7. The fitting

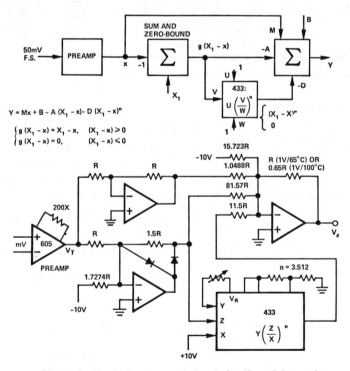

Figure 6. Block diagram and circuit for linearizing, using smooth approximation

process is aided by using slightly different coefficients for the linear portion: (12.395) (mV) + 41.35 for the exponential case, to reduce error, and (12.424) (mV) + 39.88 for the piece-wise-linear case, to reduce the number of breakpoints.

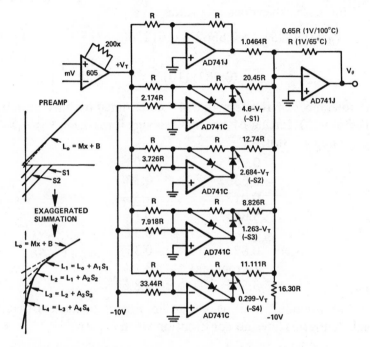

Figure 7. Circuit for linearizing using piecewise-linear approximation

The values calculated for the plots were carried out to a sufficient number of places to make the computational errors negligible for the idealized configuration. However, it should be evident that, for a practical circuit, the tolerances can in most cases be considerably looser than the numbers in (2) and (3) imply. The functions were fitted using the techniques discussed in Chapter 2-1. Then, as noted there, the next steps are to derive electrical scaling, nominal circuit values, and allowable device tolerances.

Equations (4) and (5) are electrically-scaled equations for the two cases, assuming that a gain-of-200 preamplifier is used, providing 1V/5mV at the input of the linearizer. The output scale factor is

1V/65°C. *The resistance values in Figures 6 and 7 are the nominal values required to provide the needed gain relationships. The table following equation (4) lists the tolerances necessary to embody that equation with less than 0.13°C (i.e., 2mV) error contribution by each term. V_θ is the scaled output voltage, and V_T is the scaled thermocouple voltage.

$$V_\theta = 0.9535V_T + 0.6362 - 0.01226(8.6835 - 1.5V_T)$$

$$- (0.087)(10)\left[\frac{8.6835 - 1.5V_T}{10}\right]^{3.512} \quad (4)^\dagger$$

The tolerances of the terms in (4), determined by differentiation, $(\partial V_\theta/\partial A_i = S_i)$, setting $\Delta V_\theta = 2mV$, solving for $\Delta A_i/A_i = 2/A_i S_{i_{max}}$, and rounding down, are:

0.9535	0.02%
0.6362	0.3%
0.01226	1.5%
8.6835	0.1%
1.5	0.3%
0.87	0.35%
10 (denom)	0.1%
3.512	0.8%

The "ideal-diode" limiting circuit in Figure 6 ensures that the bracketed terms have no contribution when negative.

Equation 5 is the scaled equation for the piecewise-linear case. Though the error is "lumpier" than that of the exponential approximation, and there are more circuit details to attend to, the circuitry is repetitive, and the tolerances are somewhat looser.

---

*This scaling was chosen to obtain the benefits of using the full output range. Though not making use of the full-scale range, a scale-factor of 100°/1V would permit direct readout of temperature on a 3 or 4-digit panel meter. It can be obtained without further modification by appropriately attenuating the output.

†This equation is derived from equation (2) by normalizing it, then setting the normalized equation equal to a normalized electrical equation, thus arriving at the constant voltages and coefficients. A further step was to recognize that it would be advantageous to use the major portion of the full-scale range of nonlinear devices. To do so, the difference terms were multiplied and divided by 1.5, resulting in larger input voltages and smaller overall coefficients. The constants in the exponent term were manipulated to provide the 10V denominator and 10V input multiplier desirable for a $Y(Z/X)^m$ device, in an application where Z is the only active input.

$$V_\theta = 0.9557V_T + 0.6135 - 0.0489(4.6 - V_T) - 0.0785(2.684 - V_T)$$

$$- 0.1133(1.263 - V_T) - 0.09(0.299 - V_T) \qquad (5)$$

The tolerances of the terms in equation (5) are:

0.9557	±0.02%
0.6135	±0.3%
0.0489	±0.75%
4.6	±0.8%
0.0785	±0.9%
2.684	±0.9%
0.1133	±1.2%
1.263	±1.2%
0.09	±5%
0.299	±5%

for less than ±2mV error from any term, or 6.3mV (0.41°C) root-sum-of-squares error (allowing for ±0.6°C of theoretical error). Tolerances, as applied to circuit elements, should take into account resistance-ratio mismatch, and the drift variations of amplifiers, resistances, and references, with time and ambient temperature, as well as scale factor, drift errors, and shape errors of the exponentiating device (e.g., the Model 433, if applied as the $Y(Z/X)^m$ in Figure 6).

It is interesting to note, as an exercise in function fitting, the value of plotting a curve. While it would appear natural to fit a function by seeking the best numerical fit, starting with a linear slope from the origin, this case proves the contrary. From the plot, it is immediately obvious that the departure from linearity is *greatest* at the origin, and that the most rewarding approach is to offset and reverse the "origin of nonlinearity."

AMPLITUDE COMPRESSION

If the result of an analog measurement, having a modest frequency content and a wide range of variation, must be made available at some distance, with an intervening noisy medium that is likely to result in pickup and loss of amplitude information, the designer has a number of possible options. Popular ones include:

1. Transmission as a frequency-modulated signal

2. Conversion to digital form and transmission either serially or in parallel

3. Logarithmic compression and analog transmission

4. Logarithmic compression and digital transmission

Some general comments can be made about these options:

1. Frequency modulation calls for wide bandwidth, depending on the dynamic range and frequency content of the signal, and highly-linear modulation and demodulation. If the DC level is important, a precise phase or frequency reference must be made available.

2. A/D conversion requires adequate resolution (16 bits for less than 30% error for the smallest signal in a 10,000:1 dynamic range). Adequate sampling rate and bandwidth, and a stable clock are necessary for serial (2+ wires) transmission; many wires are required for parallel transmission. All alternatives are costly, but SERDEX (see Page 88) is more convenient than most, if its bit-rate is adequate.

3. Logarithmic compression can be implemented at low cost (Figure 8). The signal-to-noise ratio of a compressed signal depends only on the noise level and the choice of log scaling; it is essentially independent of the signal level over a wide dynamic range. Bandwidth requirements are those of the analog signal, in its compressed

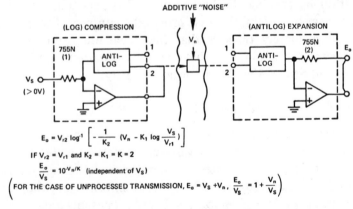

$$E_o = V_{r2} \log^{-1} \left[-\frac{1}{K_2} \left(V_n - K_1 \log \frac{V_s}{V_{r1}} \right) \right]$$

IF $V_{r2} = V_{r1}$ and $K_2 = K_1 = K = 2$

$$\frac{E_o}{V_s} = 10^{-V_n/K} \quad \text{(independent of } V_s\text{)}$$

$$\left(\text{FOR THE CASE OF UNPROCESSED TRANSMISSION, } E_o = V_s + V_n, \quad \frac{E_o}{V_s} = 1 + \frac{V_n}{V_s} \right)$$

Figure 8. Log compression used for improving dynamic range of transmitted signal

form. Though suitable for transmitting small or large signals impartially, the compression process is inherently insensitive to small signal components riding on larger signals. That is, the signal-to-noise ratio, even when mediocre, is independent of amplitude.

4. Logarithmic compression, combined with digital transmission (Figure 9), results in greatly-increased signal-to-noise, as long as the induced noise is below the logic thresholds. A further advantage of compression is the reduction of the required digital resolution: a 10,000:1 dynamic range can be comfortably resolved to within 1% of the actual value at any level using a 12-bit converter (cf. 2 above). Besides the obvious cost savings, there is also a slight reduction of the number of wires (parallel transmission) or an improvement in speed (serial transmission).

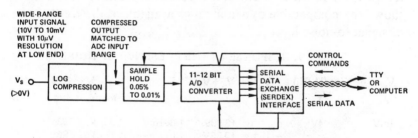

Figure 9. Log compression allows signal having wide dynamic range to be converted to digital at moderate resolution, and transmitted digitally via standard twisted-pair 20mA current loop with high noise immunity

The logarithmic compression process involves a logarithmic operator, such as the Model 755N, which computes $-K\log_{10}(V_s/V_r)$, where K may be 1V or 2V (per decade), and $V_r = 0.1V$. If the signal is transmitted in this logarithmic form (K = 2), a span of 10,000:1 of V_s is translated to a span of 8V at the compression output. An input swing of 1-10V will produce a 2V output change; so will an input swing of 1mV to 10mV. Thus, high-level signals are attenuated (average gain = 2/9 = 0.22) and low-level signals are amplified (average gain = 2/0.009 = 222). Noise picked up or induced in transmission will add to the logarithmic version of the signal. That signal-to-noise ratio can be greatly improved for small signals should be evident.

At the receiving end, the Model 755N is used as an antilog operator (a difference "instrumentation" amplifier may be used to reject common-mode errors, if appropriate), producing the inverse operation: $V_r (10)^{V_{in}/-K}$. Table 1 shows what happens to an instantaneous voltage V_s —assuming ideal (or matched) log conformance, that the V_r's are matched, and that the K's are adjusted for net unity gain— in the presence of a spurious instantaneous "noise" voltage, V_n. The "signal-to-noise" ratio with logarithmic compression is compared to what it would be without compression.

It can be easily seen that the signal-to-noise ratio depends only on the noise level, and that 10mV of noise is rejected in the same ratio, whether the signal is 1mV or 10V. While linear transmission does a much better job at high levels, it is virtually useless at low levels. Table 2, which is extracted (and interpolated) from Table 1, shows the comparable dynamic range available at different choices of signal-to-noise level.

TABLE 1. RESPONSE IN THE PRESENCE OF NOISE

V_s sig.	$-K\log(V_s/V_r)$ V_T	V_n	$V_n + V_T$	E_o	$\|E_o - V_s\|$ error	S/N log	S/N lin.
1mV	+4V	− 1mV	+3.999V	1.001mV	1.15µV	868	1
		− 10mV	+3.99V	1.012mV	11.6 µV	86	0.1
		+100mV	+4.1V	0.891mV	0.11mV	9.2	−
		−100mV	+3.9V	1.12 mV	0.122mV	8.2	−
		+ 1V	+5V	0.32mV	0.7mV	1.5	−
		− 1V	+3V	3.16mV	2.2mV	0.5	−
10mV	+2V	− 1mV	+1.999V	10.01mV	11.5µV	868	10
		− 10mV	+1.99V	10.12mV	116 µV	86	1
		−100mV	+1.9V	11.22mV	1.22mV	8.2	0.1
		− 1V	+1V	31.6mV	22mV	0.5	−
100mV	0V	− 1mV	−0.001V	0.1001V	115 µV	868	100
		− 10mV	−0.01V	0.101V	1.16mV	86	10
		−100mV	−0.1V	0.112V	12.2mV	8.2	1
		− 1V	−1V	0.316V	0.22V	0.5	−
1V	−2V	− 1mV	−2.001V	1.001V	1.15mV	868	10^3
		− 10mV	−2.01V	1.012V	11.6mV	86	10^2
		−100mV	−2.1V	1.122V	122mV	8.2	10
		− 1V	−3V	3.16V	2.2V	0.5	1
10V	−4V	− 1mV	−4.001V	10.01V	11.5mV	868	10^4
		− 10mV	−4.01V	10.12V	116mV	86	10^3
		−100mV	−4.1V	11.22V	1.22V	8.2	10^2
		− 1V	−5V	31.6V!!	21.6V	0.5	10

TABLE 2. DYNAMIC RANGE VS. SIGNAL-TO-NOISE RATIO

	|Noise|	Dynamic Range Log Channel	Dynamic Range Linear Channel
A. S/N > 865	1mV	10V:1mV	10V:865mV
	10mV	—	10V:8.65V
B. S/N > 85	1mV	10V:1mV	10V:85mV
	10mV	10V:1mV	10V:850mV
	100mV	—	10V:8.5V
C. S/N > 8.5	1mV	10V:1mV	10V:8.5mV
	10mV	10V:1mV	10V:85mV
	100mV	10V:1mV	10V:850mV
	1V	—	10V:8.5V

To determine the effects of errors in the log devices (especially variations of the coefficients with temperature), the complete relationship may be used:

$$E_o = V_{r_2} \cdot 10^{\left[\frac{K_1}{K_2} \log \frac{V_s}{V_{r_1}} - \frac{V_n}{K_2} \right]}$$

$$= \left[\frac{V_{r_2}}{V_{r_1}^{K_1/K_2}} \right] \cdot V_s^{K_1/K_2} \cdot 10^{-V_n/K_2} \tag{6}$$

If K_1 and K_2 are equal and track one another, and if V_{r_1} and V_{r_2} are equal and tracking, $E_o = V_s \cdot 10^{-V_n/K}$, giving the results in column 5 of Table 1. If they differ, equation (6) provides a means of exploring the errors. Errors of log conformance can be treated as additive values of V_n.

Since a logarithmic function is inherently unipolar (the logarithm is real only for positive values of the argument —positive signals require a 755N, negative signals a 755P), it is far from ideal for signals that are inherently zero-centered. While it may be useful to bias some types of input signals into a single polarity, functions that demand symmetrical treatment may be badly distorted by the wide variation, in both resolution and speed, between zero and full-scale input. Such functions would profit by a type of precise compression that is symmetrical about zero. An example of an easily-obtained form is a $\sinh^{-1}$ function (Figure 10), which involves two complementary antilog transconductors (752P and 752N) in the feedback path of an operational amplifier. The resulting function is logarithmic for larger values of input, but it passes through zero essentially linearly.

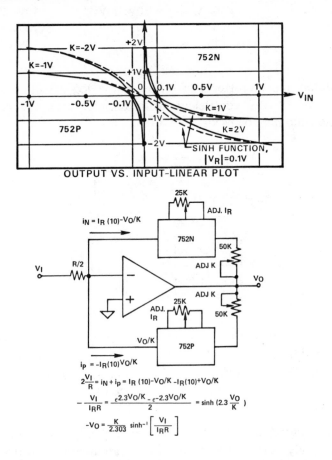

Figure 10. Bipolar signal compression using complementary logarithmic transconductors to synthesize $\sinh^{-1}$ function

Used in the forward path, the pair provides an inverse function, proportional to the hyperbolic sine. Assuming appropriate symmetry, matching, and tracking, the overall response, in the presence of an added noise voltage, is (instant by instant)

$$E_o = V_r \sinh\left[\sinh^{-1}\frac{V_s}{V_r} - \frac{\ln(10)}{K}\right]V_n \qquad (7)$$

If the magnitudes of V_r and K are, respectively, 0.1V and 2V,

$$E_o = 0.1 \sinh(\sinh^{-1} 10V_s - 1.1513V_n) \qquad (8)$$

A table similar to Table 1 may be derived to compare signal-to-noise and dynamic ranges

TABLE 3. IDEAL RESPONSE OF BIPOLAR COMPRESSION/EXPANSION IN THE PRESENCE OF A NOISE VOLTAGE

V_s signal magnitude	$\sinh^{-1}10V_s$	V_n "noise"	E_o output	$E_o - V_s$ error	S/N nonlin.	S/N linear
+10V	5.2983	−0.001V	10.0121V	0.0121V	827	10,000
		−0.01	10.1164	0.116	86	1,000
		−0.1	11.221	1.221	8.2	100
+1V	2.9982	−0.001V	1.00123V	0.00123V	810	1,000
		−0.01	1.0117	0.0117	85	100
		−0.1	1.123	0.123	8.2	10
+0.1V	0.8814	−0.001V	0.10017V	0.00017V	600	100
		−0.01	0.10164	0.0016	61	10
		−0.1	0.117	0.017	5.9	1
+0.01V	0.09983	−0.001V	0.01012	0.00017	85	10
		−0.01	0.01116	0.0012	8.6	1
		−0.1	0.0217	0.012	0.9	−
+0.001V	0.010	−0.001V	0.00112	0.00012	8.7	1
		−0.01V	0.0022	0.0012	0.9	−
		−0.1V	0.0125	0.012	−	−

While not quite as impressive as the logarithmic function, because of limited gain through the origin, the hyperbolic compression/expansion can be improved by extending the logarithmic portion of the range. If, for example, V_r is reduced to 0.01V, the signal-to-noise ratio for $V_s = V_n = 0.01\,V$, is increased to 60. The penalty that is paid is in the maximum speed through zero, since the effective feedback resistance and gain in the vicinity of zero are multiplied tenfold.

A final word: The $\sinh^{-1}$ function can also be used as a high-accuracy tapered null meter with calibrated (approximately-equal) intervals off-null for equal ratios of change. It has great sensitivity at null, wide dynamic range, and continuous indication of direction of approach to the null.

EXTRACTING A MEASURE

We have seen how nonlinear analog-computing circuits may be used to compensate for transducer nonlinearity and to reduce the effects of noise in transmitting data. A historically important factor in data reduction (especially in the days before the use of oscilloscopes), still in widespread use, and likely to remain so forever, is the use of the meter (i.e., *measure*) to "boil down" information to a simple reading or trend of readings that can be interpreted by the human eye instantly, and serve as the basis for a decision. Metering is also used in computer-control and measuring systems, where the computer seeks to reduce a large number of measurements to a few significant indications of the present status and trend of an element or a process, as ingredients of a series of decisions.

There is a large universe of computational meters, and it would be presumptuous in these pages to seek to accomplish more than to skim the surface lightly, picking out those operations that have great usefulness, universal validity, and specific appropriateness to the devices under discussion here. These include, to begin with,

> Mean
> Mean Absolute Value
> Root Mean-Square
> Peak (or Valley)

While one ordinarily thinks of free-running devices, with fixed averaging times, it might be profitable to at least consider, in addition, single-shot measurements and variable-period measurements.

In addition, there are a number of measurements that involve two or more variables. A few interesting and popular ones include

> Power (instantaneous, peak, and average)
>
> Energy and energy-per-cycle
>
> Power factor (and phase angle)
>
> Vector Sum (Root-sum-of squares)
>
> Ratio and log ratio (dB)

The circuitry that produces these functions will usually operate on the input after it has been preamplified and scaled, and perhaps linearized, but quite often it can be implemented with devices having sufficient stability, input impedance, common-mode rejection, or what-have-you, to operate directly on the transducer output.

MEAN AND MEAN ABSOLUTE-VALUE

Figure 11a shows the usual circuit employed for a running average, a simple unit-lag. In Figure 11b, an inverting version is shown; the averaging time-constant can be increased, without resorting to high-value resistors, by employing "T" networks (Figure 11c) at the cost of increased voltage drift because of the attenuation-and gain. This circuit is adequate for determining averages of signals having high-frequency fluctuations and relatively-slow mean variations. The settling time for a step change is 4.61RC to 1%, 6.91RC to 0.1%, and 9.21RC to 0.01%. High-frequency attenuation is modest: 3db at $f_0 = 1/2\pi RC$, 20dB at $10f_0$, 40dB at $100f_0$, etc.

If the average must respond more quickly to changes of non-stationary functions, one needs a filter having a response continuously approximating the ideal average response over a period τ,

$$E_0 = \frac{1}{\tau} \int_{t-\tau}^{t} v(t)dt, \text{ or operationally } \frac{1 - \epsilon^{-\tau p}}{\tau p} \qquad (9)$$

A reasonable approximation is the transfer function

$$\frac{e_0}{v_i} = \frac{1 + \overline{RCp}^2}{(1 + 1.2RCp + 1.6\overline{RCp}^2)\,(1 + 2RCp)} \tag{10}$$

(where p is the Heaviside derivative operator). It can be stably embodied with a unit-lag and a second-order state-variable (integrator-loop) band-reject filter or, less stably but more compactly, with a single operational amplifier, 8 precision resistors and 6 precision capacitors.[3]

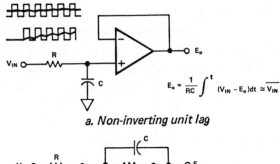

$$E_0 = \frac{1}{RC}\int^t (V_{IN} - E_0)dt \cong \overline{V_{IN}}$$

a. Non-inverting unit lag

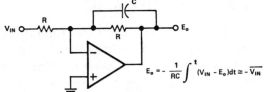

$$E_0 = -\frac{1}{RC}\int^t (V_{IN} - E_0)dt \cong -\overline{V_{IN}}$$

b. Inverting unit lag

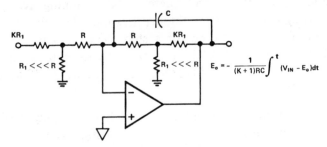

$$E_0 = -\frac{1}{(K+1)RC}\int^t (V_{IN} - E_0)dt$$

c. Long-period unit-lag, using T-network to avoid extremely-large R and C values

Figure 11. Classical unit-lag running-average circuits.

Another kind of "running average" is the average over each cycle of a train of events having differing periods, for example, the blood pressure averaged over each heartbeat, or the volume of CO_2 averaged over each breath. This can be accomplished with two integrators, a divider, and a sample-hold (Figure 12). At the beginning of each cycle (determined independently), the two integrators (*signal* and *period*) are gated to *run*. At the conclusion of the cycle, the integrators are momentarily placed in *hold*, the divider output (accumulated signal divided by period) is *sampled*, and the integrators are then dumped in preparation for the next cycle, which may start immediately (or after a wait of arbitrary duration). Meanwhile, the sample-hold retains the last average reading

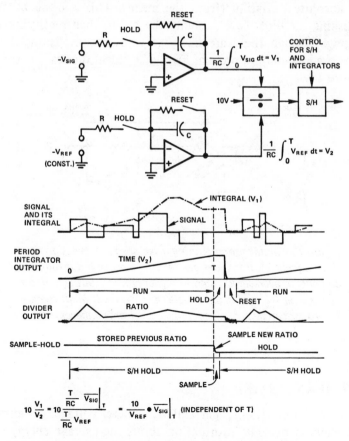

Figure 12. Averaging signals over variable periods

and holds it until updated. The divider-and-sample-hold could be a fast variable-reference A/D converter, especially desirable if long intervals are required between integrations.

Although averaging usually is considered to be a linear, or time-varying process (except where ratios are involved, as above), it has a place in the discussion of applying nonlinear devices in data reduction and measurement, because averaging is one of the most universally-used forms of data reduction.

The average, or mean, is not always directly appropriate, as a measure of a signal. For example, AC measurements seek to ignore the DC level (or *mean*), and instead concern themselves with the mean absolute deviation (from the mean). This is done by first establishing a "zero" level, usually the mean, then rectifying and averaging. Though there are many ways of accomplishing this, a widely used op-amp approach, involving "ideal diodes", is shown in Figure 13.

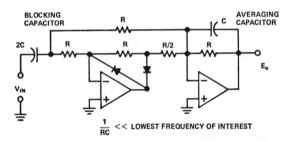

Figure 13. Circuit for computing mean absolute deviation at low frequencies. For wider bandwidth, external input follower and output averager will permit higher currents through diodes and greater RC's using reasonable capacitance. If input average is zero or can be zeroed, coupling capacitor is unnecessary.

ROOT MEAN-SQUARE

For many applications, particularly where voltage or current measurements provide information about the average energy generated, transmitted, or dissipated, the root-mean-square (rms) is a more useful measure. Straightforwardly, it involves squaring an input, taking the average, and obtaining the square root.

$$E_{rms} = \sqrt{\frac{1}{T}\int_0^T (V_{in})^2 \, dt} \qquad (11)$$

Classically, it has been measured by meters sensitive to the heating effect of an rms level. Electronically, because of the difficulty of tailoring a general-purpose instrument to fit the characteristics of a given thermal device, most "rms" meters for many years didn't

WAVEFORM		RMS	MAD	$\dfrac{RMS}{MAD}$	CREST FACTOR
	SINE WAVE	$\dfrac{V_m}{\sqrt{2}}$ 0.707 V_m	$\dfrac{2}{\pi}V_m$ 0.637 V_m	$\dfrac{\pi}{2\sqrt{2}}$ = 1.111	$\sqrt{2}$ = 1.414
	SYMMETRICAL SQUARE WAVE OR DC	V_m	V_m	1	1
	TRIANGULAR WAVE OR SAWTOOTH	$\dfrac{V_m}{\sqrt{3}}$	$\dfrac{V_m}{2}$	$\dfrac{2}{\sqrt{3}}$ = 1.155	$\sqrt{3}$ = 1.732
	GAUSSIAN NOISE CREST FACTOR IS THEORETICALLY UNLIMITED. q IS THE FRACTION OF TIME DURING WHICH GREATER PEAKS CAN BE EXPECTED TO OCCUR	RMS	$\sqrt{\dfrac{2}{\pi}}$ RMS = 0.798 RMS	$\sqrt{\dfrac{\pi}{2}}$ 1.253	C.F. q 1 32% 2 4.6% 3 0.37% 3.3 0.1% 3.9 0.01% 4 63ppm 4.4 10ppm 4.9 1ppm 6 2x10⁻⁹
η: "DUTY CYCLE"	PULSE TRAIN η MARK/SPACE 1 ∞ 0.25 0.3333 0.0625 0.0667 0.0156 0.0159 0.01 0.0101	$V_m\sqrt{\eta}$ V_m 0.5V_m 0.25V_m 0.125V_m 0.1V_m	$V_m\eta$ V_m 0.25V_m 0.0625V_m 0.0156V_m 0.01V_m	$\dfrac{1}{\sqrt{\eta}}$ 1 2 4 8 10	$\dfrac{1}{\sqrt{\eta}}$ 1 2 4 8 10

Figure 14. RMS, MAD, and crest factor of some common waveforms. See also Table 1, Chapter 3-7, for additional waveforms.

measure rms at all. They measured the mean absolute deviation ("ac average") but indicated it on an "rms" scale calibrated to the ratio of rms-to-mean for sine waves. This was all right as long as sine waves were being measured (if they weren't badly distorted). It was even acceptable if signals having a more-or-less constant ratio of rms/mad, such as symmetrical square waves, Gaussian noise, or symmetrical triangular waves, were being measured, so long as a calibration was provided (see Figure 14). But, for unpredictable waveforms, variable-width pulse trains, and SCR'd sine waves, average-measuring rms meters were useless.

Accurate, wide-range "true-rms" circuits are made possible at reasonable cost by the availability of transconductance multiplier/dividers (XY/Z), such as the AD531, and by stable log-antilog circuits, such as the 433, and (more recently) by the 440 rms module. Examples of practical rms circuits can be seen in Chapter 3-7. The basic scheme is shown in Figure 15. It employs squaring, averaging, and *implicit* square-rooting. The crucial dynamic-range characteristic of rms devices is *crest factor*, the ratio of peak input to the rms value of the waveform. For example, an input signal having a dynamic range of rms of 20:1 and a crest factor of 5, calls for a device having substantially greater-than-100:1 dynamic range.

Straightforward open-loop schemes —square, average, root— call for excessive dynamic range internally; for example, a 100:1 ratio of maximum to minimum input, when squared, becomes 10,000:1, placing near-impossible demands on an open-loop square-rooter. When a transconductance-type multiplier-divider is connected for *implicit* square-rooting, the squarer's gain is controlled by the output, reducing the dynamic range, in the steady state, to first-order. Types employing logarithmic circuitry are even more effective, if slower, because they can reduce wide dynamic ranges to equal per-decade internal voltage swings.

For simplicity, filters are almost always first-order unit-lags. However, it is not unfeasible to specify filters having a more-nearly ideal "running-average" response.

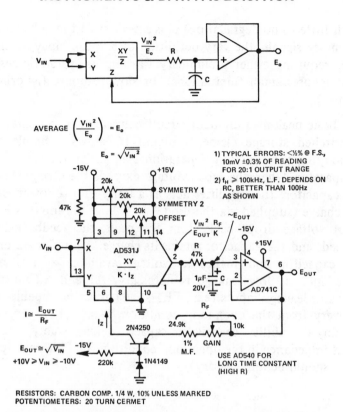

Figure 15. Block diagram of rms circuit and a practical circuit employing an I.C. multiplier-divider. See also Chapter 3-7.

PEAK AND VALLEY

For some purposes, averaging-type measurements do not provide adequate information about a waveform. Examples include periodic signals with rapidly-changing amplitudes, signals with variable crest factors, and amplitude-modulated waveforms. There are also applications in which it is necessary to determine the largest peak (or valley, or p-p spread) of a waveform over a given time interval.

Peak-detection-and-measurement circuits are numerous, and the choice (and cost) depends on the characteristics to be optimized. Such characteristics include accuracy, speed, leakage rate, sensitivity, and complexity. For free-running applications, a built-in leak is necessary; for one-shot applications, very long *hold* time (with little or no degradation) plus a *reset* circuit may be required. For noisy signals, the very definition of a "peak" may be in question, requiring either preliminary filtering, a hysteretic response (that ignores small fluctuations), or slow response (ignoring fast "blips").

The basic peak-measurement circuit consists of a comparator and a switched storage element. Figure 16 shows a simple circuit embodying the function. Operational amplifier A1 serves as the comparator. When the input voltage exceeds the charge stored on the capacitor, the amplifier acts as a unity-gain follower, causing the charge (supplied via the diode) to follow the input. When the input voltage drops back from the peak, the feedback loop is opened, and the capacitor retains its charge. Amplifier A2 unloads the capacitor and makes its voltage available at low output impedance. If low leakage is necessary, A1 and A2 must both have low-leakage inputs (e.g., FET's). A1 must be capable of fast recovery from the open-loop condition, it must have large phase margin, the ability to drive a capacitive load stably, and high input impedance in the open-loop condition (i.e., internal "protection" should be unnecessary).

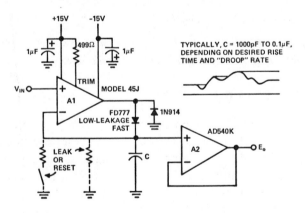

Figure 16. Peak-follower circuit

The capacitor determines both the charging rate and the "droop:" $dE_o/dt = I/C$. If 10mA are available for charging the capacitor (1000pF), the slewing rate is $10^{-2}/10^{-9} = 10V/\mu s$. If the total leakage current is 100pA, the droop rate will be $10^{-10}/10^{-9}$ = 0.1V/s. If the capacitance is increased to $0.1\mu F$, the droop rate will be reduced to 1mV/s, and the maximum charging rate will be $0.1V/\mu s$. If the circuit is to free-run, a leak must be provided to allow downward variations of the peak level to be followed. A resistor will provide exponential decay (proportional to the last peak), and a current sink will provide linear decay at a fixed rate. The output of the peak follower may be averaged to determine the average variation of the peaks. If, on the other hand, the circuit is to provide a one-shot measurement of the highest of a series of peaks, a *reset* switch must be provided to discharge the capacitor before the next series of readings.

The negative-going edge at the output of A1 can be used to indicate that a peak has just occurred. If fast following and long *hold* are necessary to "catch" a single fast peak, two of these circuits may be cascaded, the first using a small capacitor (paralleled by a reverse-biased diode to the negative supply to ensure a small downward leak), and the second using a large capacitor for leisurely acquisition and long *hold*.

The "valley" follower is essentially the same circuit, but the diodes are reversed. It will track negative-going voltages that are below the stored level, and *hold* the lowest level experienced. For peak-to-peak measurements, the outputs of the followers can feed a simple subtractor-connected op amp. Alternatively, the A1-diode-capacitor portions of a peak and a valley circuit may be used, with the capacitor voltages applied to the inputs of a differential instrumentation amplifier, such as the AD520, the 603, or the 605. If the peak-to-peak circuit requires a leakage path to enable it to follow an envelope, the capacitors can feed directly a subtractor-connected FET input op amp, with resistors of appropriate magnitude for the desired leakage rate. As an added bonus, capacitors may be connected across the feedback resistors to filter out the cyclical swings of the peak measurements.

Usually, peaks are above ground, and valleys are below ground. However, if it is desired to measure peaks or valleys of widely-ranging signals anywhere in the range, this can be done by connecting the capacitor and the leak resistor (or the *reset* switch) to a voltage lower than the lowest peak (peaks) or higher than the highest valley (valleys), instead of to ground; typically, the negative and positive supply voltages serve the purpose. The *reset* switch should, of course, always have resistance in series for protection.

As indicated earlier, there are many circuits for peak-following. They include single op amps with diode-capacitor inputs (outside the feedback loop), multiple-op-amp loops, sample-holds with comparators (input is compared with the S/H output, and the comparator operates the S/H control logic, often in synchronism with a clock to avoid oscillation), and A/D converters.

Converters used for peak-following are typically the counter-DAC-comparator type. A D/A converter continuously provides an output voltage proportional to the state of a digital counter. The converter output is compared with the signal input. If the input is the lesser, the comparator continuously inhibits the count. If the input is the greater, the counter accumulates clock pulses until the comparator threshold is crossed, and the count is again inhibited with the next clock pulse. Though slow, the A/D converter types have the advantage of essentially "infinite" *hold* times, since retention of data does not depend on the charge stored in a capacitor. The A/D converter is an ideal second stage of a two-stage peak follower (note that the DAC output, corresponding to the digital count, is an analog quantity). It is also obvious that if the reduced data must be converted to digital form, this is an ideal way to "kill two birds with one stone", if peak information is a suitable measure of the input.

POWER MEASUREMENT (Figure 17)

Analog multipliers are well-suited to the accurate measurement of *instantaneous* power ($e \cdot i$). Their outputs can be averaged to obtain *average* power, applied to peak-detectors to obtain *peak* power, and integrated to obtain *energy*. Furthermore, the energy output can be computed and divided by the period, in the manner indicated in Figure 12, to obtain *energy per cycle*.

When power is measured, voltage can be picked off, differentially,

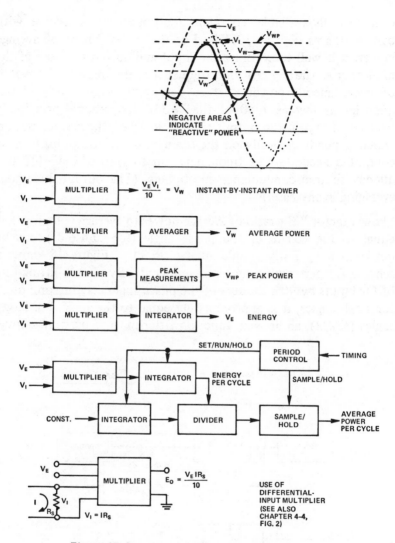

Figure 17. Power and energy measurement

if necessary, by a differential or isolation amplifier, and scaled to the multiplier input. Current can be measured by a differential pickoff across a shunt. If the passively-scaled voltage and current happen to fall within the common-mode and amplitude limitations of available analog multipliers, it is worthwhile to consider the use of *differential-input* multipliers, such as the low-cost monolithic AD532.

Commercially-available complete multipliers are available with bandwidths as high as 10MHz. For the measurement of average power, it is well to consider the input and output response of the multiplier separately. At high frequencies, the phase relationships of instantaneous power start to deteriorate significantly at frequencies as low as 1/50 of the "–3dB frequency," principally because of lags in the output stages. However, the *average* power, which depends critically on the *input* phase relationships, can be computed accurately at frequencies up to 1/10 of the –3dB frequency in transconductance multipliers. (The output undergoes averaging in any case.)

"Power factor," the ratio of average power to average volt-amperes, equal to the cosine of the phase angle (for sinusoids), can be determined by fairly simple analog circuitry. Figure 18 shows a scheme for performing such measurements. By phase-shifting one of the inputs by 90°, the *sine* of the phase angle may be computed; for small angles, it is approximately equal to the angle. For larger angles ($< \pi/2$), an arc-sine function fitter may be used, if a direct

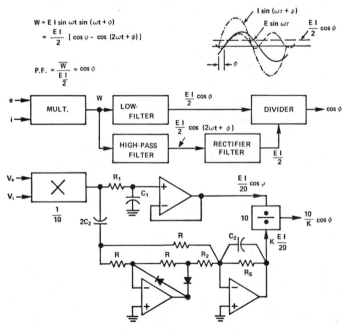

Figure 18. Power-factor measurement

measurement of the angle is required.

Impedance (magnitude) may be measured by computing the scaled ratio of the average or rms voltage to the average or rms current, using a divider (Figure 19). It is important not to fall into the trap of seeking to take the ratio of two ac quantities by an instantaneous measurement. Conceptually, the measurement will not be finite for zero denominator unless the two signals are $n\pi$ apart in phase (n = 0 or any integer). As a practical matter, analog dividers call for unipolar denominators; with the added complication of polarity-switching, bipolar denominators may be handled, but the vicinity of zero is ordinarily forbidden.

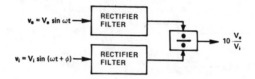

Figure 19. Impedance-magnitude measurement

VECTOR SUM

The vector sum of any number of mutually orthogonal voltages may be obtained by a circuit that solves the equation

$$E_0 = \sqrt{V_1{}^2 + V_2{}^2 + V_3{}^2 + \ldots V_n{}^2} \tag{12}$$

As noted in the introductory chapter, and confirmed elsewhere,[4] the straightforward approach (squaring each input, summing, then taking the square-root of the sum) can be expensive, and may lead to poor results over wide dynamic ranges of E_0, because of the expansion of dynamic range inherent in the squaring operation.

An implicit approach, using a ZY/X device, such as the 433, solving the equation

$$E_0 = V_2 + \frac{V_1{}^2}{E_0 + V_2} \tag{13}$$

[4]See *Analog Dialogue*, Vol. 6, No. 3, page 3.

for two variables, and, in general

$$E_o = V_n + \frac{V_1{}^2}{E_o + V_n} + \frac{V_2{}^2}{E_o + V_n} + \frac{V_3{}^2}{E_o + V_n} + \dots \quad (14)$$

is far more satisfactory, because each nonlinear term, $V_i^2/(E_o + V_n)$, is net first order, with no external manifestation of square-law dynamic range (Figure 20).

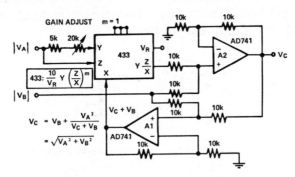

a. Preferred circuit to compute $\sqrt{V_A{}^2 + V_B{}^2}$

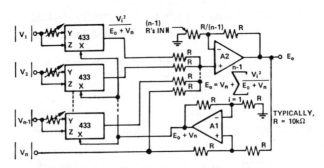

b. Extension of the technique to n input signals

Figure 20. Square-root of sum of squares

Performance, adjustment, and choice of components are straightforward. An important consideration, not immediately obvious, is the need to scale down inputs and outputs to avoid overdriving amplifier A1 or the 433's. If all inputs can have the same maximum value simultaneously, for the maximum output of A1 to be less than an arbitrary level, E_{max}, the maximum input value, $V_i = E_{max}/(1 + \sqrt{n})$. For $E_{max} = 10V$, the corresponding values of n and V_{max} are:

n	V_{max}
2	4.14V
3	3.66V
4	3.33V
5	3.09V

If n = 2, and A1 can swing to 12.1V, V_{max} is 5V.

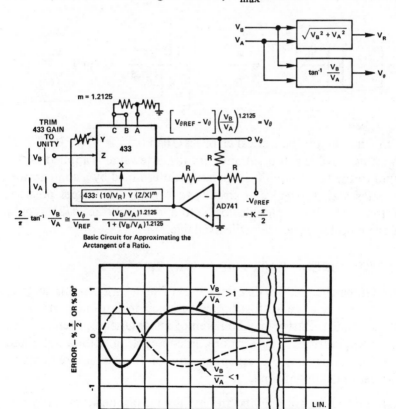

Basic Circuit for Approximating the Arctangent of a Ratio.

Theoretical Errors of Arctan Approximation

Figure 21. Arctangent circuit, with error plot

Magnitude is one aspect of vector composition, but not the whole story. In addition to magnitude, phase-angle is often desired. If the phase angle, θ, is equal to the arctangent of the ratio V_B/V_A, it can be approximated by function fitting. An excellent first-quadrant fit ($V_B, V_A \geqslant 0$) can be obtained simply, to within 0.75% (theoretically), using a single 433 and an operational amplifier, in an implicit feedback circuit. It maintains its accuracy over an extremely wide range of ratios, because the ratio never appears explicitly — only as a difference of logarithms within the 433. The circuit of Figure 21, which embodies the approximation, solves the normalized equation:

$$\theta = \left[\frac{\pi}{2} - \theta\right] \left[\frac{V_B}{V_A}\right]^{1.2125} = \frac{\pi}{2} \cdot \frac{\left[\dfrac{V_B}{V_A}\right]^{1.2125}}{1 + \left[\dfrac{V_B}{V_A}\right]^{1.2125}} \cong \tan^{-1}\frac{V_B}{V_A} \quad (15)$$

with a maximum theoretical error less than $0.75\% \frac{\pi}{2}$ (or $0.68°$). If V_B is negative (IVth quadrant), its absolute value is applied as the input to the $\tan^{-1}$ circuit. Its polarity, determined by a comparator, operates a sign/magnitude circuit, to furnish the proper polarity of the angle. With suitable logic (to add or subtract $\pi/2$), ranges of angle up to $\pm\pi$ can be made available.

RATIO AND LOG RATIO, dB (Figure 22)

Dividers can be used for direct readout of such ratios as efficiencies, losses or gains, % distortion, impedance magnitudes, elasticity (stress/strain). Ratios may be taken of instantaneous, average, rms, or peak quantities. Furthermore, in conjunction with sample/hold devices, ratios may be taken of any of these measurements at different instants of time.

Ratiometric measurements are by no means new, but the low (and still-decreasing) cost of analog dividers (and of variable-reference A/D converters) should serve as an encouragement to designers to consider employment of the technique as a realistic alternative (or adjunct) to tightly-regulated reference supplies for measurements, ultra-stable light sources, etc.

To eliminate the effects of a common parameter, whether physical or electrical, many measurements can profitably involve the use of ratio techniques. For example, in bridge measurements, variations of the power supply directly affect the scale factor. But if the output is divided by the bridge-supply voltage, the scale-factor stability depends only on the stability of the divider. This scheme can be combined with linearization, as shown in Figure 3b. Naturally, the divider should be at least as stable as the bridge-reference voltage if the ratiometric compensation is to be useful.

Compensation for reference-voltage variations is an example of reducing the effects of a common *electrical* parameter. However, ratios can also be used to eliminate the effects of a common *physical* parameter. For example, in light-transmission measurements, it is common to compensate for variations in light intensity by transmitting two beams, one through a reference medium, the other through the medium being measured, and to take the ratio of the two measurements.

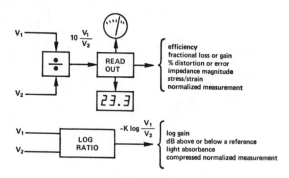

Figure 22. Ratio and log-ratio measurements

Often, logarithmic ratios are more useful than linear ratios. There are two broad categories of such measurements. The first is the measurement of phenomena covering a wide dynamic range, with reference to a normalized level, with log-compression, and either the display of the results on a limited-range meter scale, or the transmission of the measurement through a noisy medium. The second category consists of those measurements that are performed linearly but are normally characterized (or thought about) in terms of logarithmic ratios. One example is light transmission measurements. For another, electrical gain or attenuation may be measured as a ratio of output to input; by the use of a log ratio device, such as Model 756, a direct measurement of the log ratio, or "dB,"* may be performed.

CONCLUSION

In this chapter, we have suggested a number of uses of analog nonlinearities in the reduction of data for display or transmission. It is not unlikely that the thoughtful reader's experience and needs will suggest many more.

---

*The decibel, one-tenth of a bel (B), is the logarithm of an electrical power ratio of 1.259. The number of dB corresponding to a power ratio is $10 \log_{10} (P_2/P_1)$. If resistance is constant, the number of dB also is equal to $20 \log_{10}(V_2/V_1)$ or $20 \log_{10}(I_2/I_1)$, since power is proportional to the square of voltage or current. The term has been widely corrupted to express log ratios of any two quantities (even engineers' salaries), by the definition $dB = 20 \log_{10}(Q_2/Q_1)$. Though confusing (some would say deplorable), it is almost universally understood.

II

Communications & Signal Processing

Chapter 4

Nonlinear devices have always been used in audio signal communications to stabilize or modulate oscillator amplitude and frequency, achieve automatic gain control, and demodulate the received signal. Classically, diode, transistor, and thermionic-device characteristics have been used. Effectiveness and stability of such operations have generally depended on the designer's skill and ingenuity in circuit design and on the availability of components having suitable stability, parameter match, "linearity" (i.e., *parametric conformance*), and low cost.

Now with the availability of operational amplifiers, multiplier-dividers, and logarithmic elements, with their tightly-specified (guaranteed) parameters, convenient (modular or black-box) packages, and low (or decreasing) cost, the designer has a set of new options to make his job easier and more fruitful. In addition to standard signal-processing circuits, he can now consider new approaches to waveform synthesis and control, and the design and uses of such tools as voltage-controlled amplifiers, filters, and oscillators (VCA, VCF, VCO) with uniform behavior and predictable characteristics. Combining these operations with some of the "hybrid" techniques described in Chapters I-4 and I-5B of the *Analog-Digital Conversion Handbook*, one becomes aware of a formidable arsenal of signal-manipulating possibilities, virtually at one's fingertips.

Just a few additional examples of the applications of nonlinear analog devices in signal-handling include compression and expansion, phase-sensitive detectors, phasemeters, phase-locked loops, low-noise recording systems, correlators, spectrum analyzers, speech and music synthesizers, and so on.

AUTOMATIC GAIN CONTROL

An analog multiplier or divider is inherently a gain controller (Figure 1a) since the signal applied to one of its inputs can be considered a dependent variable, either multiplied or divided by a second input that controls its gain (or attenuation). Since the gain-setting voltage can be derived from any source, there is a wide range of possible applications. For example, a DC voltage applied from a remote manually-adjustable source can cause the multiplier to act as a potentiometer with a "long shaft." The control voltage can be derived as a measure of one or more other voltages in a system and used to control the gain in response to their variation. A useful special case is *automatic gain control* (Figure 1b).

The circuit of Figure 1c is a practical example illustrating the application of the low-cost AD531 I.C. multiplier-divider in an AGC application. It maintains 3V peak-to-peak output for inputs ranging from 0.1Vp-p to more than 12Vp-p, with better than 2% regulation from 0.4Vp-p to 6Vp-p, and distortion well below 1%. Input frequency can range from 30Hz to 400kHz (–3dB). The set point is adjustable either manually or by an external DC reference voltage. The input signal can be either single-ended or differential.

The feedback circuit works in a straightforward manner: if the input signal increases, the output will tend to increase. Its negative peaks, as recognized by the diode and stored on the 1µF capacitor, tend to increase, causing the output of the inverting integrator to increase. This, in turn, causes the denominator to increase, reducing the gain of the AD531 multiplier-divider (an XY/I device), and tending to keep the output level constant.

In the steady state, the average voltage at point A must be ideally equal to $\frac{1}{2}V_B$, but of opposite polarity, making the net input to the integrator equal to zero, and holding the output of the integrator at whatever constant level is necessary to keep the loop in balance. In that state, the negative peak value of E_{out} is approximately one diode-drop below V_A, so $|E_{out} \text{ (peak)}| \cong \frac{1}{2}V_B + \text{diode drop}$

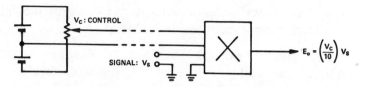

a. Analog multiplier as gain controller

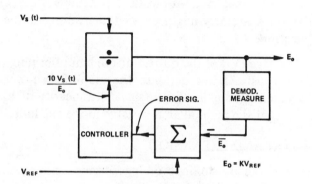

b. Typical AGC loop using divider

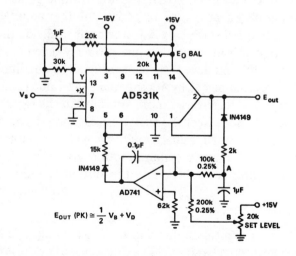

c. Practical AGC loop using IC multiplier-divider

Figure 1. Gain control with multipliers and dividers

In practice, the *set level* potentiometer would be adjusted empirically to calibrate the output at the desired level.

In the simple practical example given here, to illustrate the principle, an unembellished half-wave diode-and-capacitor circuit reads the peak level of the waveform. Naturally, other measures of the waveform, such as mean absolute-value or RMS, might be used; in addition, somewhat more-sophisticated temperature-compensated rectification circuitry might be used, depending on the needs of the application.

The control voltage (V_c) at the output of the amplifier ranges from about $-2V$ (lowest AD531 gain) to the amplifier's lower limit, $-13.5V$ (to handle the smallest input signals). Linearity of V_c is not important, since it is a manipulated variable inside the loop.

COMPRESSION AND EXPANSION

In Chapter 2-3, the possibilities of logarithmic compression and expansion in transmitting small voltages safely through a noisy medium were touched upon. Though it operates instant-by-instant, a drawback of the scheme is that the logarithmic gains must be matched to ensure linear overall response. Another approach, that can be applied to quasi-stationary waveforms, is to divide the signal by a voltage that is a measure of some property, such as squared-peak, transmit the modified signal through the medium, then multiply the received signal by its squared peak-value (Figure 2a). Since the control voltage varies more slowly than the signal (essentially DC), it does not affect the signal's shape, only its amplitude. The high gain for small signals and low gain for large signals produces a predictable compression function. At the receiving end, the inverse function is applied, and the output amplitude variation is recovered. Mismatches affect only the overall gain, without introducing distortion.

An example of a typical application of this technique is in high-fidelity tape recording systems. The Burwen Laboratories Model 2000, outlined in Figure 2b,[1] has a 110dB dynamic range when used with a 15ips tape-recorder.

[1] "Design of a Noise Eliminator System," by R.S. Burwen, Audio Engineering Society Preprint No. 838(B-8), October, 1971.

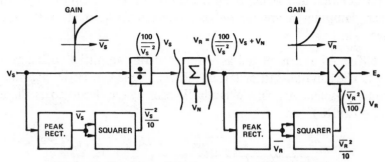

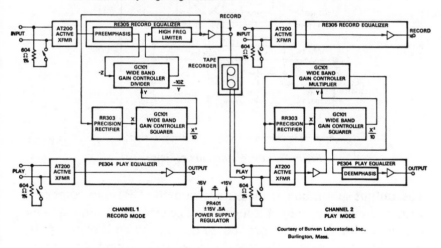

Figure 2a. Gain compression-expansion. Gain function is nonlinear, but signal is transmitted throughout essentially without distortion. Small signals are greatly amplified before transmission. Noise is either suppressed by squaring or masked by high signal levels.

Courtesy of Burwen Laboratories, Inc., Burlington, Mass.

Figure 2b. A commercial wide-range record-playback system having 110dB dynamic range.

SIGNAL GENERATION

A number of schemes for signal generation are discussed in Chapter 2-2, including a variable-frequency two-phase oscillator. As noted there, nonlinear elements can be used to control frequency, phase, amplitude, etc. As a further example, Figure 3 is a schematic diagram of a very low distortion (0.01%) fixed-frequency (1kHz) single-phase phase-shift sine-wave oscillator. Its amplitude (about 7Vrms) is controlled by an AGC loop that applies linear damping in greater or lesser degree without affecting the waveform.

Amplifier A1 is connected as a non-inverting amplifier with a gain of +3. The band-pass filter R1, C1, R5, C2, tuned to 1kHz, provides frequency-selective positive feedback, causing the circuit to oscillate at $f_o = (2\pi RC)^{-1}$.

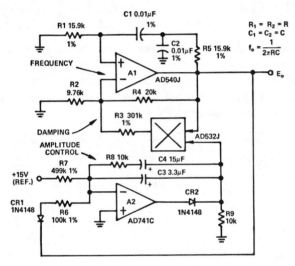

Figure 3. Low-distortion oscillator

The output amplitude is measured via diode CR1 and compared with a reference current through R7. The error is accumulated by the integrator (A2) and, applied to one of the multiplier inputs, increases or decreases the negative feedback around A1, appropriately affecting its gain and the damping of the oscillator. In the steady state, the net input to the integrator is zero, its output is constant, and R4 is in effect paralleled by a large trim resistance of exactly the right magnitude to keep the oscillation stable at a constant amplitude.

Since the multiplier output is essentially linear and is attenuated to provide a "vernier" gain adjustment on the oscillator amplifier, its distortion has a negligible effect on the output. The distortion is affected primarily by the nonlinearity of operational amplifier

A1 at the frequency of oscillation. The AD540J FET-input op amp provides distortion in the neighborhood of 0.01%. If distortion of 0.04% is tolerable, an AD741C may be used.

Capacitors C1 and C2 may be changed to obtain other frequencies of oscillation. The amplitude reference (the +15V supply in Figure 3) can be provided by a zener reference diode (for a 9V diode, reduce R7 to 301kΩ).

MODULATION

The terms "multiplier" and "modulator" are closely related. The modulation process almost invariably either uses or creates a multiplication operation. To illustrate this, Figure 4a shows that the "balanced modulator" is simply an analog multiplier; Figure 4b shows the block diagram of a "pulse-height-pulse-width" multiplier—one variable modulates the amplitude, the other modulates the duty cycle, and the area (measured by an averager) is proportional to the product of the two inputs. Historically, modulation was used in the design of multipliers far more frequently than multipliers were used for modulation. But now, with the coming of low-cost IC transconductance multipliers, the pendulum is swinging the other way. Analog multipliers are considered for a variety of modulation applications, from amplitude modulators (Figure 5) to frequency-modulated triangular, square, and sine waves (Chapter 2-2, Figures 3, 8, and 9).

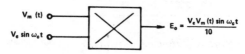

a. Multiplier as balanced modulator

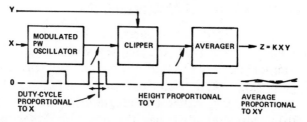

b. Pulse-height, pulse-width-modulation multiplier, first quadrant

Figure 4. Modulation and multiplication

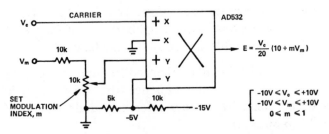

Figure 5. Multiplier as amplitude modulator

Voltage-symmetrical (but not necessarily time-symmetrical) triangular waves may be used to produce duty-cycle-modulated square pulse trains by biasing the triangular waves with the modulating waveform and detecting zero crossings with a precision comparator (Figure 6).

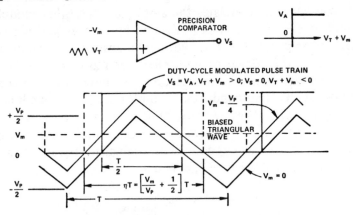

Figure 6. Duty-cycle-modulated triangular wave

FREQUENCY DOUBLING AND n-TUPLING

A multiplier, connected as a squarer, can be used to obtain low-distortion sine waves of twice the frequency of an input sine wave. The DC component of the output can be removed with a high-pass filter (Figure 7a). Alternatively, one of the inputs can be phase-shifted by 90°, using either an integrator or an all-pass filter (Figure 7b). This alternative has the advantage that amplitude variations do not result in large transient "bounces" at the output; however, its performance is somewhat sensitive to frequency,

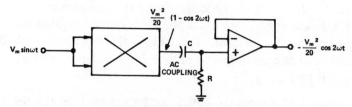

a. Multiplier as frequency doubler

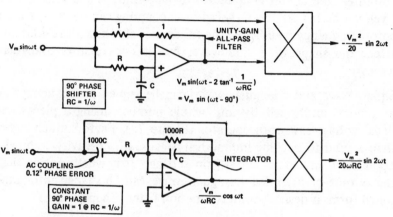

b. Multiplier as "low-bounce" frequency doubler – two approaches

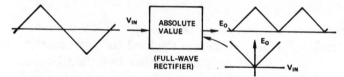

c. Absolute-value as triangular-wave frequency doubler

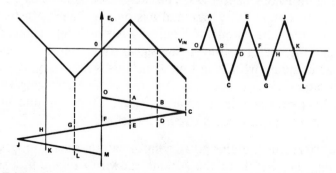

d. Triangular-wave frequency tripler using 3 piecewise-linear segments

Figure 7. Frequency multiplication (see chapters 2-1, 2-3, 3-5)

whereas that of Figure 7a is wideband for frequencies well above the filter's crossover. Typically, phase error of the double-frequency signal becomes significant at 1/100 of the multiplier's −3dB frequency, and the envelope amplitude loses accuracy above 1/10 of the −3dB frequency.

Frequencies of triangular waves can be doubled by the use of an absolute-value circuit (Figure 7c). If amplitude is constant, the dc level can be biased out. Otherwise, ac coupling can be used, with a cutoff frequency well below the fundamental (phase shift does not affect the shape of a sine wave, but it does distort triangular waves).

Square-wave and triangular waves can be tripled in frequency, or in general multiplied by any whole number, using a piecewise-linear voltage sawtooth operator (Figure 7d). Factors much larger than 3 tend to become impractical because of sensitivity to breakpoint drift and incremental gain settings. It is worth noting that the output of a tripled triangular wave can be shaped into sinusoidal form, if desired, using a function fitter (Chapter 2-1).

DEMODULATION

We have touched on peak, average, and RMS measurements in Chapter 2-3. Similar techniques are used for demodulating amplitude-modulated signals. Figure 8 shows two "ideal diode" high-accuracy full-wave rectifier circuits that are perhaps less well-known than Figure 13 of Chapter 2-3. The circuit of Figure 8a uses 5 equal resistors to obtain unity gain and has only a single path from the input source. The circuit of Figure 8b has high input impedance, an especially useful feature if the signal source must be unloaded or if ac coupling with long time constants is necessary. A possible disadvantage in both cases is that the output-averaging filtering must be performed in a separate stage. These circuits may be followed by a peak-reading circuit, if desired.

If the waveform contains polarity information, *synchronous-detection* may be useful. In the scheme shown in Figure 9, a square-wave reference signal multiplies the alternate half-cycles by positive and negative constant voltages. If signal and reference are in

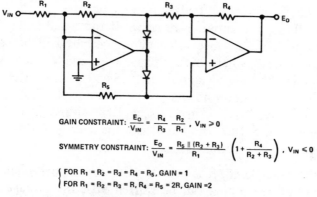

GAIN CONSTRAINT: $\dfrac{E_O}{V_{IN}} = \dfrac{R_4}{R_3}\,\dfrac{R_2}{R_1}$, $V_{IN} \geqslant 0$

SYMMETRY CONSTRAINT: $\dfrac{E_O}{V_{IN}} = \dfrac{R_5 \parallel (R_2 + R_3)}{R_1}\left(1 + \dfrac{R_4}{R_2 + R_3}\right)$, $V_{IN} \leqslant 0$

$\begin{cases} \text{FOR } R_1 = R_2 = R_3 = R_4 = R_5,\ \text{GAIN} = 1 \\ \text{FOR } R_1 = R_2 = R_3 = R,\ R_4 = R_5 = 2R,\ \text{GAIN} = 2 \end{cases}$

a. Full-wave rectifier circuit

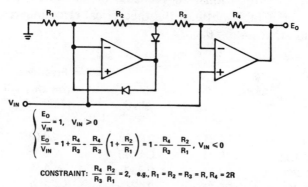

$\begin{cases} \dfrac{E_O}{V_{IN}} = 1,\quad V_{IN} \geqslant 0 \\[2mm] \dfrac{E_O}{V_{IN}} = 1 + \dfrac{R_4}{R_3} - \dfrac{R_4}{R_3}\left(1 + \dfrac{R_2}{R_1}\right) = 1 - \dfrac{R_4}{R_3}\,\dfrac{R_2}{R_1},\ V_{IN} \leqslant 0 \end{cases}$

CONSTRAINT: $\dfrac{R_4}{R_3}\,\dfrac{R_2}{R_1} = 2$, e.g., $R_1 = R_2 = R_3 = R$, $R_4 = 2R$

b. High-input-impedance full-wave rectifier circuit

Figure 8. Absolute-value (full-wave rectifier) circuits

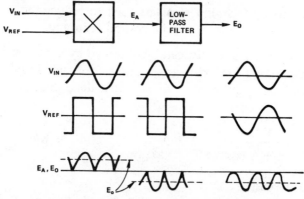

Figure 9. Synchronous (phase-sensitive) detection

phase, the full-wave-rectified output is positive; if they are in opposite phase, the output is negative. If the signal and reference are sinusoidal, the average value of the output will be equal to $(V_{rm} V_{sm}/20) \cos\theta$, where θ is the phase angle and V_{rm} and V_{sm} are the reference and signal amplitudes.[2] Small phase shifts do not greatly affect detection accuracy; for example, $0.8°$ gives 0.01% error, $2.56°$ gives 0.1%, $8°$ gives 1%, and $18°$ gives 5%. If the signal and reference are $180°$ out of phase, the average output will be negative, with the same ideal phase tolerances.

If, on the other hand, it is desired to measure small *phase* deviations, one of the inputs can be shifted $90°$; the average output will then be proportional to the sine of the phase angle. The following brief table outlines the theoretical error inherent in the assumption that $sin\ \theta = \theta$.

Angle		Sine	Fractional Error (% 1rad)
θ_{rad}	$\theta°$	$\sin \theta$	$\|\sin\theta - \theta_{rad}\|$
0.084	4.813	0.0839	< 0.01%
0.180	10.31	0.1790	< 0.1%
0.390	22.35	0.380	< 1.%
0.490	28.1	0.471	< 2.%
0.670	38.4	0.621	< 5.%

Function-fitting techniques can be used to reduce the error if the range of angle is too large for the desired accuracy.

Greatly-improved linearity can be obtained by combining sine and cosine demodulation with an implicit feedback loop to obtain "tan-lock" demodulation[3]. A tan-lock demodulator solves the equation

$$-E_O = \frac{B\ \sin\theta}{1 + A\ \cos\theta} = B\ \sin\theta\ +\ AE_o\ \cos\theta \cong K\theta \qquad (1)$$

[2]See Figure 18, Chapter 2-3.

[3]"Use this Tan-Lock Demodulator," by R.P. Hennick, *Electronic Design* No. 25, December 6, 1970, pp74-75.

as shown in Figure 10. In addition to the improved linearity over a wide range of angle, as shown in the error plot, with its attendant reduction of distortion, one might expect to realize improvements in noise threshold, hold-in range, and pull-out frequency (see discussion below).

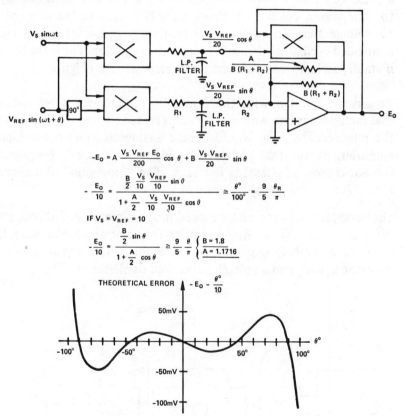

Figure 10. "Tan-lock" demodulator circuit, scaling, and theoretical error

Phase demodulators are often preceded by AGC or limiting circuits to ensure constant ac input amplitude and avoid amplitude modulation of the output. Wideband multipliers, such as the 429, have less than 1° of differential phase shift at 1MHz. The output double-frequency phase shift of about 24° @1MHz is unimportant, since only the dc component of the output is used; the dc level depends critically only on the input frequency characteristics.

PHASE-LOCKED LOOPS

A phase detector may be used as the "summing point" of a feedback loop that generates a frequency that is compared with the average input frequency and "locked in" to that frequency with a fixed (e.g., 90°) phase relationship (in the steady state, ideally, for sine waves, cos θ = 0). Phase error is usually in the form of a dc voltage that drives the local frequency generator (a voltage-controlled oscillator) through a high-gain amplifier. Thus, the loop, if stable, seeks to maintain the phase error at zero (Figure 11).

To anyone familiar with the principles of feedback (and today, that includes anyone who uses op amps creatively and successfully), the phase-locked loop would appear analogous to an operational amplifier, except that phase is the input variable and frequency (rate-of-change of phase) is fed back. The "loop gain" of a phase-locked loop is expressed in terms of %Δf/radian.

The basic elements of a phase-locked loop, as mentioned above, are the phase detector, a filter-amplifier (to remove ac components from the dc voltage that represents the phase error and to amplify the error signal), and a voltage-controlled oscillator (VCO).

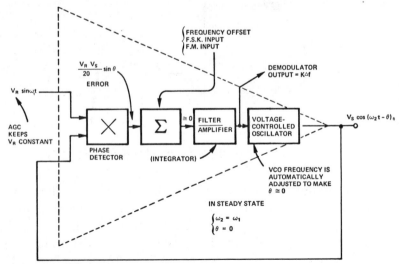

Figure 11. Phase-locked loop, or phase follower, with response to sinusoidal signals.

Applications of phase-locked loops are in two basic classes: frequency re-creation and multiplication, and narrow-band filtering, based on the ability to respond to an input frequency; and frequency modulation and demodulation, based on the ability of a well-designed phase detector and VCO to respond accurately, stably, and linearly to a dc voltage.

In the first class, a received signal may be noisy, distorted, and actually jittering in frequency. The job of the phase-locked loop is to generate a clean waveform of appropriate shape that is locked-in to the average signal frequency. The filter prevents the output frequency from responding to rapid fluctuations of phase; it tends to null out the average phase error. The VCO need not be very linear (the same may be said for the phase detector), but the range of frequencies, or phase, within which the loop is captured (i.e., under control) must be sufficiently wide to embrace the expected fluctuations, either at the signal frequency or a harmonic (if the job of the loop is frequency multiplication).

In the second class, the VCO and/or the phase detector should have a linear relationship over a dynamic range corresponding to the maximum deviation present in the modulated signal. The filter should be slow enough to filter out carrier, but fast enough to follow the modulation. If the loop is acting as a demodulator for a frequency-modulated signal, the output of the VCO will track the modulation, and the "dc" input of the VCO will be the demodulated output voltage. If the loop is acting as a modulator, the modulating signal is added at a voltage summing point after the phase detector. The output frequency will change to the degree necessary to create a phase-error voltage that will continuously balance out the disturbance caused by the modulating signal, while remaining locked-in to the input frequency.

An intermediate class is "frequency-shift keying" (FSK), in which the modulating signal is a step change of voltage (such as a change of binary logic levels), which produces a step-change of frequency. On the receiving end, the loop responds to a step-change of frequency with a step-change of voltage. This form of operation is often referred to as "modem" (modulate-demodulate). A linear relationship is not required, but the phase-voltage-frequency band should be adequate.

Signals and VCO outputs may have any waveshape, as long as one can ensure that the loop will not lock in on a harmonic (unless so desired). Since a phase-locked loop is an active closed-loop system, it must be designed not to be dynamically unstable (i.e., to run away or oscillate, rather than behaving as desired). Two ranges of frequency are usually important in governing lock-in: *capture* (pull-in) *range,* the band of frequencies within which a lock-in condition can be acquired, and *lock* (dropout) *range* (tracking or holding), the band of frequencies within which lock-in can be maintained, always wider than the capture range.

Some ideas about VCO design may be found in Chapter 2-2 (and in many other places in the literature). The low cost of modular and IC multipliers makes high-performance medium-frequency phase-locked loops quite practical today. For the future, one may expect to find integrated-circuit phase-locked loops that transcend in performance today's elementary IC devices, requiring considerably fewer external components, and more suitable for high-precision applications.

VOLTAGE-CONTROLLED FILTERS

A "state-variable" active filter is one in which an analog-computing feedback loop (or loops), involving one or more integrators, is used to simulate the desired transfer function. Though less compact than the usual active-filter circuit, in terms of the number of op amps required to obtain a transfer function characterized by an n-degree polynomial, it has an important advantage: If an integrator is preceded (or followed) by an analog multiplier (or divider), the overall characteristic frequency ω_0 (or characteristic time T_0) will be directly proportional to the multiplying or dividing input voltage. This makes it possible to build filters in which the capacitors are, in effect, adjustable, either directly or inversely, by a control voltage. For a filter that does not involve inductors, is thus possible ideally to manipulate the frequency scale of a filter by means of a single voltage, without affecting any other parameters. The cost is one multiplier per capacitor, formerly impractical, but now quite feasible, because of the low price of multiplier/dividers.

Figure 12 shows how a multiplier (a) or a divider (b) can be used to adjust the "break" frequency or time constant of a first-order lead or lag (for a lead-lag (c), the outputs are summed in an external adder-subtractor with appropriate polarities and coefficients). Note that the multiplier is used ahead of the integrator, with passive summation at its input, while the divider follows the

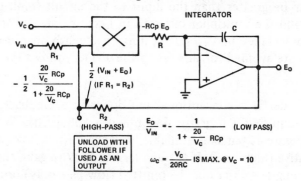

a. Multiplier adjusts break frequency of unit-lag circuit.

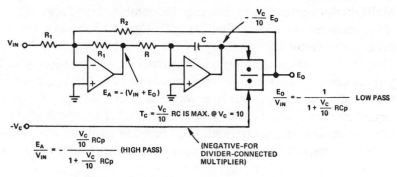

b. Divider adjusts time-constant of unit-lag circuit

c. Adder circuit combines outputs of (b) to form lead-lag response (normalized)

Figure 12. First-order variable filters using multipliers or dividers

integrator. The reason for this can be seen if one considers the consequences of, for example, using the multiplier following the integrator. If the multiplier's output is 10V, then for *any* value of V_c less than 10V, the integrator output must be greater than 10V. Placing the multiplier ahead of the integrator solves the out-of-range problem, because the integrator output, in the closed loop, can never be greater than the input to the circuit (multiplied by R_2/R_1), and the multiplier output can never be greater than 10V because of its inherent $V_1 V_2/10$ scaling. In like fashion, out-of-range problems are avoided for this case if the divider *follows* the integrator.

If the divider is a conventional multiplier in a feedback configuration, requiring a negative denominator voltage, the configuration of Figure 12b makes available both the output (low-pass) and its derivative (high-pass). If the divider has *positive* gain, the inverting summing amplifier may be omitted (low-pass only) or replaced by a *non*-inverting summer (high pass and lead-lag).

Multiplier-integrator elements can be combined to form higher-order state-variable filters. For example, Figure 13 shows a second-order filter (note the similarity to the oscillator circuit of Fig. 9, Chapter 2-2). Depending on which output or combination of outputs is used, it can serve as a high-pass, low-pass, band-pass, band-reject, all-pass, etc. Again, it is important to note that, for ideal circuit elements, the control voltage V_c affects only the frequency scale; damping and coefficient weightings, normalized frequency-response characteristics, and normalized time response are all unaffected.

This feature is especially useful when first- and second-order filter responses are cascaded to obtain nth order Butterworth, Chebyshev, Bessel, or other response characteristics. Variation of V_c to adjust cutoff frequency does not affect the coefficient weightings, once the relative-frequency-and-damping relationships have been set.

Where digital control is desired, the multiplier blocks could be embodied by multiplying D/A converters, such as the IC AD7520.

Variable time-constant integrators, and the tunable filter networks that they make possible, can be usefully and profitably employed

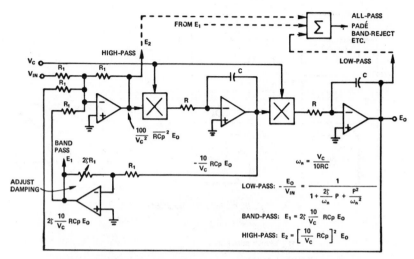

Figure 13. 2nd Order filter with variable natural frequency, using two multipliers. If desired, damping could also be controlled via a multiplier

in a number of ways. Examples include adaptive control (adjustment of time constants to achieve an automatically-minimized control-loop error function), spectrum analysis (variable-frequency sweeps to obtain amplitude spectra of stationary waveforms), variable analog delay lines, variable-bandwidth systems for audio "hiss" and "rumble" noise reduction, tunable-carrier transmitters and receivers for narrow-band signals, programmable filters, etc. If the filter is controlling an oscillator frequency, it may be used in FM detection with a phase-locked loop, where the input to the filter is the phase-detector output voltage required for tracking, proportional to the modulating signal.

SPECTRUM ANALYZERS

There are many ways of analyzing and plotting a signal spectrur A few that are relevant to the techniques and devices described in these pages are "spot" measurements (or frequency "combs"), band measurements, and swept measurements. The first two types can be achieved either with fixed-frequency (narrow-band or bandpass) filters in parallel or with a single stepped (narrow-band or bandpass) filter (Figure 14). If many frequencies or bands are to be measured, the serially-stepped-filter approach is considerably

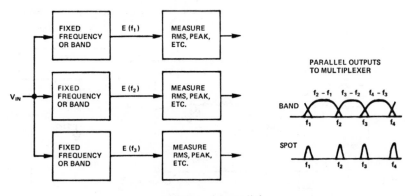

a. Filters in parallel

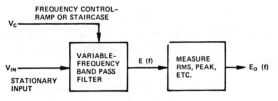

b. Single variable-frequency filter

Figure 14. Spectrum analyzers

more economical in terms of equipment, but consumes more time for the measurement. The swept filter may provide a continuous "spot" measurement; however, the sweep must be slow enough to not introduce substantial errors as a result of its rate of variation. The dc measure can be obtained in terms of rms, peak, average, or "one-shot-per-step" integral measurement.

MUSIC SYNTHESIZERS

These versatile instruments serve as a tonal palette embracing an extremely wide range of audio waveforms and sounds for the ministrations of the musical composer, performing artist, and special-effects creator. They permit a wide range of pitches, tones, attack-decay-sustain-release sequences, amplitudes, and combinations, both linear, and nonlinear. They tend to use the whole gamut of waveform processing trickery, including voltage-controlled amplifiers, voltage-controlled oscillators, voltage-controlled filters, modu-

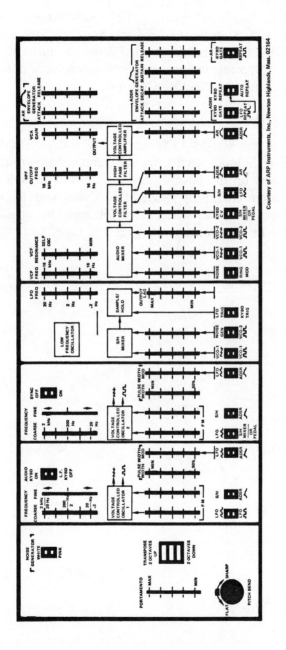

Figure 15. Typical music-synthesizer functions

lators and demodulators, phase-locked loops, sample-holds, noise generators, and pressure-sensitive transducers. Both analog and digital (ROM) functions are used. Figure 15 shows the control panel of a typical commercially-produced moderately-priced keyboard instrument, the ARP Odyssey.

CONCLUSION

This chapter has sought to touch briefly and suggestively on a number of techniques used in audio communications and signal-processing, and on the possible contributions of today's low-cost, compact, comprehensively-specified modular and IC nonlinear devices. It is hoped that the reader will consider omissions and elisions (due to the pressures of space and time) as a challenge to creativity.

Computing & Control

Chapter 5

In this final chapter of the Applications section, we discuss a few ways that nonlinear analog computing techniques are used in industry, suggest a few additional ones, review further applications of ideas suggested earlier, and, in effect, present a modest list of topics that, possibly landing in fertile ground, may be fruitful in terms of the ideas that are inspired in thoughtful readers. It is always important to bear in mind that improvements in device performance, reductions in cost and size, and ready availability from multiple sources, have brought many of these ideas from the status of "merely interesting" to feasibility as everyday tools of the designer.

FILTER-FREE THREE-PHASE POWER MEASUREMENT

Figure 1 shows the block diagram of a simple scheme for

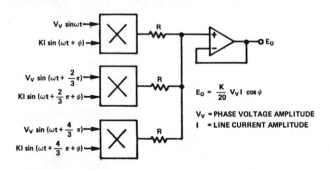

$$E_0 = \frac{K}{20} V_V I \cos \phi$$

V_V = PHASE VOLTAGE AMPLITUDE
I = LINE CURRENT AMPLITUDE

Figure 1. Three-phase average-power measurement without filters

computing the average power in a three-phase system.[1] Voltages proportional to the phase voltages and the corresponding line currents are multiplied individually in three multipliers, and the outputs are summed. The output is ripple-free for a balanced system, with the three average levels averaged and the double-frequency ac components cancelled.

Since low-pass filtering, with its essential delays, is not needed, rapid measurement or detection of the power level is made possible. If one input of each multiplier is phase-shifted by $90°$, the output will be a continuous measurement of *reactive* power, $KV_V I \sin\phi$. The $90°$ phase shift can be obtained by measuring the *line-to-line* voltages, and absorbing the stray $\sqrt{3}$ factor in the analog circuitry.

The improved speed of response makes possible faster-responding, more-stable control loops, and clean, easily-metered monitoring signals. An interesting application is in the excitation control of synchronous motors.

Besides real and reactive power, other useful output signals may be obtained by combining the real and reactive power measurements in various ways. For example, the square-root of the sum of the squares (Figure 20, chapter 2-3) may be used to compute total volt-amperes. The ratio of the power to volt-amperes is the *power factor* ($\cos\phi$), while the ratio of reactive power to volt-amperes is a good approximation to the phase angle for small angles. The ratio of real power to reactive power, $\tan\phi$, a nonlinear function of the power factor, is particularly useful as a control signal because of its high sensitivity.

Finally, it is often useful to control the excitation of a synchronous motor so that it is overexcited at low loads, with reduced excitation as load increases, to avoid exceeding the normal current limitations of the motor at full load. The control criteria for this operation can be established by setting a simple linear combination of the reactive and real power equal to a constant.

[1] "Detection and Measurement of Three-Phase Power, Reactive Power, and Power Factor, with Minimum Time Delay," by I. R. Smith and L. A. Snyder, *Proc. IEEE*, November, 1970, p. 1866.

RATIOMETRIC MEASUREMENTS – LIGHT TRANSMISSION

Figure 2 shows a scheme commonly employed to measure transmittance or absorbance of light by an unknown medium, independently of variations of the light-source level. The light is transmitted through a reference medium (which might be air or vacuum) and through an unknown. Both samples are transduced to current by a pair of matched photosensitive detectors. The 756 log-ratio module converts the output currents (at essentially zero input impedance) to an output voltage proportional to the logarithm of their ratio.

Since the measurement is ratiometric, it is independent of the source intensity. Since it is logarithmic, it can deal accurately with a wide range (4 decades) of unknowns, and furthermore it can be read out directly in logarithmic absorbance or transmittance units.

Always useful, ratiometric measurements (with analog dividers) in general, and log-ratio measurements in particular, are becoming increasingly feasible and accessible for an ever-wider variety of applications as cost decreases and availability and performance increase.

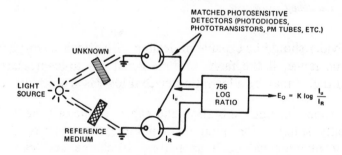

Figure 2. Measuring light transmission independently of light-source variations. Log-ratio gives direct reading of relative transmittance or absorbance

EXPONENTIAL DECAY TIME-CONSTANT

Figure 3 shows a circuit that can be used to rapidly measure, compute, and display continuously the time constant of an exponential decay. For example, a 10-minute time constant can be measured within seconds.

The operating principle is simple: the time-derivative of $\epsilon^{-t/\tau}$ is equal to $-(1/\tau)\epsilon^{-t/\tau}$. Therefore, if we divide the argument by the negative of its time-derivative, the result is the time constant τ, available immediately after the startup transient has died away.

In order that the differentiator be stable and not have excessive noise at high frequencies, it has a second-order rolloff ($R_c C$ and RC_c). Naturally, these time constants should be short compared to the shortest time constants being measured, but they should be no shorter than is necessary on that account.

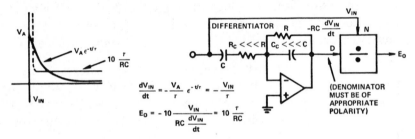

Figure 3. Circuit for determining time constant of exponential decay

The divider should be capable of dealing with signals having a wide dynamic range, if the range of time constants to-be-measured is substantial. It may be a log-ratio device if $\log \tau$ is acceptable.

Applications include calibration and capacitor measurements. It is especially suitable for obtaining rapid measurements of slowly-varying phenomena, such as battery discharge and capacitor-dielectric "soakage." By recording the measurement continuously, or sampling the waveform from time-to-time, the "quality" of the time constant can be investigated (e.g., is the response truly exponential?). The differentiator must not introduce substantial errors; therefore, though the initial tolerance is not critical, the capacitor should be the highest grade available (polystyrene, teflon, polycarbonate, etc.), and the amplifier should have low leakage current and low noise.

MASS GAS FLOW COMPUTATION

This is an example of the use of low-cost nonlinear analog circuit elements in the conditioning of transducer outputs to obtain an essentially direct measurement of a quantity that depends on a number of variables.

The measurement of gas flow through a resistive element, such as a nozzle, venturi, or an orifice, requires that we know the absolute pressure, the absolute temperature, and the pressure-drop. An equation typically used to relate the gas flow to these variables is*

$$F = K_1 \left(1 - K_2 \, \frac{\Delta P}{P} \right) \sqrt{\frac{P \Delta P}{T}} \qquad (1)$$

If ΔP is small compared to P, this expression simplifies to the frequently-used

$$F = K \sqrt{\frac{P \Delta P}{T}} \qquad (1a)$$

Figure 4a shows how a divider and a multiplier-divider can be used to compute equation (1a). For a fixed value of K, the electrical inputs are scaled in the preceding preamplifiers so as to utilize the full output range (E_1), and as much of the input ranges as is consistent with the various combinations that produce full output. If the input divider is a multiplier-divider, the third input can be a constant voltage with the effect of adjusting K^2.

Figure 4c shows how logarithmic circuits can be used to embody equation (1a). If all three variables can vary widely, the logarithmic approach is the more useful, because it allows the scaling to be flexible, without fear of overranging, as long as the output is properly scaled.

If equation (1) is used, it can be embodied by feeding the output of equation (1a) into the circuit of Figure 4b. It amounts to subtracting from E_1 a correction term proportional to $E_1 V_{\Delta P}/V_P$,

*NASA Tech Brief 71-10407, Lewis Research Center, J. Watson, D. Noga, J. Dolce, and J. Gaby, Jr., "Low-Cost Logarithmic Mass Flow Computer"

with a coefficient K_2, determined by the resistor ratio R_1/R_2.* The output of the flow circuit can be integrated to determine the total mass transferred over a period of time, or it can be averaged by a unit-lag or other averaging filter.

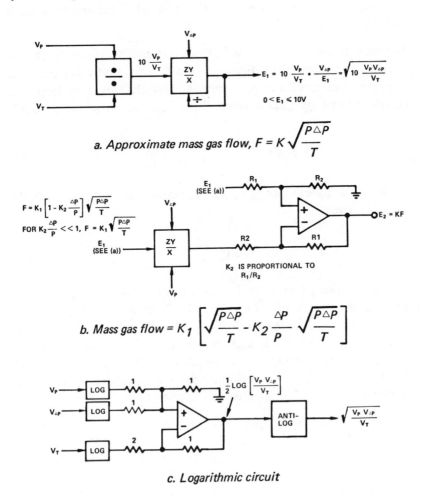

a. Approximate mass gas flow, $F = K\sqrt{\dfrac{P\triangle P}{T}}$

b. Mass gas flow $= K_1\left[\sqrt{\dfrac{P\triangle P}{T}} - K_2\dfrac{\triangle P}{P}\sqrt{\dfrac{P\triangle P}{T}}\right]$

c. Logarithmic circuit

Figure 4. Mass gas flow configurations

*The subtractor configuration of Figure 4b may be found useful for implementing many of the circuits in this book (and elsewhere) that call for differences of the form $(x - Ay)$. The gain or attentuation, $-A$, is determined by R_1/R_2, and the gain at the plus input will be unity if the resistor ratio is matched as shown.

OXYGEN CONCENTRATION WITH ANTILOGS

Electrical measurements of ion concentrations are logarithmic. For solutions at $29^{\circ}C$, a relative concentration change of one decade (i.e., $\times 10$ or $\times 0.1$) produces a 60mV change, of appropriate polarity, at the measuring electrode.

If it is desired to measure the actual (relative) concentration, the output of the detector is preamplified and applied to the input of an antilog circuit. A circuit for performing this job is shown in Figure 5.[2]

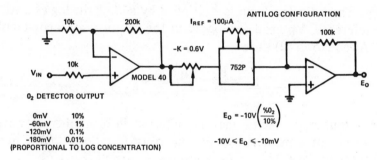

Figure 5. Linearizer for oxygen detector (see Chapter 4–3)

TRANSIENT-FREE RANGING PICOAMMETER

The conventional electrometer circuit using an inverting operational amplifier requires large-value feedback resistors to convert the input current to an output voltage. If the input covers a wide range, either manual or automatic range-switching may be needed, involving several large-value resistors.

There are a number of discomforting factors to consider when designing such circuits. First of all, resistances in the $10k M\Omega$ region, and greater, are difficult to obtain with tight tolerances and good stability vs. time and temperature. Stray capacitance tends to make the response of these circuits quite slow. Range switches tend to have leakage and capacitance; besides the inherent steady-state errors, the settling time after switching can

[2]See also Chapter 4–3.

be of the order of many seconds, certainly inappropriate for autoranging. In addition, there are all the inherent problems of low-level current measurement by *any* means: cable problems, stray capacitance and leakage, and amplifier input-circuit problems.[3,4]

A workable answer to this problem (Figure 6) involves the use of log-antilog circuitry, similar to that employed in the Model 434 multiplier-divider, but with one of the inputs designed specifically for electrometer-level current-handling. The input current is "logged" in the feedback circuit of A1, and referred to an input I_{REF}. The ratio I_{IN}/I_{REF} is multiplied by adding the log of a voltage reference, V_R, and antilogged in the circuitry associated with A4. As equation (2) shows,

$$E_o = \frac{R_2}{R_1} \cdot \frac{I_{IN}}{I_{REF}} \, V_R \qquad (2)$$

the output scale factor may be adjusted in a number of ways, separately or concurrently: by the resistor ratio, R_2/R_1, by a reference voltage, V_R, or by a reference current, I_{REF}. I_{REF}, in turn, may be determined by a stable low-current source.

Since the scale factor is proportional to V_R, the gain may be set directly by a voltage, without the need for switch circuitry in automatic ranging. The saturation current of the log transistors is quite low, typically well below 10^{-13} A at 25°C. Since the saturation currents of the two transistors in each pair are monolithically matched, temperature affects the ratio negligibly. Because the monolithic dual transistors are essentially at the same temperature (in close proximity), the kT/q terms cancel, and the performance of the circuit is essentially independent of temperature. When ranges are switched, the circuit recovers quickly (from milliseconds to microseconds), because the switching is remote from the picoampere-level circuitry.

---

[3] See "The World of fA—Op Amps as Electrometers," *Analog Dialogue*, Volume 5, No. 2.

[4] "High-Performance Flame-Ionization Detector System for Gas Chromatography," *Hewlett-Packard Journal*, Volume 24, No. 7.

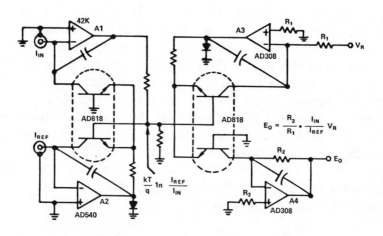

Figure 6. Temperature-compensated wide-range picoammeter with normal resistance values, non-interactive range scaling and voltage-adjustable scale factor

CORRELATION AND CONVOLUTION

These topics involve equations of the typical form

$$F(\tau) = \int_0^T f(t) \cdot g(\tau - t)dt \tag{3}$$

While there is simply not enough space available in the present volume even to touch (however inadequately) on these topics, they must nevertheless be mentioned, because the high speed and low cost of multipliers and multiplying D/A converters, and their small space requirement, makes analog or partially-analog approaches more competitive with digital techniques than has been the case in the past. A typical circuit that embodies (3) is shown in Figure 7.

Correlation is used as a means of recovering information in the presence of noise or unrelated signals. If the information is sinusoidal, and the "noise" is an out-of-phase component at the same frequency, equation (3) can be recognized as a phase-sensitive detector, where τ is the delay corresponding to the phase-shift ϕ. For signal waveshapes having less-predictable properties, τ is the delay of an adjustable delay line. The integration is performed a number of times (depending on the desired resolution), for various values of τ up to the full period, and each integration reconstructs one point on the correlation function. Adjustable delay lines are available, for short delays, in analog form (e.g., "bucket-brigade" types), and for arbitrary delays, in digital form.[5] The most-popular forms of correlation are auto-correlation and cross-correlation, determined by the relationship between the functions f( ) and g( ).

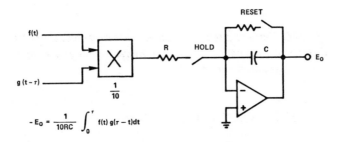

Figure 7. Basic analog correlation circuit

Convolution of time functions corresponds to multiplication of their transforms in the s or $j\omega$ domains. Multiplying the transform of an analog signal by the transform of the indicial (step, pulse, etc.) response of a linear circuit (i.e., its complex transfer function) provides the transform of the time response of the circuit to the analog signal. It is therefore possible to *model* the time response of a circuit to a time waveform without actually building the circuit by a series of convolutions of the input waveform with an independently-generated waveform that has been fitted* to conform to the desired indicial time response. This is an especially useful technique if the desired indicial response requires an unreasonable or not-physically-attainable transfer function.

[5]*Analog-Digital Conversion Handbook*, Analog Devices, Inc., 1972
*See chapters 2-1 and 2-2 on function fitting and function generation.

ALARM CIRCUITS

In a system, there are usually a number of variables the magnitudes of which are unimportant from the standpoint of their contribution to the equations of the performance or efficiency of the on-going process, but must nevertheless be maintained within given tolerances. While it is possible to convert and record or observe these variables, it is usually cheaper and simpler to take note of them only when they have deviated beyond one or more sets of thresholds.

Figure 8 shows three circuits that can be used to activate alarms if

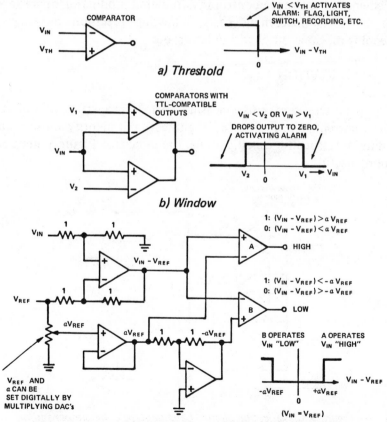

a) Threshold

b) Window

c) Fractional deviation ($\pm aV_{REF}$) from preset reference (V_{REF})

Figure 8. Alarm circuits

the input exceeds or falls below a preset threshold (a), if the input departs from a prescribed range of operation (b), or if the input departs from a reference value by more than a prescribed percentage (c).

These are the simplest kinds of deviations requiring alarm. Most others can be constructed using them as basic elements. Using the Serdex system (see page 88), it is possible to obtain remote information on the state of both alarm-only and measured variables at the same time via the same twisted-pair, in the form of a coded printout.

Naturally, these same alarm functions can be recognized as key elements of tolerance control in automated production operations of all sorts: machining parts, adjusting precision resistances, keeping machine speed within limits, etc.

CLASSIFICATION

For processes that must measure certain properties of objects (size, current-gain, resistance, brightness) and identify those units falling into specific classifications, the circuit of Figure 9 may be found useful.

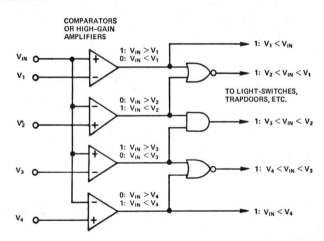

Figure 9. Classification circuit. Latching-type comparators and/or hysteresis may be used to reduce ambiguities and "cycling" due to noise

The input voltage, corresponding to the measurement of the property of interest, is compared with a series of graduated reference levels. The outputs of the comparators are processed by simple logical operations, the outputs of which indicate uniquely into which grade the object whose property is being measured should be placed. These outputs activate the appropriate trapdoor, indicator light, etc.

MEDIAN CIRCUIT

An analog signal may be transmitted along several redundant paths to improve reliability or reduce noise. While simple summation of the outputs will reduce uncorrelated noise, the output may be greatly in error if one path has failed in a saturation mode. One means of combination that secures both noise reduction and

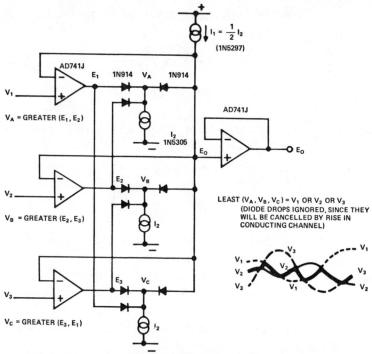

Figure 10. Median circuit continuously selects middle value among three inputs for noise reduction or improved reliability through redundant circuitry

protects against failure of one path involves computing the median signal. That is, the output is always the signal the value of which is between the other two signals.

Figure 10 shows one form of circuit that can compute the median of three input signals V_1, V_2, and V_3. It computes the greater of each pair (V_1, V_2), (V_2, V_3), (V_3, V_1), and then follows the least of the three "greaters." It will therefore follow continuously the signal which is neither the greatest nor the least, irrespective of which one it happens to be at a given instant.

It has been claimed possible, using similar circuitry, to design a circuit that will follow the mth in magnitude among n input signals.[6]

TRIGONOMETRIC FUNCTIONS AND COMBINATIONS

Rectangular-to-polar conversion involves computations of the form (vector composition)

$$\left. \begin{aligned} r &= \sqrt{x^2 + y^2} \\ \theta &= \tan^{-1}(y/x) \end{aligned} \right\} \tag{4}$$

and polar-to-rectangular (vector resolution) involves the inverse operation,

$$\left. \begin{aligned} y &= r\sin\theta \\ x &= r\cos\theta \end{aligned} \right\} \tag{5}$$

Chapter 2-1 has discussed in great detail the manner of fitting $\sin\theta$. Since the cosine of an angle θ is the sine of $(90° - \theta)$, circuits that fit $\sin\theta$ can also be used to fit $\cos\theta$ (Figure 11). Usually, two sine-function-fitters are needed per angle, but if the signal varies slowly, multiplexing may be used to allow one function fitter to share sine and cosine (of a number of different angles, if necessary).

[6]"Analog Sorting Network Ranks Inputs by Amplitude and Allows Selection," *Electronic Design* 2, January 18, 1973, and sequel, *Electronic Design* 17, August 16, 1973, p. 7.

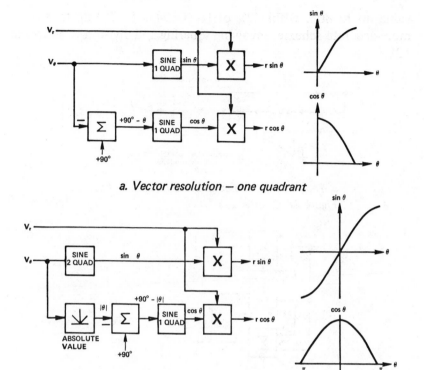

a. Vector resolution — one quadrant

b. Vector resolution — two-quadrant

c. Cosine approximation — 4 quadrants

Figure 11. Cosine approximations using sine function fitters

Vector composition (4) is discussed in the text that accompanies Figures 20 and 21 in Chapter 2-3.

If it is necessary to compute the tangent of an angle, the two schemes outlined in Figure 12 may be of interest. The first, based on the arctangent scheme, is the simpler but somewhat less accurate (within 1.4% of tan 45° up to 50°, within 2.5% of ideal value up to 80°, within 2% of 10 (i.e., tan 84.3°) up to 86°. A more-accurate scheme, involving squaring and division, is shown in 12b.[7]

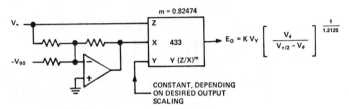

a. Tangent approximation based on arctan circuit of Figure 21, Chapter 2-3

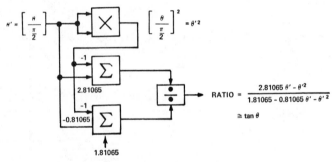

b. Tangent approximation with multiplication, division, and summing. Theoretical absolute error less than 0.0001 to 63°, relative error less than 0.1% to 72°, less than 0.5% to 81°, less than 1.2% for all values of θ < 90°.

Figure 12. Approximations to the tangent

Figure 13 is a circuit for computing the *difference* between two orthogonal quantities, using an XY/Z device (e.g., the AD531 multiplier-divider) for implicit square-rooting:

$$E_o = \sqrt{V_a{}^2 - V_b{}^2} = (V_a + V_b)(V_a - V_b)/E_o \qquad (6)$$

To minimize the use of external amplifiers, the sum-term is obtained passively and the difference is inherently available at the

[7]The equation that this scheme embodies is similar to that used (in a different guise) in a patented digital-to-synchro converter designed by F. H. Fish, of the U.S. Naval Avionics Facility, Indianapolis, Indiana.

differential "x" inputs. The AD741J feedback amplifier converts the output voltage to a current. When properly calibrated, and adjusted for less than 100mV error at full-scale, the output of this circuit will differ from the theoretical value by less than ±100mV for any pair of input voltages over an output dynamic range between 10V:0.3V and 10V/0.1V. Bandwidth is dc to 100kHz for best accuracy and 600kHz for -3dB error.

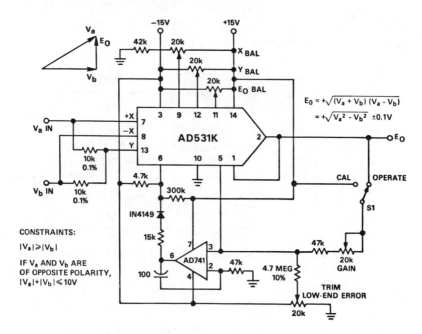

CALIBRATING THE CIRCUIT

Step	Condition	Adjust	For
1.	CAL, $V_a = V_b = 0V$	E_O BAL	$E_O = 0V$
2.	CAL, $V_a = 20Vp\text{-}p$, 10Hz, $V_b = 0V$ Pin 13 (AD531) grounded Scope sensitivity 50mV/cm(V), 100ms/cm(H)	Y BAL	Min. E_O swing
3.	CAL, $V_a = V_b = 0$, 20Vp-p to pin 13 Scope sensitivity as in (2)	X BAL	Min. E_O swing
4.	OPERATE, $V_{ia} = 10.00V$, $V_b = 0V$	GAIN	$E_O = 10.00V$
5.	OPERATE, $V_a = 1.00V$, $V_b = 0V$	"Low end"	$E_O = 1.00V$
6.	OPERATE, $V_a = V_b = 0V$	E_O BAL	$E_O = 0V$

Figure 13. Vector-difference circuit using AD531

ADAPTIVE CONTROL (Figure 14)

It has been noted, in Chapter 2-4, that the multiplier is essentially a remotely-operated gain control. Because it is free from the reliability and speed problems of servoed potentiometers, and is several orders-of-magnitude lower in cost, it is a natural choice for variable gains in adaptive control systems. If the gain criteria are determined by analog computation, analog multipliers (or dividers) are used; if they are determined digitally, multiplying D/A converters, such as the monolithic AD7520, are used.

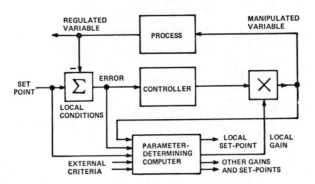

Figure 14. Adaptive control loop: multiplier as loop-gain adjuster

LINEARIZATION

Besides transducer linearization (discussed at length in Chapter 2-3), another interesting applications area for nonlinear devices, especially high-speed multipliers, lies in linearizing inherently nonlinear displays. For example, aside from other sources of nonlinearity and error, the spot-position on the face of a non-spherical cathode-ray tube is subject to "pincushion" distortion and defocusing as a consequence of the inherent geometrical relationships.

As a result of such distortion, the x-coordinate, the y-coordinate, and the spot width are multiplied by a factor of the form

$$\sqrt{(a_1 x)^2 + (a_2 y)^2 + 1} \qquad (7)$$

where x and y are the deflection voltages.

In order to correct for this distortion, the deflection voltages must, in effect, be divided by this term. Although it is feasible, speed and accuracy limitations make it preferable to use an approach in which a correction term (usually small) is *added* to the deflection voltages at the deflection amplifier. In this way, introduction of additional nonlinearity is minimized, and additional delay through the correction circuit applies only to a small correction rather than the entire deflection signal.

The choice of a suitable additive function is not a matter agreed upon by all designers. Examples of functions that have been employed are:

$$kX = AV_x + BV_x(CV_x + DV_x^2 + EV_xV_y) \qquad (8)[8]$$

$$kX = AV_x + BV_x(V_x^2 + V_y^2) \qquad (9)[9]$$

$$kX = AV_x + BV_x + CV_x\sqrt{V_x^2 + V_y^2 + D} \qquad (10)[10]$$

The X-correction is shown in the above examples, but the Y-correction is similar in form, as is the focus correction.

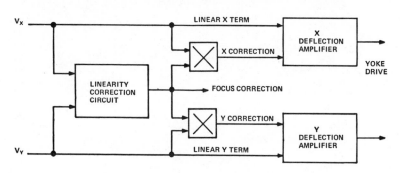

Figure 15. Linearity correction for cathode-ray tubes

[8] "Linearize Your CRT Displays," *Electronic Design* 17, August 16, 1970.

[9] "IC Op Amps Straighten Out CRT Graphic Displays," *Electronics*, January 4, 1971.

[10] *Distortion Correction in Precision Cathode-Ray Tube Display Systems,* Intronics, Inc., 1970.

CONCLUSION

We have shown in these chapters many of the ways that nonlinearity can be, is being, and will be used by system designers to do jobs in a practical, economical, and (often) uncomplicated manner. We have considered function fitting, function generation, instrumentation and measurement, signal-processing, and sundry other applications, as examples of the near-universality of the analog approach.

In the next section, we shall inspect the devices that are used for these applications more closely, with an eye to learning more about how they are designed and understanding their important properties, characteristics, and specifications.

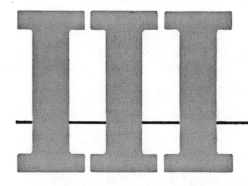

Logarithmic Circuits

Today's logarithmic circuits almost invariably use the inherent logarithmic properties of silicon junction devices. Readily available in monolithically-matched pairs, they are easily compensated for temperature variation, low in cost, and characterized by wide dynamic range, typically 10^{-2}A to 10^{-11}A.

In this chapter, we discuss their basic properties, techniques for obtaining both thermal and dynamic stability, some commonly-used circuits, specifications and definitions, and means of adjustment and test.

BASIC CONSIDERATIONS

An "ideal logarithmic diode" would be characterized by the current-voltage relationship

$$I = I_0 \; (\epsilon^{qV/kT} - 1) \tag{1}$$

Connected in the feedback path of an operational amplifier (Figure 1), it would constrain the output voltage to be

$$E_0 = \frac{kT}{q} \ln(I/I_0) = \frac{kT}{q} \ln(10) \cdot \log(I/I_0) \tag{2}$$

as long as $I/I_0 \gg 1$.

q   is a constant equal to the unit charge 1.60219×10^{-19}C
k   is Boltzmann's constant, 1.38062×10^{-23}J/°K
T   is absolute temperature, °K = °C + 273.15
I_o is the extrapolated current for E_o (= V) = 0

Typical round-number values in the vicinity of room temperature are:

°C	T(°K)	$\dfrac{kT}{q}$	$\dfrac{kT}{q} \ln(10)$
24.21	297.36	25.62mV	→59. mV
→25.	298.15	25.69mV	59.16mV
26.85	→300.	25.85mV	59.52mV
28.58	301.73	→26. mV	59.87mV
29.25	302.4	26.06mV	→60. mV

Thus, at 25°C, a 10:1 change of I would produce a 59.16mV change of E_o; an ϵ:1 change (2.7183) of I would result in a 25.69mV change of E_o.

Such a diode would be very useful. Since it is a 2-terminal device, it can be used for currents of either polarity; a number of them could be "stacked up" in series to obtain greater voltage; it can operate away from ground. Unfortunately, most diodes that are available as two-terminal devices have a limited range of logarithmic behavior. At the high end, ohmic bulk resistance produces an additional voltage drop:

$$V = \frac{kT}{q} \ln(I/I_o) + IR_B \qquad (3)$$

At the low end, the slope undergoes one or more changes of a multiple m* ($1 \leqslant m \leqslant 4$) to

$$V = m \frac{kT}{q} \ln(I/I_o) \qquad (4)$$

Since both the magnitude of m and the value of voltage at which the slope changes are functions of the individual device (within a family), general-purpose diodes are impractical for accurate

*This factor has been attributed to diffusion current flow in extended regions such as surface inversion layers or channels and to generation-recombination mechanisms in space-charge regions.

logarithmic operations over more than 1 or 2 decades, even though it is possible to devise a circuit to balance out the ohmic resistance. And special-purpose diodes are hardly price-competitive with diode-connected monolithic dual transistors.

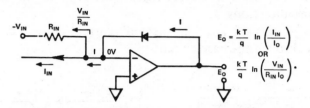

Figure 1. Ideal-log-diode circuit

THE TRANSDIODE CONFIGURATION

Figure 2a shows a transistor connected in the feedback path of an operational amplifier. The collector current is determined by the input current or voltage. The operational amplifier ideally maintains the collector current equal to the input current and holds the collector voltage at zero. Since the base is grounded, the collector and base are at the same potential, though the base current flows independently. The amplifier output voltage, which is also the emitter-to-base voltage, must be whatever value is necessary to meet the collector constraints, while furnishing any necessary amount of emitter current.

Let us now investigate the relationships governing this circuit. The modified Ebers and Moll equations[1] for emitter and collector currents of a grounded-base bipolar transistor are

$$I_E = I_{ES}(\epsilon^{qV_E/kT} - 1) - \alpha_I I_{CS}(\epsilon^{qV_C/kT} - 1) + \Sigma I_{ES_i}(\epsilon^{qV_E/m_ikT} - 1) \quad (5)$$

$$I_C = -\alpha_N I_{ES}(\epsilon^{qV_E/kT} - 1) + I_{CS}(\epsilon^{qV_C/kT} - 1)$$

$$+ \Sigma I_{CS_j}(\epsilon^{qV_C/m_jkT} - 1) \quad (6)$$

[1]"Multiplication and Logarithmic Conversion by Operational-Amplifier-Transistor Circuits," by W. L. Paterson, *The Review of Scientific Instruments*, 34-12, December 1963.

where

V_E and V_C are the emitter-base and collector-base voltages
I_{ES} and I_{CS} are the emitter and collector saturation currents
α_N and α_I are the current-transfer ratios in the normal and reverse directions
$m_i > 1$, $m_j > 1$ are "uncollected" current components that flow through the base circuit.

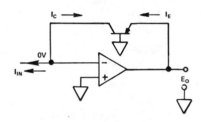 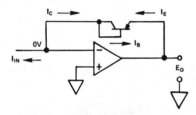

a. Transdiode (PNP)

b. Diode-connected transistor. As a two-terminal device, only one kind (NPN or PNP) will serve for either polarity of input current.

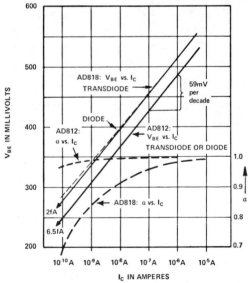

c. Transdiode and diode V_{BE} and α vs. I_C for two dual transistor types

Figure 2. Transdiodes and diode-connected transistors

For the circuit of Figure 2a, since V_C is held at zero, the relationship between collector current and emitter voltage (6) becomes

$$I_C = -\alpha_N I_{ES}(\epsilon^{qV_E/kT} - 1) \tag{6a}$$

Since the operational amplifier holds the collector current equal and opposite to the input current, the output voltage, V_E must be

$$V_E = \frac{kT}{q} \ln(I_{IN}/I_{ES}) - \frac{kT}{q} \ln \alpha_N{}^* \tag{6b}$$

for $I_{IN}/I_{ES} \gg 1$. Typically, I_{ES} is of the order of 10^{-13} A or less for most of the silicon planar transistor types used for log operations; therefore the relationship of (6b) should be valid over a very wide range of current. α_N is very nearly unity and essentially constant over the range of current for which (6b) is valid; therefore the ($\ln \alpha_N$) term is negligible (if $\alpha_N = 0.99$, its contribution is about ¼mV of constant offset). α_N in this equation should not be confused with the commonly-used grounded-base current gain $\alpha = I_C/I_E$: Since the emitter current includes both the collector current and the $m_i > 1$ terms, $\alpha = I_C/I_E$, always less than α_N, is a function of emitter voltage that decreases substantially at low values of current. Figure 2c shows plots of V_{BE} and α for two dual-transistor types widely used in logarithmic circuitry. Note that fidelity to logarithmic response is excellent, even at currents for which α is much less than 1.

If the collector and base of a transistor are connected together, a two-terminal diode is created (Figure 2b). Since the current that flows through it is the emitter current, its behavior, according to this model, is governed by (5). The first term is very nearly equal to the collector current, the second term is zero, and the sum of

---

*For PNP transistors, the input current I_{IN} is positive in the direction shown. If the input current is positive in the opposite direction, an NPN transistor is used. The polarities of the currents and voltages in equations (5) and (6) are reversed.

the $m_i > 1$ terms is therefore equal to the base current. Because

$$I_{IN} = -I_C - I_B = -I_C \left[1 + \frac{1}{h_{FE}} \right]$$

$$= \alpha_N I_{ES} (\epsilon^{qV_E/kT} - 1) \left[1 + \frac{1}{h_{FE}} \right] \qquad (7)$$

it is reasonable to consider that $1/h_{FE}$ is a measure of the $m_i > 1$ terms. From (7),

$$V_E = \frac{kT}{q} \ln(I_{IN}/I_{ES}) - \frac{kT}{q} \ln \left[\alpha_N (1 + \frac{1}{h_{FE}}) \right] \qquad (7a)$$

$1/h_{FE}$ is equal to $(1 - \alpha)/\alpha$, therefore the error term is equal to $+kT/q \ln(\alpha/\alpha_N)$. Typical values of error, according to this model, are

h_{FE}	α/α_N	$-\frac{kT}{q} \ln (\alpha/\alpha_N)$
$(\alpha_N \cong 1)$		mV @ 25°C
∞	1	0
1000	0.999	0.03
200	0.995	0.13
100	0.99	0.26
50	0.98	0.51
19	0.95	1.32
11.5	0.92	2.14
9	0.9	2.7
4	0.8	5.7
3	0.75	7.4
1	0.5	17.8

It is clear that a transistor to be used as a log diode should have high h_{FE} and maintain it over a wide range of emitter current. Figure 2c shows a comparison of AD812 (a high-h_{FE} dual monolithic transistor) and AD818 (a large-geometry dual-monolithic transistor having low bulk resistance), connected as transdiodes and as 2-terminal diodes. Though the AD818 would appear to have poor performance as a log diode at low current, its low series-resistance makes it suitable for log operation at currents exceeding 1mA, more than an order-of-magnitude better than the AD812 at high current.

OTHER SOURCES OF ERROR

If $V_{CB} \neq 0$, the other terms of equation (6) will contribute error currents that may significantly affect V_E, especially for low values of input current. From (6), in the forward conducting region,

$$V_E = \frac{kT}{q} \ln \left[\frac{I_{IN}}{\alpha_N I_{ES}} + \frac{I_{CS}}{\alpha_N I_{ES}} (\epsilon^{qV_C/kT} - 1) \, + \, \Sigma \, (\text{etc.}) \right] \qquad (8)$$

For grounded-base applications, the amplifier offset voltage V_{os} will bias the collector voltage, as will any common-mode input voltage. For applications with the base driven, the designer should ensure that the expected swing of V_{CB} and the desired low-current range are compatible for the device employed in the application. The magnitude of the effective collector-current errors may be determined experimentally by connecting emitter and base together, and applying a voltage V_C.* The collector current will consist only of the $V_C \neq 0$ terms, since $V_E = 0$. This measurement with the collector-base diode forward-biased is worst-case, since error current is extremely low with reverse bias. Care should be taken to avoid applying excessive currents or voltages.

Amplifier bias current, I_b, will cause linearity errors, referred to the input, and log-conformance errors at the output (Figure 3).

---

*V_{CB} positive for PNP transistors, negative for NPN's.

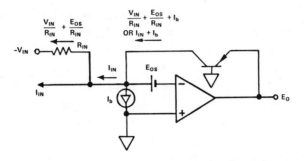

Figure 3. Offset voltage and bias-current errors

INPUT	ERROR
$\dfrac{I_{IN} + I_b}{I_{IN}}$	$\dfrac{kT}{q}\left(\ln\left[\dfrac{I_{IN} + I_b}{\alpha I_{ES}}\right] - \ln\dfrac{I_{IN}}{I_{ES}}\right)$
0.9	− 2.7 mV
0.99	− 0.26mV
0.999	−26. μV
1.000	0
1.001	26. μV
1.01	0.26mV
1.1	2.45mV

Amplifier offset voltage E_{OS} will develop an error current in the feedback path, as a function of the input resistance: E_{OS}/R_{IN}. This current will have the same effect as bias-current error. For measurements of current sources, where $1/R_{IN} \to 0$, the major contribution of E_{OS} will be through its effect on V_{CB}.

Bias current errors may be reduced in several ways. The most effective and obvious is the choice of an amplifier having appropriate specifications. However, an amplifier with more modest performance (and price) may be used. In the configuration of Figure 4a, a compensating current is summed, nulling the bias current error at one temperature. In the configuration of 4b, a compensating resistance in series with the positive input provides tracking bias-current compensation (if the amplifier inputs track)

$$-I_C = I_{IN} + I_{b_1} - I_{b_2} \frac{R_2}{R_1} \qquad (9)$$

If, for example, $I_{b_1} = I_{b_2}$ and $R_1 = R_2$, $-I_C = I_{IN}$. $I_{b_2} R_2$ should not be large enough to cause V_C effects to be significant. This is seldom a problem for current levels that permit the use of bipolar-transistor-input op amps.

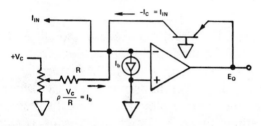

a. Nulling the effects of I_b with a compensating current

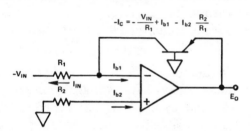

b. Nulling the effects of I_b through symmetry

Figure 4. Bias-current nulling

Offset-voltage-caused current errors are reduced at one temperature (and—to some degree—over the temperature range) by zeroing the amplifier. Otherwise, either the ambient temperature should be controlled or an amplifier having appropriate performance should be specified. Because I_{ES} may be at 10^{-14}A or less, it is important that careful consideration be given to the selection of an operational amplifier, and to sources of summing-point leakage current, since the lower limit of the log-performance range is usually determined by the amplifier input characteristics and the designer's skill in circuiting.

There are two additional sources of error, inherent in the use of single transistors and diodes, that can be minimized by the circuit techniques to be discussed later in this chapter. They are the · temperature variation of I_{ES} (doubling per ~10°C increase) and the proportionality of kT/q to temperature (0.33%/°C at 25°C), amounting to about 2mV/°C, an intolerable 8%/°C (per ε) change.

CLOSED-LOOP STABILITY

In operational amplifier circuits, a necessary condition for stability is that phase shift around the loop be less than 180° at the frequency at which the loop gain Aβ drops through unity. On a Bodé plot for a circuit using minimum-phase (RC) networks, this implies that A and β have slopes differing by less than 40dB/decade when they cross at unity loop gain (Figure 5). In operational amplifier circuits with passive feedback components, 1/β is never less than unity. Therefore, if the amplifier gain rolls off at 20dB per decade to unity, the circuit must be stable with resistive feedback.

However, in the transdiode connection, the feedback path, both active and nonlinear, may have voltage gain at the higher input-current levels, and even purely-resistive feedback (if possible) would not insure stability, since the unity-gain crossover may occur at a frequency at which the *amplifier* gain is considerably less than unity, accompanied by large phase shift. Moreover, the fact that gain is a function of signal level may force a choice between stability at high levels and bandwidth at low levels.

The effective feedback admittance, for small changes of the emitter voltage, is

$$\frac{dI_C}{dV_E} = \frac{q}{kT}I_C \cong \frac{I_C}{0.026} = \frac{1}{r_E} \tag{10}$$

Since the emitter and collector currents are nearly equal at the high end, the resistance looking into the emitter circuit is also $0.026/I_C$.

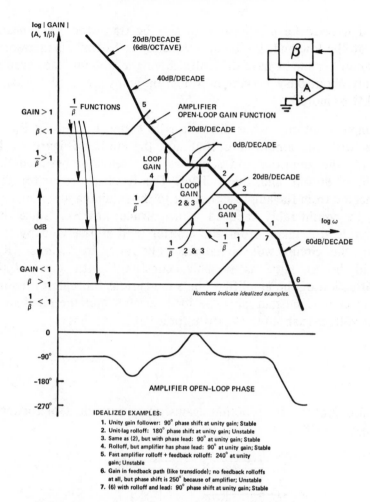

Figure 5. Bodé-plot stability analysis showing stable and unstable loop-gain ($A\beta$) situations. $A\beta = A/(1/\beta)$. On log scale, loop-gain can be estimated graphically (log $A\beta$ = log A − log $(1/\beta)$), 0 dB at crossing. Numbers indicate idealized examples.

The range of r_E is quite large: for example, it is 26 ohms at 1mA, and 26 megohms at 1nA. Therefore, it is impractical to seek to stabilize the circuit by the conventional tactic of connecting capacitance directly across this feedback element. For example, to obtain a break frequency of 1.6MHz at the high end, $0.039\mu F$

would have to be paralleled with the log transistor. This means that at the low end, the break frequency is 1.6Hz! Furthermore, the amplifier might have difficulty driving a 26-ohm load, even if to only about 0.6V maximum. (Most op amps are rated for loads of 1kΩ or more.)

A simple solution to this dilemma is to connect a resistor R_E in series with the amplifier output and the emitter (Figure 6). It unloads the amplifier and serves as an attenuator between the amplifier output and the emitter. The feedback capacitor C_c, connected from the amplifier output to the summing junction, can now be considerably reduced in magnitude; however, since the output is still taken from the emitter (and still servo'd by the loop), the circuit will be considerably faster in response. R_E should be as large as possible, consistent with the output specification of the amplifier. Since the current through it is equal to the emitter current plus the load current, and the maximum diode voltage is about 0.7V, then, for a 10V amplifier,

$$R_E \cong \frac{9.3V}{I_C + I_L} \tag{11}$$

R_E also protects the junction against excessive values of forward voltage.

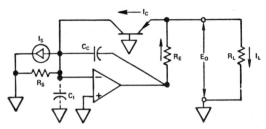

Figure 6. Transdiode circuit with stabilizing elements R_E and C_C

The choice of feedback capacitance depends on the summing-point capacitance and the maximum and minimum current levels. Its value can be determined from the Bodé plot (Figure 7), obtained as follows:

The small-signal response of the feedback portion of the loop, $\beta = \Delta V_f / \Delta E_A$ can be obtained from

$$\frac{\Delta V_f}{R_s}(1 + R_s C_I p) = (\Delta E_A - \Delta V_f)C_c p + \Delta I_c \qquad (12)$$

where

$$\Delta I_c = \frac{\Delta E_o}{r_E} = \frac{\Delta E_A}{R_E + r_E} \qquad *$$

Solving for β,

$$\beta = \frac{\Delta V_f}{\Delta E_A} = \frac{R_s}{R_E + r_E} \frac{1 + (R_E + r_E)C_c p}{1 + R_s(C_I + C_c)p} \qquad (13)$$

If the input is a current source ($R_s \rightarrow \infty$)

$$\beta = \frac{1 + (R_E + r_E)C_c p}{(R_E + r_E)(C_I + C_c)p} \qquad (14)$$

At high frequencies ($p \rightarrow j\omega \gg 2\pi f_T$)

$$\beta = \frac{C_c}{C_I + C_c} \qquad (15)$$

At low frequencies, for the voltage case (R_s finite)

$$\beta = \frac{R_s}{R_E + r_E} \qquad (16)$$

Noting that r_E is inversely proportional to I_C (from 10), the time constants containing r_E will be proportional to r_E for low values of I_C and constant ($\cong R_E$) for high values of I_C.

---

*The effect of load resistance can be included by adding $r_E R_E / R_L$ to $(R_E + r_E)$ wherever it appears.

In order to achieve small-signal stability, the numerator break frequency $\omega_c = 1/(R_E + r_E)C_c$ should be at least 1 octave less than (i.e., ½) the frequency at which $1/\beta = 1 + C_I/C_c$ crosses the amplifier's open-loop gain plot, at the highest value of I_C.

For example, if $R_E = 2.2k\Omega$, $\omega_t = 10^7$ rad/s, $C_I = 10pF$, r_E (@1mA) $= 26\Omega$,

$$\frac{1}{2200C_c} = \frac{1}{2}\frac{\omega_t}{1 + C_I/C_c} \qquad (17)$$

Solving for C_c gives: 88pF; hence, 100pF would be a reasonable value.

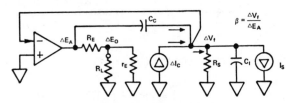

a) Model for stability analysis

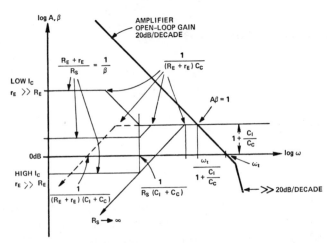

b) Bodé magnitude plot

Figure 7. Bodé plot stability analysis of transdiode circuit

PRACTICAL CIRCUITS

The basic circuits that we have considered so far are of little practical value because of their temperature sensitivity. Also, the output level depends on the value of the reference current, αI_{ES}, which differs from device to device, and is in any event quite sensitive to temperature, approximately doubling for each $10°C$ increment. The scale factor, kT/q, changes in proportion to absolute temperature, $0.33\%/°C$ in the vicinity of room temperature ($27°C$).

For two matched transistors (V_{BE} match for constant collector current and temperature) the ratio of the αI_{ES} terms tends to be constant with temperature. For this reason, log transistors are nearly always used in pairs, in order to compensate for variations of αI_{ES} with temperature. Compensation is achieved by performing the subtraction:

$$\frac{kT}{q} \ln \frac{I_1}{\alpha I_{ES_1}} - \frac{kT}{q} \ln \frac{I_2}{\alpha I_{ES_2}} = \frac{kT}{q} \left[\ln \frac{I_1}{I_2} + \ln \frac{\alpha I_{ES_2}}{\alpha I_{ES_1}} \right] \quad (18)$$

The error term is a constant very nearly equal to $\ln(1) = 0$; if it cannot be ignored, it can be biased out by a fixed value of voltage or current in a subsequent stage.

The subtraction may be performed with a subtractor, as in Figures 8 and 9, or by connecting the log elements in series opposing, as shown in Figures 10, 11, and 12.

In Figure 8, the outputs of A1 and A2, shown with NPN transistors,

$$E_{o_1} = -\frac{kT}{q} \ln \frac{I_1}{\alpha I_{ES_1}} , \quad E_{o_2} = -\frac{kT}{q} \ln \frac{I_2}{\alpha I_{ES_2}} \quad (19)$$

are subtracted in the circuit of A3 to obtain the output

$$E_o = \frac{R_2}{R_1} \frac{kT}{q} (\ln \left[\frac{I_1}{I_2} \right] + const.) \quad const. \rightarrow 0 \quad (20)$$

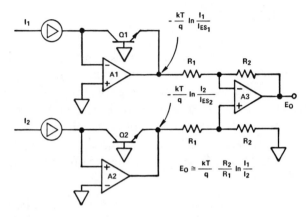

Figure 8. Log ratio circuit with temperature-compensated I_{ES}

Performance of this log ratio circuit is now independent of I_{ES}, if the transistors are adequately matched. For a single input, I_1, the ratio reference I_2 may be fixed at whatever value is desired to normalize I_1. That is, if $I_2 = I_1$, $\ln(I_1/I_2) = 0$. I_2 may be set, for example, at the upper or lower end of I_1's range, or at the geometric mean, for symmetry.

Since kT/q is not usually considered a convenient value of voltage, R_2/R_1 can be scaled to provide an appropriate value of gain. If, for example, it is desired that the output of the circuit have a scale factor of 1V/decade

$$E_o = K \log_{10}(I_1/I_2) = 1.0 \log_{10}(I_1/I_2) \qquad (21)$$

then $R_2/R_1 = q/(kT \ln 10) = 16.903$ at $25°C$.

If the temperature sensitivity of this circuit ($0.33\%/°C$) is too great for the desired stability and range of temperature variation, it may be followed by a gain stage having an equal-and-opposite temperature coefficient. In Figure 9, the unity-gain subtractor is followed by a follower-with-gain circuit. The resistor R_{TC} is chosen so that the gain equation

$$G = 1 + \frac{R_3}{R_{TC}} \qquad (22)$$

has a sensitivity of $-0.33\%/°C$. For example, if $G = 16.9$, and R_{TC} = $1k\Omega$ at $25°C$, R_3 = $15.9k\Omega$, and the temperature coefficient of $R_{TC} \cong +0.35\%/°C$.

For the convenience of the circuit-designer, the Model 751 logarithmic circuit element contains a pair of matched transistors (751P: PNP's, 751N: NPN's) and a resistive divider designed to provide a temperature-compensated gain in log transistor circuits (See Chapter 4-1).

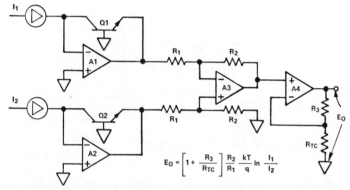

Figure 9. Log ratio circuit with compensation for both I_{ES} and kT/q

The circuits of Figure 8 and Figure 9 were shown without the dynamic stabilization elements R_E and C_c, for clarity in presentation. However, they would be used in the manner of Figure 6 in the circuits of both A1 and A2.

While the circuits of Figures 8 and 9 are workable, they tend to be expensive to implement and are rarely used by designers of constant-reference logarithmic converters. The circuits of Figures 10 and 11 are somewhat more representative. With minor modification, they can be used for antilog operations. They differ from the circuits of Figures 8 and 9 in that they perform the subtraction by a series-opposing connection of the log diodes.

Figure 10 demonstrates the principle. I_1 is the input current; it may be furnished by a current source, or it may be developed through an input resistor R_{IN} by an input voltage V_{IN}. I_2 is either a reference or a second input current furnished by a current source. The emitter-to-base voltage of Q1 is $-kT/q \ln (I_1/\alpha I_{ES1})$. Assuming that Q2 has high h_{FE} (and that the base current is therefore negligible), the emitter-base voltage of Q2 is $-kT/q \ln (I_2/\alpha I_{ES2})$.

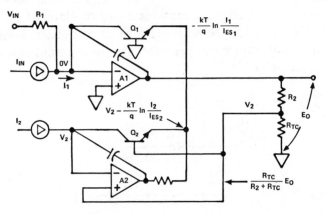

Figure 10. Temperature-compensated log circuit

Inasmuch as the base voltage of Q2 is $E_o R_{TC}/(R_2 + R_{TC})$, the base voltage of Q1 is 0, and both emitters are at the same voltage,

$$V_2 - \frac{kT}{q} \ln \frac{I_2}{\alpha I_{ES_2}} = - \frac{kT}{q} \ln \frac{I_1}{\alpha I_{ES_1}} \qquad (23)$$

$$E_o = \left[1 + \frac{R_2}{R_{TC}}\right] V_2 = - \left[1 + \frac{R}{R_{TC}}\right] \frac{kT}{q} \ln \left[\frac{I_1}{I_2} \frac{\alpha I_{ES_2}}{\alpha I_{ES_1}}\right] \qquad (24)$$

The "bootstrap" connection of V_2 to the reference input of A2 ensures that the collector-base voltage of Q2 is held at zero, since the negative input of A2 follows V_2. However, it also requires that I_2 be furnished from either a current source, a voltage source referenced to V_2, or a high voltage source in series with a large value of resistance.

The resistive divider compensates for the temperature variation of kT/q and provides a magnified scale factor. If $(1 + R_2/R_{TC}) = 16.9$, at $25°C$,

$$E_o = - 1V \cdot \log_{10} \frac{I_1}{I_2} \tag{25}$$

Figure 11 shows a similar circuit, with a fixed current reference, for accurate log conversion of a single current or voltage input signal (or sum). The reference current is equal to V_{Z_1}/R_3.

$$E_o = K \log_{10} \frac{I_{IN}}{I_{REF}} = K \log_{10} \frac{V_{IN}}{E_{REF}} \tag{26}$$

where

$$E_{REF} = V_{Z1} \left[\frac{R_{IN}}{R_3} \right]$$

and

$$K = \left[1 + \frac{R_2}{R_{TC}} \right] \frac{kT}{q} \ln 10$$

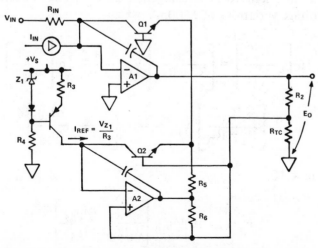

Figure 11. Temperature-compensated log circuit with internal current reference

Resistor R6 allows the high end of the dynamic range to be extended beyond 1mA; its negative-resistance effect tends to cancel the voltage drop of the bulk resistance of Q1. It is calculated by the formula of (27)

$$R_6 = \frac{R_5}{R_B} \frac{R_{TC} R_2}{R_{TC} + R_2} \tag{27}$$

where R_B = Bulk Resistance.

INVERSE OPERATION

If the positions of the input resistor and the log element are interchanged, the same basic circuit configuration may be used to obtain the antilog

$$E_o = -E_{REF} \, \epsilon^{-V_{IN}/K} = -E_{REF} \, (10)^{-V_{IN}/K_{10}} \tag{28}*$$

In Figure 12, assuming that Q2 operates with a value of reference current sufficient to ensure logarithmic operation unaffected by base-voltage variations of ±600mV,

$$-\frac{kT}{q} \ln\left[\frac{E_o}{R_1 \alpha I_{ES_1}}\right] = \left[\frac{R_{TC}}{R_2 + R_{TC}}\right] V_{IN} - \frac{kT}{q} \ln\left[\frac{I_{REF}}{\alpha I_{ES_2}}\right] \tag{29}$$

$$\frac{q}{kT}\left[\frac{R_{TC}}{R_2 + R_{TC}}\right] V_{IN} = -\ln\left[\frac{E_o}{R_1 I_{REF}} \frac{\alpha I_{ES_2}}{\alpha I_{ES_1}}\right] \tag{30}$$

or, to base 10,

$$= -(\ln 10)\log_{10}\left[\frac{E_o}{R_1 I_{REF}} \frac{\alpha I_{ES_2}}{\alpha I_{ES_1}}\right] \tag{31}$$

*$E_{REF} = I_{REF} R_1$

If $\alpha I_{ES_2} = \alpha I_{ES_1}$, and

$$\frac{kT}{q}\left[1 + \frac{R_2}{R_{TC}}\right] \ln 10 = K_{10}$$

then

$$E_o = R_1 I_{REF} (10)^{-V_{IN}/K_{10}} \tag{31}$$

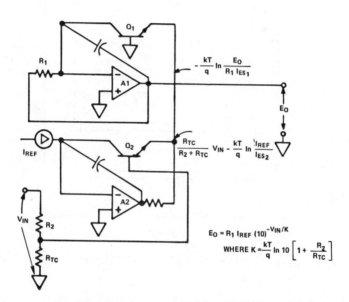

Figure 12. Antilog (exponential) circuit

LOG MODULES

The circuit of Figure 11 is similar to the circuit used for Models 755 (N and P) and 752 (N and P). The principal difference between the two families is that the 755 is a complete, self-contained log circuit, including output amplifier A1, with fixed choices of K (2/3, 1, and 2) and fixed (but modifiable) I_{REF}, specified for operation over 4 decades of voltage and 6 decades of current, while the 752 has a 7-decade range of input current, requires an external amplifier, and has an approximately 10:1 range of adjustment for both K and I_{REF} (Figure 13). Since the major source of error in the 755 at the low end is the amplifier, the 752 can be used with a choice of operational amplifiers to obtain a wider current or voltage range. It also permits greater flexibility of parameter choice and can be used in complementary pairs to assemble sinh or $\sinh^{-1}$ ("bipolar log") functions.

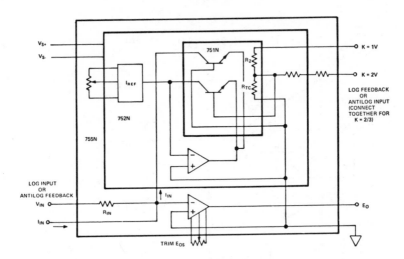

Figure 13. Comparative block diagrams of log/antilog modules (simplified)

NOMENCLATURE AND CHARACTERISTICS (N vs. P)

For all logarithmic devices that offer a choice between an "N" and a "P" version, N signifies that an NPN transistor is used as the basic log element; P signifies a PNP transistor.

For N versions (Figure 14),

● Input voltage or current for the log connection is always positive.

● Output voltage for the antilog (exponential) connection is always positive.

● Output voltage for the log connection is negative for $V_{IN} >$ E_{REF} or $I_{IN} > I_{REF}$, positive for $V_{IN} < E_{REF}$ or $I_{IN} <$ I_{REF}.

● Output voltage for the antilog connection is more positive than E_{REF} with negative inputs, less positive than E_{REF} with positive inputs, equal to E_{REF} for zero input.

- In the log connection, output approaches positive limits as V_{IN} or I_{IN} approach zero.

- In the antilog connection, output approaches zero when the input has large positive values.

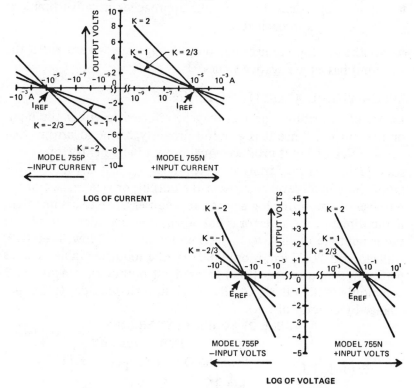

Figure 14. Output vs. input of Model 755N & 755P
in log connection (log input scales), showing voltage,
and polarity relationships

For *P* versions,

- Input voltage or current for the log connection is always *negative*.

- Output voltage for the antilog (exponential) connection is always negative.

- Output voltage for the log connection is negative for $V_{IN} > E_{REF}$* or $I_{IN} > I_{REF}$, positive for $V_{IN} < E_{REF}$ or $I_{IN} < I_{REF}$.

- Output voltage for the antilog connection is more positive (less negative) than E_{REF} with negative inputs, less positive than E_{REF} with positive inputs, equal to E_{REF} for zero input.

- In the log connection, output approaches negative limits as V_{IN} or I_{IN} approach zero.

- In the antilog connection, output approaches zero when the input has large negative values.

SPECIFYING LOGARITHMIC DEVICES

Errors of logarithmic devices may be referred to either the input or the output. Since it is a useful property of the logarithm that equal ratios of input produce equal output increments (for a given scale factor) we may translate percentage errors at any level of the input to millivolt-level changes at the output, or vice versa. For the purpose of specifying log devices, it is conservative to assume that, if the direction of an error is unknown, it be considered to reduce the magnitude of the argument (i.e., log 0.8, 20% low, is −0.097, while log 1.2, 20% high, is only 0.079). The following table† relates percentage input errors (low) to millivolt output increments for 3 commonly-used values of K. K is in volts/decade (volts per change-by-a-factor-of-ten).

TABLE OF EQUIVALENT ERRORS

Error R.T.I.	Output error (mV)		
	Error RTO = $-K \log_{10} (1 - RTI/100)$		
% low	K = 1V	K = 2V	K = 2/3V
0.1	0.43	0.87	0.29
0.5	2.18	4.35	1.45
1.0	4.36	8.73	2.91
3.0	13.2	26.5	8.82
4.0	17.7	35.5	11.8
5.0	22.3	44.6	14.9
10.0	45.8	91.5	30.5

*i.e., V_{IN} more positive or less negative than E_{REF}
†See also Table 4, Chapter 4-1

For intermediate values, linear interpolation is quite adequate.

It is not considered good practice to state (log) output error in percent unless it is clearly stated that the error is in percent of nominal K or of full-scale (= nK) or of some other such measure.

When applying devices such as the Model 755 (used as our example here), a firm understanding of the error sources associated with log amplifiers is beneficial for achieving best results. The principal error sources are of two kinds:

1. Parametric errors, due to tolerances and changes in the constants of the ideal log equations, including offsets.

2. Log conformity error, the error that remains when all parametric effects have been removed by nulling and calibration.

Parametric errors are stated separately for voltage and for current operations, as defined in the equations

$$E_o = -K \log_{10} \frac{V_{IN} - E_{os}}{E_{REF}}$$

and

$$E_o = -K \log_{10} \frac{I_{IN} - I_{os}}{I_{REF}} \tag{32}$$

Scale Factor (K) is the voltage change at the output for a decade (i.e., a 10:1) change at the input, when connected in the log mode. Error in scale factor is equivalent to a change in gain, or slope, and is specified in percent of the nominal value. K is positive for "N" types and negative for "P" types. Its specification for Model 755 is 1% maximum tolerance and 0.04%/°C maximum change with temperature (0°-70°C).

Offset Voltage (E_{os}) depends on the operational amplifier used for the log operation. Its effect is that of a small voltage in series with the input resistor. For current logging operations with high-impedance sources, its error contribution is negligible. However, for voltage logging, it modifies the value of V_{IN}. Though it can be adjusted to zero at room temperature, its drift over the temperature range should be considered. In the 755, E_{os} is zero $\pm 400\mu V$, with a maximum drift of $\pm 15\mu V/^{\circ}C$.

Reference Voltage (E_{REF}) is the effective internally-generated voltage to which all input voltages are compared. It is related to the internally-generated reference current I_{REF} by the equation: $E_{REF} = I_{REF}R_{IN}$, where R_{IN} is the value of input resistance. Typically, I_{REF} is considerably less stable than R_{IN}; therefore, practically all the tolerance is due to I_{REF}. In the 755N, E_{REF} is nominally $+0.1V \pm 3\%$ (3mV *max*), with a maximum temperature coefficient of $0.1\%/^{\circ}C$. In the 755P, $E_{REF} \cong -0.1V$, same tolerances.

Offset Current (I_{os}) is the bias current of the amplifier, plus any stray leakage currents. This parameter can be a significant source of error when processing signals in the nanoampere region. For this reason, it is held to within 10pA (doubling per $10^{\circ}C$ increase) in such devices as the 755.

Reference Current (I_{REF}) is the internally-generated current source output to which all values of input current are compared. I_{REF} tolerance errors appear as a DC offset at the output. For the 755N, I_{REF} is $+10\mu A \pm 3\% \pm 0.1\%/^{\circ}C$ *max* (polarity of I_{REF} is negative in 755P). From the table, $\pm 3\%$ tolerance of this input parameter corresponds to $\pm 13.2mV$ at the output, an offset that is independent of input signal and removable by adjusting the reference current (where possible), adding a voltage to the output by injecting a current into the scale-factor attenuator, or simply by adding a constant bias at the output's destination.

In addition to the parametric errors, *log-conformity error* must be considered. When the parameters have been adjusted to compensate for offset, scale-factor, and reference errors, the output will be found to still deviate from ideal logarithmic behavior (principally at the extremes of the range). Since the behavior of an ideal log device is linear on a semi-log plot, *log conformity error* is the deviation from a straight line on a semi-log plot over the range of interest. For Model 755, the best linearity of the log relationship is found in the middle 4 decades of the current range (10nA to 100µA). For this range, the log-conformity error is ±0.5% RTI or 2.18mV RTO (K = 1).

It should be obvious that the large number of degrees of freedom, both parametrically, and in terms of the user's variables, would make it difficult to summarize a log devices's specifications in one overall number. As an alternative, sufficient information is provided to calculate performance to fit the desired range. While the Model 755 specifications were cited for the sake of example, the reader should naturally seek to acquaint himself with the properties of devices available at the time he initiates a design effort, via data sheets, the *A D Product Guide*, and other media. Complete specifications of the 755 are listed at the end of this chapter, to aid the reader's understanding of device behavior. Applications information about specific devices is available in great detail on their data sheets.

LOG AND ANTILOG DEVICES

In logarithmic circuits of the type we have been discussing, we can view the causal explanation of what basically happens as follows:

1. Current is applied at the input of the circuit, developing an amplifier input voltage.

2. The amplifier output voltage changes in the opposite direction.

3. The amplifier output voltage, applied to the input (V_{BE}) of the log diode, causes a collector current to flow that balances the amplifier's input current, and holds the input voltage at zero.

4. The amplifier's output is proportional to the log of the input current; but the log diode's output current is proportional to the antilog of *its* input voltage.

Thus, we can consider any log device as consisting of an operational amplifier and an antilog (exponential) circuit. As noted earlier, if we interchange the input resistor and the feedback element, we will therefore have a circuit that develops an antilog input current at the summing point, and a corresponding antilog voltage at the output.*

In the antilog connection, the same sources of error are present and can be considered in the characteristic equation

$$E_o = E_{REF}(10)^{-V_{IN}/K} \pm E_{os} \tag{33}$$

Errors appearing as constant increments of input will give rise to constant percentage errors at the output.

The 755 can be connected for either log or antilog operation. The 752 can be connected for either log or antilog operation when used with an external operational amplifier. The chart (Figure 15) compares the operating ranges of the 752, with a variety of op amp types, and the 755, subject to the constraint of ±2% error over a ±10°C range. A chopper-stabilized amplifier (233J) maximizes the voltage range; a low-bias-current FET (42J) maximizes the current range; a general-purpose FET (40J) minimizes cost.

LOG MODULE	755	752 OPERATING WITH OP AMP		
		233J	42J	40J
Op Amp Type ➤	Internal-High Performance FET	Chopper Stabilized	Electrometer FET	Economy FET
Input Range for ±2% Error, Over ±10°C				
V_{in}[1]	3.5mV to 10V	500μV to 10V	37.5mV to 10V	25mV to 10V
E_{OS} Drift	150μV	10μV	750μV	500μV
I_{in}	1nA to 1mA	3.5nA to 1mA	50pA[2] to 1mA	5nA to 1mA
$I_{OS} + I_{OS}$ Drift	20pA	70pA	1pA	100pA
Selection Criteria	Complete log amplifier, high performance, trimmed internally	Extends lower limit of voltage range. Minimum drift and offset errors, long term stability	Extends lower limit of current range	Lowest cost for a complete log amplifier
Relative Costs (1–9)	100%	140%	104%	80%

[1] Values selected are consistent with a 10kΩ input register.
[2] Log conformity error restricts the lowest input signal to 100pA.

Figure 15. Comparison of capabilities of 755 complete log/ antilog module and 752 log/antilog transconductor

*It is also possible to consider the log diode as a *current* follower, isolated from the amplifier's output voltage by R_E, that develops the log voltage as an *output* (by-product) across r_E.

For applications calling for a variable reference, or the log of a ratio, the 756 has been designed. It has log ratio conformity of 1%, typically over 7 decades (4 decades of numerator change, 3 decades of denominator). Figure 16 shows typical % error, referred to the ratio, as a function of I_N, for 4 values of I_D.

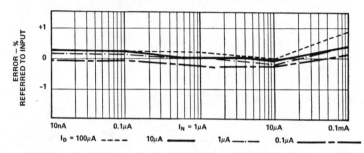

Figure 16. Log-conformity errors of Model 756 log-ratio module

DYNAMIC ERRORS

Speed and frequency response of logarithmic devices depend on scaling, the signal level and the direction of change. Typically, above $1\mu A$, the response is dominated by the integrator time constant, changing little with signal level. Below $1\mu A$, $r_E C_c$ dominates the response, reducing speed in proportion to the input current. A tabulation of typical 755 response time follows. It is interesting to note that response time is shorter for increasing signal magnitude than for the same decreasing increment, because the new current value determines the speed.

I_{IN} (Increasing)	Time	I_{IN} (Decreasing)	Time
1nA to 10nA	1ms	10nA to 1nA	4.5ms
10nA to 100nA	100μs	100nA to 10nA	400μs
100nA to 1μA	7μs	1μA to 100nA	30μs
1μA to 1mA	4μs	1mA to 1μA	7μs

The logarithmic response will of course distort wide-ranging sinusoids, as will the asymmetric and nonlinear delay times. Frequency response is therefore given in terms of *small-signal* response at differing current levels. Typical frequency response (−3dB) of the 755 is

I_{IN}	−3dB frequency
1nA	80Hz
1μA	10kHz
10μA	40kHz
1mA	100kHz

TESTING LOG DEVICES

The following equipment (or its equivalent) will be found useful in performing the basic tests to be discussed.

Picoampere Current Source	Keithley 261
Precision dc Voltage Standard	Electronic Development Corp. 100N
Function Generator	Hewlett-Packard 3310A
Digital Voltmeter	Hewlett-Packard 8300A
Oscilloscope	Tektronix 543B
	with Type 1A5 preamplifier

In this section, circuits and techniques will be discussed for the evaluation of some of the basic log-device parameters, such as Scale Factor, Log-Conformity Error, Reference-Current accuracy, Response Time, Bandwidth, and Input Offset (to the degree that they apply to a given log device). For the 755, which is a complete, self-contained "black box," all of these parameters apply; they will be discussed in the context of measurements of a 755.

For brevity, pin-number connections will be referred to. The relationship between pin connections and circuit functions can be seen in Figure 17, a simplified functional diagram of the 755N. The principles, however, are applicable to all similar devices. The results will depend on the care with which measurements are performed.

Scale Factor (K) has been defined as the change in output voltage

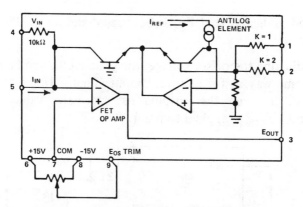

Figure 17. Connection diagram of model 755

for a decade (10:1) change at the input, when connected in the log configuration. It is the slope of a semi-log plot of the output.

Scale factor is most-easily measured using a current source (Figure 18). Apply a value of current I_1 to pin 5 (input summing point) and measure the output. Increase the current by a factor of exactly 10 to $10I_1$ and again measure the output. The scale factor K is simply the difference between the two measurements.

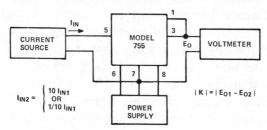

Figure 18. Scale-factor measurement with current input (shown for k = 1)

If it is desired to measure scale factor in the voltage mode, it is first necessary to carefully adjust the input voltage offset to a value near zero. A sensitive adjustment can be performed using the circuit of Figure 19, with the voltage input (pin 4) grounded. Using an external 100kΩ 10-turn trim pot, the output is adjusted (K = 1) to a value between +4V and +5V, for N-type devices. (For 755P, the value would be between −4V and −5V.) Since E_{REF} is 10^{-1}V, an output voltage between 4V and 5V, with the input grounded, indicates that E_{os} is between 4 and 5 decades lower, or $1-10\mu V$.

After adjusting the offset voltage, the scale factor can be measured in the same way as for a current input (Figure 18), by taking the difference between the outputs for successive values of input V_{IN_1} and $10V_{IN_1}$, applied to pin 4.

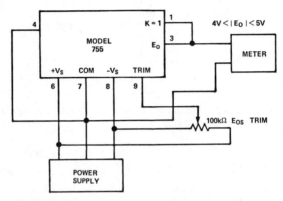

Figure 19. Trimming E_{OS} in the log mode

Reference Voltage is the input voltage at which the ratio of input-to-reference becomes unity; the log output at that value should be zero. The value of reference voltage for a given unit is thus measured by empirically applying precisely-determined voltage to the input (4) in the neighborhood of E_{REF} (0.1V for 755N), and adjusting for zero output. The reference voltage can be modified either by connecting an external resistance of appropriate value to the current input (5) and using it as the input resistor, by modifying the current reference, or by adding a constant in an external summing amplifier at the output. In any of these instances, the new value of reference voltage can be calibrated by applying precisely the value desired at the input, and adjusting for zero output.

Reference Current is the value of input current at which the ratio of input-to-reference becomes unity; the log output at that value should be zero. It is measured similarly to reference voltage, by applying a precisely-measured adjustable input current and adjusting for zero output. I_{REF} for 755N is about $10\mu A$. I_{REF} can be modified by adding a constant in an external summing amplifier at the output or by modifying the reference source. It is an input variable in the 756 log ratio unit, adjustable in the 752 log/antilog transconductor, and fixed in the 755. However, it can be modified in the 755 by applying a current from a current-source to an unused K pin (1 or 2). The value of current required is $66\mu A$ per decade of shift. If it is required for correction only, $0.29\mu A$ per 1% change is the approximate sensitivity; for small shifts, the current may be developed by voltage in series with a high value of resistance ($\geqslant 2.2M\Omega$).

Log Conformity error is the difference between the actual output voltage and the output voltage predicted by the log transfer equation, with the effects of offsets, reference shifts, and scale factor either nullified or taken into account. A plot of output vs. input on semi-log paper should be a straight line. Any deviation from a straight line is log conformity error. (Output is measured on the linear scale, input on the log scale.) Apply enough input voltage or current values over the range of interest to allow a smooth curve to be drawn connecting the plotted values of output. The input should be accurately determined, the output accurately measured, and the paper sufficiently large to permit tolerances of the magnitude of interest to be observed. A "best straight" line may then be drawn; the deviations from it represent log-conformity error.

A more sensitive way of plotting is to subtract the expected value of output from the actual value and plot that result (the total error) on the linear scale vs. input on the log scale. A "best straight line" will then show the average slope and offset error, and the deviations from it the log-conformity error. Output error may be referred to the input via the Table of Equivalent Errors (p. 188).

Dynamic Measurements are performed by observing small changes from or about an appropriate bias level. The current input may be used to sum a dc bias and a voltage input (in series with a large external resistance) for incremental changes (Figure 20).

For example, to measure the response time for I_{IN} increasing from 10nA to 100nA or decreasing from 100nA to 10nA, apply 10nA to pin 5 (of the 755N). Through a 10MΩ resistor, apply a 0.9V square pulse, starting from 0V. This will produce an incremental current step of 90nA, and an overall input swing of 10-100nA. Assuming that the pulse is of adequate width, response to both increasing and decreasing steps can be seen at the same time. At the output (K = 1), the increasing-input step response will swing from +4V to +3V; the decreasing-input response will swing from +3V to +4V.

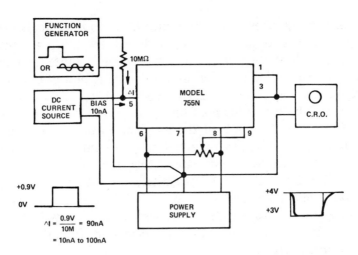

Figure 20. Performing incremental response measurements

Frequency response is measured typically by using a ±5% sinusoidal deviation about a fixed input value. For example, at the 1µA level, the deviation will be ±50nA. Apply 1µA dc to the current input (5), and sum it with 100mVp-p through a 1MΩ resistor. The output swing at low frequencies should be 43.5mVp-p. Increase the frequency, maintaining the input amplitude constant, until the output amplitude is 30.5mVp-p. This is the frequency at which the output amplitude is down 3dB from its low-frequency value.

CONCLUSION

Logarithmic devices have been summarized in Part 1, and applications have appeared throughout Part 2. This chapter has discussed the basic properties of logarithmic devices, techniques for obtaining thermal and dynamic stability, some commonly-used circuits, specifications and definitions, and means of adjustment and test. Chapters 4-1, 4-2, 4-3, "Aids for the Designer," will provide further information, as well as guidelines for selection and use of devices for log, log ratio, and antilog applications.

APPENDIX TO CHAPTER 3-1
COMPLETE SPECIFICATIONS OF A TYPICAL LOG/ANTILOG
MODULE (MODELS 755N & 755P)

(typical @ +25°C and ±15VDC unless otherwise noted)

TRANSFER FUNCTIONS	DYNAMIC RANGE OF INPUT
Log of Current	120dB

$$E_o = -K \log_{10} \frac{I_{in} - I_{OS}}{I_{REF}}$$

1nA to 1mA (755N)
−1nA to −1mA (755P)

Log of Voltage

80dB

$$E_o = -K \log_{10} \frac{E_{in} - E_{OS}}{E_{REF}}$$

1mV to 10V (755N)
−1mV to −10V (755P)

Antilog of Voltage

$$E_o = E_{REF} \, 10^{-E_{in}/K} \pm E_{OS}$$

$-2 \leqslant E_{in}/K \leqslant 2$

TRANSFER FUNCTION PARAMETERS

Symbol	Value	Tolerance	Drift	Note
K	2/3, 1, 2V	1% max	±0.04%/°C max	1, 2
E_{REF}	0.1V	3% max	±0.1%/°C max	2
I_{REF}	10μA	3% max	±0.1%/°C max	2
E_{OS}	0 ± tol.	±400μV	±15μV/°C max	3
I_{OS}	0 ± tol.	+0, −10pA max	2x/10°C	

LOG CONFORMITY ERROR REFERRED TO INPUT

Input Current Range	Conformity Error	Input Voltage Range	Conformity Error
1nA to 10nA	±1% max		
10nA to 100μA	±0.5% max	1mV to 1V	±0.5% max
100μA to 1mA	±1% max	1V to 10V	±1% max
1nA to 1mA (Total Range)	±1% max		

RESPONSE TIME

I_{in} (increasing)	Time	I_{in} (decreasing)	Time
1nA to 10nA	1ms	10nA to 1nA	4.5ms
10nA to 100nA	100μs	100nA to 10nA	400μs
100nA to 1μA	7μs	1μA to 100nA	30μs
1μA to 1mA	4μs	1mA to 1μA	7μs

SMALL SIGNAL FREQUENCY RESPONSE

I_{IN}(Level)	3dB Down At
1nA	80Hz
1μA	10kHz
10μA	40kHz
1mA	100kHz

NOISE REFERRED TO INPUT, 10kHz BANDWIDTH

Noise Voltage	2μV rms
Noise Current	2pA rms

RATED OUTPUT (Note 4)

±10V, ±5mA

POWER REQUIREMENTS (QUIESCENT)

±15V, regulated ±1%, 7mA

TEMPERATURE

Operating	0°C to +70°C
Derated	−25°C to +85°C
Storage	−55°C to +125°C

MECHANICAL

Case Size	1.5″ × 1.5″ × 0.4″
Weight	38.1 × 38.1 × 10.2 mm
	1 oz. (28.3g)

PRICES

(1−9)	$55.00
(10−24)	$49.00

NOTES:

1. Use terminal 1 for K = 1V, terminal 2 for K = 2V, terminals 1 and 2 (shorted together) for K = 2/3V.
2. Parameter is + for 755N, − for 755P.
3. Externally adjustable to zero.
4. No device damage due to any pin being shorted to ground.
5. Specifications subject to change without notice.

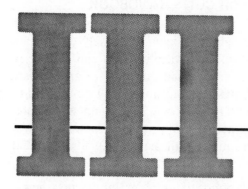

Multipliers

Chapter 2

An analog multiplier is a device that produces an output voltage or current that is proportional to the product of two or more independent input voltages or currents

$$E_o = V_x V_y / V_r = K V_x V_y \tag{1}$$

The proportionality constant, $1/V_r$, has the dimension V^{-1}. V_r may be identifiable with a specific voltage or current in the circuit, or it may be independently determined. It is usually fixed at 10V.

The operating range of a multiplier may be defined in terms of its inputs. For two inputs, and a possibility of two polarities for each input, there are four combinations of polarity. They can be visualized as the four quadrants of the X-Y plane (Figure 1).

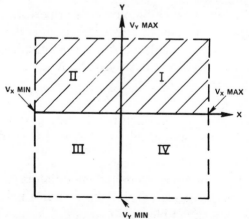

Figure 1. Multiplier operating coordinates

A pair of inputs within the operating region uniquely determines the multiplier's output voltage. A multiplier that can accept all four combinations of inputs and provide outputs of appropriate polarity is referred to as a "4-quadrant multiplier." "Two-quadrant" multipliers respond to a ± signal at one input and a unipolar signal at the other. For instance, if a multiplier responds to $\pm V_x$, but only $+V_y$, will it operate in the half-plane of quadrants I and II, indicated by shading.

One-quadrant multipliers respond to unipolar inputs in a single quadrant only. If both V_x and V_y are limited to positive values, the multiplier operates in the first quadrant. Occasionally, one will find a multiplier that responds to the appropriate number of quadrants, but with inverted output polarity. Its equation is $E_o = -KV_xV_y$. A multiplier that responds to one or more of its inputs in a single quadrant can be used for multiple-quadrant operation by being preceded by an absolute-value circuit, and followed by a sign-magnitude output circuit, with output polarity determined by input polarity (a procedure almost as cumbersome as it sounds but typical of such devices as some multiplying D/A converters). Multi-quadrant operation can also be achieved by offsetting the inputs and output (see Figure 21).

TECHNIQUES OF MULTIPLICATION

At present, the two most-common means of performing electronic analog multiplication are *variable-transconductance* and *pulse-width, pulse-height* modulation. A third method, *log-antilog*, is gaining in popularity, particularly for low-speed high-accuracy calculations.

The circuit design, and the factors affecting overall performance of these three types of multipliers will be discussed in detail in this chapter.

Many other types of multipliers and modulators have been, and still are, used in analog-computing, communications, and instrumentation circuits. Examples of these are quarter-square, diode-ring, FET, and magnetic (e.g., Hall effect). The design of these types will not be covered here; however, much of the discussion of specifications and testing could be applied to them.

CHARACTERISTICS OF MULTIPLICATION

Since the algebraic properties of multiplication are the determining factors in the design and specification of analog multipliers, a review of a few of these properties and their correspondence to physical multiplier performance will aid understanding.

One of the most salient properties of multiplication, with direct implications for design and characterization, is the fact that the product is zero for three kinds of input pairings.

Input State	Theoretical Output	Error Parameter
1	$0 \cdot 0 = 0$	Offset
2	$0 \cdot Y = 0$	Y-Null, or Y Feedthrough
3	$X \cdot 0 = 0$	X-Null, or X Feedthrough

Another important property is the relationship of the magnitude of the product to the inputs. If we assume that both products are always less than V_r (i.e., V_r is full scale), as is the case for most popular multipliers, then the product is always less than or equal to V_r

$$4 \qquad 0 \leq |V_x, V_y| \leq V_r \quad \text{Input Constraint} \qquad (2)$$

$$5 \qquad |V_x \cdot V_y/V_r| \leq V_r \quad \text{Output Constraint} \qquad (3)$$

If the two inputs are not equal, the product will be less than the smaller input, if the conditions of (2) are met, i.e., if

$$|V_x| < |V_y| \qquad (4)$$

and

$$|V_x V_y/V_r| < V_r \qquad (5)$$

then

$$|V_x V_y/V_r| < V_x \qquad (6)$$

Equations 2-6 illustrate that the output of an ideal multiplier is well-behaved for small inputs: the output goes to zero as either or both inputs are reduced to zero. The analog multiplier circuits discussed in this chapter come surprisingly close to following this ideal behavior, assuming that the first-order (or linear) errors are adjusted to zero. The reason for this is that the nonlinear component of error ($f(V_x,V_y)$) is a continuous function of V_x and V_y that goes to zero as V_x and V_y are decreased to zero. The following sections, describing the sources of multiplier errors, and the relationship to circuit design, will show why this is so.

The multiplication function can be represented by a surface in three dimensions embodying the equation

$$Z = X\,Y \qquad (7)$$

The shape of the surface can be outlined in terms of the following characteristics

1. Output (Z) is zero along the X and Y axes (zero feedthrough).

2. If one input is constant, then the output is linearly proportional to the other input, with a slope ("Gain") determined by the constant input.

3. If the two inputs are equal (X = Y, or X = –Y), then the output is proportional to the input squared. This produces two tangent parabolas of opposite polarity at right angles corresponding to the diagonals in I-III and II-IV.

The surface that fits these requirements is the hyperbolic paraboloid, or "saddle-surface," as sketched in Figure 2. The parabolic branches correspond to the intersections of the surface with vertical planes through the diagonals and parallel to the diagonals (condition 3). The straight-line elements correspond to the intersections of the surface with vertical planes parallel to the X and Y axes (condition 2), and the surface passes through the X-Y plane along the X and Y axes and their intersection of 0 (condition 1).

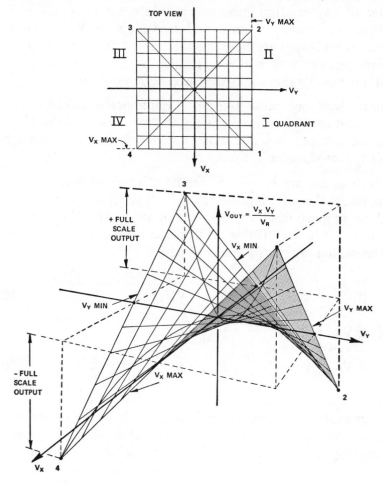

Figure 2. Four quadrant multiplier — input/output surface

Horizontal planes intersect the function in hyperbolas. That is, a contour map would show a family of right hyperbolas climbing hills along diagonal I-III and sinking into valleys along diagonal II-IV (Figure 5).

The corners of the surface labeled 1, 2, 3, 4 in the figure represent the maximum outputs of the multiplier in the respective quadrants. These maxima occur at the four combinations of $\pm X$ and $\pm Y$.

To some, the thought of a multiplier as a nonlinear device may seem paradoxical, despite the fact that it is clearly nonlinear. After all, any practical measurement of a multiplier's gain characteristics, performed at constant values of X or Y, will yield a linear output-input relationship, that is, the multiplier acts as a linear amplifier with gain KX. In fact, one thinks of a "linear" multiplier as one that obeys the ideal relationships of (1).

Clearly, with one input fixed, the multiplier is indeed a linear device and could, in concept, be replaced by a fixed-gain amplifier. The signal input can not cause the gain to vary, so it will be linearly reproduced at the output.

If both inputs are varied, the response is indeed nonlinear. For example, if the same input is applied to both X and Y, the output will be proportional to the square of the input. This is clearly nonlinear behavior, fitting neither the proportionality nor the superposition criteria for linearity (see p. 1).

$$\text{If } V_{IN} = V_1 \qquad E_{out} = KV_1^2 \qquad\qquad (8)$$

$$\text{If } V_{IN} = V_2 \qquad E_{out} = KV_2^2 \qquad\qquad (9)$$

$$\text{If } V_{IN} = V_1 + V_2 \qquad E_{out} \neq K(V_1^2 + V_2^2) \qquad (10)$$

The geometrical interpretation is that the hyperbolic paraboloid is a developed surface, a curved surface that can be constructed from straight-line elements — like the cylinder or the cone.

That the ideal analog multiplier is a linear device, if one input is maintained at a constant value, is a useful fact, one that makes it easy to characterize, adjust, calibrate, and measure the behavior of real multipliers in linear terms despite their inherently nonlinear nature.

ERRORS IN PRACTICAL ANALOG MULTIPLIERS

The output of a practical multiplier will differ from the theoretical product of its inputs by a generally unpredictable amount, ϵ, as defined in the equation

$$E_o = K V_x V_y \pm \epsilon(V_x, V_y) \qquad (11)$$

It will be quite helpful in discussing multiplier circuit properties to expand the error indicated symbolically in (11) into terms directly related to error sources in the circuit. There are four primary sources of static (or dc) error in an analog multiplier (dynamic errors are discussed later, in the section on multiplier specifications)

Error	Symbol
1. Input Offsets	X_{os}, Y_{os}
2. Output Offset	Z_{os}
3. Scale Factor	ΔK
4. Nonlinearity	$f(X,Y)$

The influences of these errors can be applied as follows

$$E_o = (K + \Delta K)\left\{(V_x + X_{os})(V_y + Y_{os}) + Z_{os} + f(X,Y)\right\} \qquad (12)$$

Multiplying out and combining terms

$$E_o = KV_xV_y + \underbrace{\Delta KV_xV_y + (K + \Delta K)\left\{V_xY_{os} + V_yX_{os} + Y_{os}X_{os} + Z_{os} + f(X,Y)\right\}}_{\epsilon(X,Y)} \qquad (13)$$

This lengthy array of error terms can be untangled by considering each separately:

Term	Description	Dependence on Input
KV_xV_y	True Product	Goes to zero as either or both inputs go to zero
ΔKV_xV_y	Scale-Factor Error	Goes to zero at V_x, $V_y = 0$

Strictly speaking, the following terms are multiplied by $K + \Delta K$, but the effect of ΔK can be ignored, since the product of ΔK and another error is a second-order error, and hence negligible.

$V_x Y_{os}$	Linear "X" Feedthrough Due to Y-input dc Offset	Proportional to V_x
$V_y X_{os}$	Linear "Y" Feedthrough Due to X-input dc offset	Proportional to V_y
$X_{os} Y_{os}$	Output Offset due to X,Y Input Offsets	Independent of V_x, V_y
Z_{os}	Output Offset	Independent of V_x, V_y
$f(X,Y)$	Nonlinearity	Depends on both V_x, V_y. Contains terms dependent on V_x, V_y, their powers and cross-products.

The error of a practical analog multiplier, $\epsilon(X,Y)$, can be visualized as a surface representing the difference between the actual multiplier output and the theoretical value. The error surface will be, in general, warped, twisted, and not level, much like a section of hilly countryside. Figure 3 shows a hypothetical error surface for a four-quadrant multiplier. The elevation, or Z coordinate of the graph, represents the error, $\epsilon(X,Y)$ defined by

$$\epsilon(X,Y) = E_0(\text{actual}) - KV_x V_y = E_0 - V_x V_y / V_r \qquad (14)$$

where

E_0 = measured value of multiplier output voltage

V_x = X input voltage

V_y = Y input voltage

$KV_x V_y$ = Ideal output voltage

$\epsilon(X,Y)$ is the measured voltage corresponding to the sum of the error terms in (13).

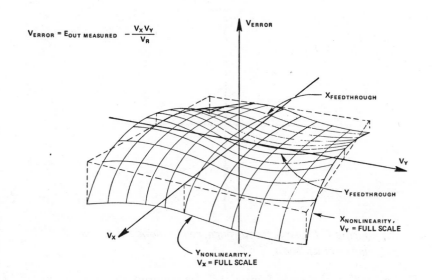

$$V_{ERROR} = E_{OUT\ MEASURED} - \frac{V_X V_Y}{V_R}$$

V_{ERROR}

$X_{FEEDTHROUGH}$

V_Y

$Y_{FEEDTHROUGH}$

$X_{NONLINEARITY,}$
$V_Y = FULL\ SCALE$

V_X

$Y_{NONLINEARITY,}$
$V_X = FULL\ SCALE$

Figure 3. Multiplier error plane

Using an error surface to describe the static error of an analog multiplier may seem awkward, but it is the easiest way to visualize the *whole* three-dimensional effects of the individual error components in (13).*

For example, consider the effect of "linear" X feedthrough $V_x Y_{os}$. Assuming that Y_{os} is a small positive quantity, then as V_x increases, the output of the multiplier will increase in proportion. As V_x goes negative, the output will go negative. This effect is clearly independent of the Y input, since equation 10 shows $V_x Y_{os}$ as an additive error. The result of the linear X feedthrough is to tilt the plane about the Y axis, as shown in an end-on view of the XY plane, Figure 4.

Similarly, the effect of X offset ("linear" Y feedthrough) is to tilt the whole error plane about the X axis. The effect of dc offset, $X_{os}Y_{os} + Z_{os}$, is to move the whole plane up or down on the Z (output) axis.

*The error surface is primarily an aid to visualization. It, and the 2-dimensional contour representation ("iso-vers": *iso*-equal, *ver(ity)*-accuracy), have been used for this purpose. However, because of wide differences in form of the error function from unit to unit, and for a given unit under various stages of adjustment and thermal environment, error surfaces are of little use as compact presentations of data for individual devices.

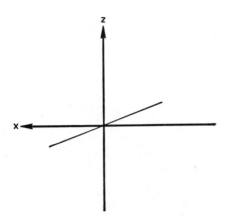

Figure 4. Linear X feedthrough, can be viewed as section of the error surface at Y = 0

The effect of the scale factor, ΔK, considered alone, is to create an error surface defined by

$$\epsilon(XY)_K = \Delta K V_x V_y \qquad (15)$$

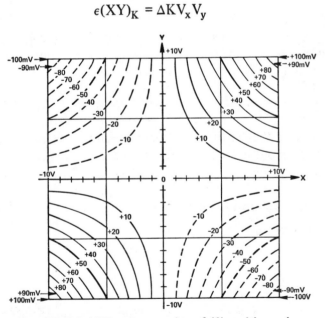

Figure 5. ISO-VER representation of 1% positive-scale-factor-error surface (all other errors zero), contour interval: 10mV

This is just a scaled-down version of the multiplier output $V_z = KV_xV_y$. Therefore, the scale-factor error surface must be a hyperbolic paraboloid, as sketched in Figure 2, but much smaller. Figure 5 is a 2-dimensional contour ("Iso-ver") representation. The effect of nonlinearity, $f(X,Y)$, is to introduce curvature on the nominally-straight-line elements parallel to the X or Y axis. That is, a section through the multiplier output surface parallel to the XZ or YZ plane is no longer a straight line (Figure 6).

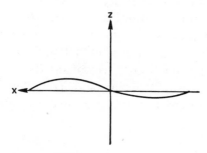

Figure 6. X nonlinearity can be viewed as section of the error surface at any value of Y

To summarize, a two-input analog multiplier has 4 sources of "trimmable" static error. Referring to (13), these are

1. X-input offset (Linear Y feedthrough)
2. Y-input offset (Linear X feedthrough)
3. Output offset
4. Scale-factor error

The effects of these four errors can be reduced to zero by introducing equal and opposite offsets for errors 1, 2, 3, and by precise adjustment of the scale factor, or gain, for 4. After the four errors are adjusted to zero, the remaining error will be due to the inherent nonlinearity of the multiplier, $f(X,Y)$. The nonlinearity is generally irreducible; however, in certain cases, a large percentage of it can be cancelled, as will be described in the *transconductance multiplier* section.

TRANSCONDUCTANCE MULTIPLIERS

The variable-transconductance multiplier is the simplest type of analog multiplier, at least in concept. One input variable controls the gain (transconductance) of an active device, which amplifies the other input in proportion to the control input.

A wide variety of active devices, such as transistors, FET's, vacuum tubes have been used, with varying degrees of success, to make "transconductance" (or "transresistance") multipliers and modulators for analog computing or communications signal-processing. However, almost all of the "transconductance" multipliers available today use silicon junction transistors as the active elements, because of the transistor's linear and consistent relationship between collector current and transconductance, given by equation (16)

$$\frac{dI_c}{dV_{be}} = \frac{q}{kT} I_c \qquad (16)$$

where

I_c = collector current in amperes

V_{be} = base-emitter voltage, in volts

q = unit of electronic charge = 1.60219×10^{-19} coulombs

k = Boltzmann's constant = 1.38062×10^{-23} joules/°K

T = absolute temperature, degrees Kelvin = °C + 273.15°

q/kT = 1/(25.69mV) at 25°C

The multiplicative property can be seen for sufficiently small increments ΔI_c, ΔV_{be} (Figure 7)

$$\Delta I_c = \frac{q}{kT} I_c \cdot \Delta V_{be} \qquad (17)$$

Figure 7. NPN transistor nomenclature

Equation (16) is derived by differentiating the simplified junction equation*

$$I_c = \alpha_N I_{ES} (\epsilon^{qV_{be}/kT} - 1) \qquad (18)$$

α_N = charge transport factor $\cong 0.99$

I_{ES} = emitter saturation current, 10^{-12} to 10^{-14} A @25°C

with the assumption that the collector-base voltage of the transistor is zero,

$$\frac{I_c}{I_{ES}} \gg 1,$$

yet the current levels are low enough so that ohmic resistances (e.g., base spreading resistance, emitter contact resistance, and bulk resistance) are negligible. For a typical monolithic dual transistor, this means collector currents of $100\mu A$ or less. At $100\mu A$, the transconductance is about 1/260mho at 26°C; parasitic resistances of about 3Ω, typical of monolithic transistors, reduce the transconductance by about 1%.

A simple, 2-quadrant variable-transconductance multiplier can be constructed from a pair of transistors and a few resistors, as shown in Figure 8. If the output of this multiplier is considered to be the difference between the collector currents of Q1, Q2, then equations 19 and 20 define the output-input relationship for this circuit

$$I_{c_1} - I_{c_2} = \Delta I_c = \frac{q}{kT} \frac{V_y + 0.6}{4.7 \times 10^3} 10^{-3} V_x \qquad (19)$$

$$\Delta I_c = 8.3 \times 10^{-6} (V_y + 0.6)V_x \quad \text{at } 25°C \qquad (20)$$

*See equation (6), Chapter 3-1.

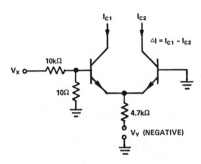

Figure 8. Simple 2-quadrant variable-transconductance multiplier

As (20) shows, the output collector-current difference is proportional to the product of the input voltages, V_x, V_y, with the following limitations:

1. The Y input has a 0.6V offset due to the assumed (constant) V_{BE}'s of Q1 and Q2. Thus, the most-positive value of V_y that can be accepted is –0.6V. Also, V_{be} is not constant. The V_{be} of Q1, Q2 increases as $|V_y|$ increases, introducing non-linearity on the Y input. Both problems can be fixed by using a more-elaborate voltage-to-current converter to replace the Y-input resistor.

2. The scale factor is a function of temperature, decreasing at –0.33%/°C near 25°C. This might be fixed by using temperature-compensating resistors on the X input, but it is hard to get precise compensation.

3. The X input is nonlinear, due to the exponential relationship between collector current and base-emitter voltage (18). The 1000:1 attenuator on the X input reduces the ±10V range to ±10mV between the bases, so that the actual X signal is less than the junction constant kT/q (25.69mV @ 25°C). However, even this small signal can result in 7% nonlinearity for X-input signals. The nonlinearity can be decreased by increasing the X-input attenuation, but at the cost of decreased signal-to-noise.

For the above reasons, the differential pair is not particularly useful or attractive as a high-level analog multiplier. However, it is quite effective as a mixer in RF applications, where the incoming signal is already quite small (millivolts or less).

There is a good, inherently simple solution to the nonlinearity, limited-dynamic-range, and temperature-coefficient problems of the simple differential pair[1]. Gilbert's circuit has rapidly gained universal acceptance, because of its low errors ($\sim$1%), wide bandwidth ($>$100MHz possible), and relative simplicity. In fact, it has become synonymous with "transconductance" multiplication. The basic circuit, shown in Figure 9, uses the logarithmic properties of diodes (or diode-connected transistors) to compensate for the exponential nonlinearity on the base inputs (hereafter labeled "X" input, for convenience) of the differential pair.

The balanced X input currents, I_{D_1}, I_{D_2}, flow through diodes D1, D2, establishing voltages V_1, V_2, which are proportional to the log-of-current ($\alpha_N \cong 1$)

$$V_1 = \frac{kT}{q} \ln \frac{I_{D_1}}{I_{ES_1}} \tag{21}$$

$$V_2 = \frac{kT}{q} \ln \frac{I_{D_2}}{I_{ES_2}} \tag{22}$$

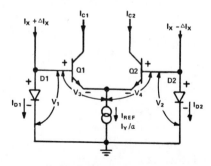

Figure 9. Linearized 2-quadrant multiplier (principle)

[1]"A New Wide-Band Amplifier Technique," by Barrie Gilbert, *IEEE Journal of Solid-State Circuits*, December, 1968, Volume SC-3, No. 4, pp. 353-365.

Since the collector currents of Q1 and Q2 are exponential functions of their base-emitter voltages (18) or the differential input voltage ($V_{BE_1} - V_{BE_2}$), it is reasonable to assume that the logarithmic input voltages provided by D1, D2 will cancel some, if not all, of the exponential nonlinearity of Q1 and Q2, resulting in a linear relationship between I_{D_1}, I_{D_2}, and I_{C_1}, I_{C_2}. In fact, the linearization is perfect in theory, and almost perfect in practice, a surprising and useful result that can be demonstrated as follows:

Assumptions:

1. Pairs Q1, Q2 and D1, D2 have zero differential offset voltage if $I_{C_1} = I_{C_2}$ and $I_{D_1} = I_{D_2}$.

2. Q1, Q2, D1, D2 obey the ideal junction equation (18)

The sum of the voltages V_1 to V_4 around the loop from the cathode of D1 to Q1, Q2, and the cathode of D2, must be zero

$$V_1 - V_3 + V_4 - V_2 = 0 \qquad (23)$$

$$V_1 - V_2 = V_3 - V_4 \qquad (24)$$

The base-to-emitter voltages of Q1, Q2 are proportional to the logarithms of their collector currents

$$V_{BE_1} = V_3 = \frac{kT}{q} \ln \frac{I_{C_1}}{I_{ESQ_1}} \qquad (25)$$

$$V_{BE_2} = V_4 = \frac{kT}{q} \ln \frac{I_{C_2}}{I_{ESQ_2}} \qquad (26)$$

Substituting for V1 through V4 in (24)

$$\frac{kT}{q} \ln \frac{I_{D_1}}{I_{ES_1}} - \frac{kT}{q} \ln \frac{I_{D_2}}{I_{ES_2}} = \frac{kT}{q} \ln \frac{I_{C_1}}{I_{ESQ_1}} - \frac{kT}{q} \ln \frac{I_{C_2}}{I_{ESQ_2}} \qquad (27)$$

Cancelling the kT/q terms and rewriting the differences of logs as logs of ratios

$$\ln \frac{I_{D_1} I_{ES_2}}{I_{D_2} I_{ES_1}} = \ln \frac{I_{C_1} I_{ESQ_2}}{I_{C_2} I_{ESQ_1}} \tag{28}$$

The constants will all be equal if the transistors and diodes are matched, as has been assumed

$$\ln \frac{I_{D_1}}{I_{D_2}} = \ln \frac{I_{C_1}}{I_{C_2}} \tag{29}$$

If the logs of the ratios are equal, then the ratios must be equal

$$\frac{I_{D_1}}{I_{D_2}} = \frac{I_{C_1}}{I_{C_2}} \tag{30}$$

This important result states that the ratio of the "output" currents I_{C_1}, I_{C_2} is linearly proportional to the ratio of the input currents, I_{D_1}, I_{D_2}, irrespective of temperature or the magnitudes of the currents! In other words, the linearization is ideally perfect, and the input-output transfer of X is constant with temperature.

The multiplier relationship can be derived directly from (30). The X input is assumed to be a difference $2\Delta I_x$ between the two diode currents I_{D_1} and I_{D_2}. The Y input controls the emitter currents I_{REF}. The multiplier output is the difference $2\Delta I_C$ in the collector currents of Q1 and Q2.

$$I_{D_1} = I_x + \Delta I_x \tag{31}$$

$$I_{D_2} = I_x - \Delta I_x \tag{32}$$

$$-I_x < \Delta I_x < I_x \tag{33}$$

$$I_{C_1} + I_{C_2} = \alpha I_{REF} \cong I_y \quad (\alpha \cong 1) \tag{34}$$

Q1, Q2 and D1, D2 are assumed matched, and Q1, Q2 have high β ($> 100, \alpha \cong 1$).

$$I_{C_1} = I_y/2 + \Delta I_c/2 \tag{35}$$

$$I_{C_2} = I_y/2 - \Delta I_c/2 \tag{36}$$

$$-I_y/2 < \Delta I_c < I_y/2, \quad I_y > 0 \tag{37}$$

Substituting for I_C and I_D in (29)

$$\frac{I_x + \Delta I_x}{I_x - \Delta I_x} = \frac{I_y/2 + \Delta I_C/2}{I_y/2 - \Delta I_C/2} \tag{38}$$

with a little bit of algebra,

$$\Delta I_C = \frac{\Delta I_x \cdot I_y}{I_x} \tag{39}$$

The output current is proportional to the product of the X input difference current ΔI_x and the Y input current, and inversely proportional to the X static current I_x, which can be seen to determine the scale factor as a linear 2-quadrant multiplier (bipolar ΔX input and unipolar Y input). The circuit can also function as a two-quadrant divider, with I_y constant, a unipolar denominator (I_x), and a bipolar numerator (ΔI_x). This linearized multiplier (Figure 9) has outstanding performance, and it represents a great improvement over the simple differential multiplier in the following ways.

1. Wide bandwidth: the circuit is basically "current-mode." At current levels of several mA, bandwidths over 100MHz can be obtained. At the lower current levels (< 1mA) normally used in multipliers, bandwidths of 1 to 10 MHz are readily achieved.

2. Excellent linearity: (39) indicates that the input-output relationship is exact for multiplication. In practice, there are

some small errors, < 1%, that will be discussed later. Nevertheless, it is greatly improved over the unlinearized multiplier.

3. Excellent temperature stability: (39) indicates that the input-output relationship is independent of temperature. In a practical circuit, there will be some slight temperature dependence due, in part, to the change in β of transistors with temperature (we have assumed β effects negligible in arriving at (39)). Changes in gain with temperature can be held to 0.02%/°C or less — more than an order-of-magnitude improvement over the simple differential multiplier (0.3%/°C).

4. Wide dynamic range: Since the X (base) input is linearized, the ratio of X input currents can be varied over a range almost equal to $-I_x < \Delta I_x < I_x$, allowing much greater input signals than the differential pair.

As a consequence of these advantages, the linearized "gain cell" has become almost universally accepted as a general-purpose multiplier building-block. With slight modification, it can be used directly as a 2-quadrant multiplier circuit.

Two-Quadrant Multiplier

The circuit of Figure 10 is an example of a workable 2-quadrant multiplier. The differential X input current is obtained from a differential pair Q6, Q7 with emitters coupled by resistor R1. Constant-current sources Q8, Q9 provide the I_x bias to the emitters of Q6, Q7. The 100kΩ emitter resistor R1 determines the differential X current, ΔI_x, per volt of input, V_x.

The X input current drives the emitters of the diode-connected transistors Q2A-B, rather than the collectors (or anodes) as in the circuit of Figure 9. This "inverted" connection is much easier to drive, since the emitters present low impedance and readily accept the current from the X input stage, Q6, Q7. The only practical difference between the "inverted" circuit and the basic current cell is that the output will also be inverted (i.e., "180° out of phase with the input"). This is easily corrected by proper phasing of the output amplifier.

The Y input current is derived from the closed-loop controlled current source A1-Q5. If Q5 has very high β (> 400), then the collector current of Q5 will be V_y/R_2, with negligible error. Diode D1 protects the base-emitter junction of Q5 from breakdown, in the event of positive Y input voltage. The differential output current (between the collectors of Q1A-Q1B) is converted to a single-ended voltage by the dynamic bridge R3, R4, R5, R6, and A2. The resistances must be very closely matched (0.1%) to minimize output voltage errors due to changing common-mode input voltage with the Y input signal.

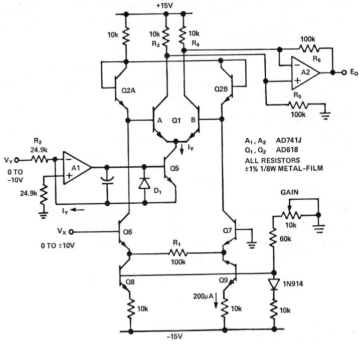

Figure 10. Practical 2-quadrant variable-transconductance multiplier

The two-quadrant multiplier is especially useful where very low feedthrough on one input is required. With the Y input at zero, the output is effectively disconnected from the input, providing at least 80 dB attenuation of any X input signal. Along with this advantage goes a disadvantage: as the Y input is reduced in magnitude, the current in Q1A and Q1B decreases, and the bandwidth of the circuit is reduced.

Four-Quadrant Multiplier

The basic two-quadrant linearized multiplier circuit can be extended to operate in 4 quadrants, accepting bipolar signals on either the X or Y inputs. This is accomplished by adding a second differential pair, Q3A-B, with bases connected in parallel with the bases of Q1, as shown in Figure 11. The collectors of the added pair are cross-connected to the collectors of Q1A-B. The single-ended Y current source of the 2-quadrant multiplier is replaced by a differential current source, Q10, R2, Q11, identical to the X current source.

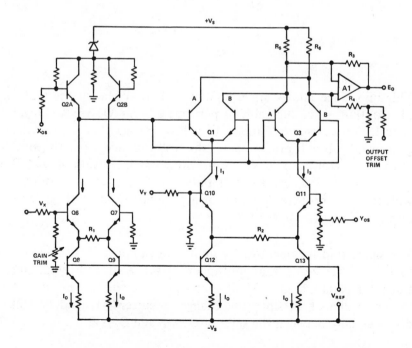

Figure 11. 4-quadrant variable-transconductance multiplier

One output of the paired Y current source is connected to the emitters of Q1A-B, the other to the emitters of Q3A-B. V_y now varies the ratio of the currents through the differential pairs, Q1A-B, and Q3A-B, and therefore controls their relative gains. For example, if $Y = 0$, then $I_1 = I_3$, and the gains of the two pairs are equal. Since their collectors are cross-coupled (bases in parallel), the outputs subtract, so there will be zero net gain for a signal on the X input. In this "balanced" condition, the X input experiences null suppression

$$E_o = V_x \cdot 0 = 0 \tag{40}$$

if $V_y = 0$.

If a non-zero voltage is applied to the Y input, then the currents I_1 and I_3 will be unbalanced

$$I_1 = I_{10} + V_y/R_2 \tag{41}$$

$$I_3 = I_{30} - V_y/R_2 \tag{42}$$

This unbalance allows an X input signal to appear at the multiplier output, since the gains of pairs Q1A-B and Q3A-B no longer cancel. For a positive Y input, I_1 will be greater than I_3, and the gain of Q1A-B will predominate; this will produce a positive output voltage (for positive X). On the other hand, the gain of Q3A-B predominates for negative Y inputs, causing a negative output voltage for a positive X input, or a positive output voltage for a negative X input.

The signal transfer operation from Y input to output is analogous to that in the 2-quadrant multiplier. If the X input is zero, and transistor pairs Q1, Q2, Q3 are matched, there will be zero output for any value of Y, since the change in currents I_1 and I_3 will divide equally between the sides of Q1A-B and Q3A-B.

$$E_o = 0 \cdot V_y = 0 \tag{43}$$

The overall output-input relationship of the multiplier can be developed from (39) as follows ($I_o = I_x$). The output of Q1A-B is

$$\Delta I_{C_1} = \frac{\Delta I_x I_1}{I_o} \qquad (44)$$

Similarly, for Q3A-B,

$$\Delta I_{C_3} = \frac{\Delta I_x I_3}{I_o} \qquad (45)$$

Since the collectors of Q1A-B and Q3A-B are cross-coupled, the output currents will subtract. The difference is ΔI_c:

$$\Delta I_c = \Delta I_{C_1} - \Delta I_{C_3} \qquad (46)$$

$$\Delta I_c = \frac{\Delta I_x}{I_o} (I_1 - I_3) \qquad (47)$$

Substituting for I_1 and I_3 from (41) and (42)

$$\Delta I_C = \frac{\Delta I_x}{I_o} (I_{10} + V_y/R_2 - I_{30} + V_y/R_2) \qquad (48)$$

Since $I_{10} = I_{30}$

$$\Delta I_C = 2 \frac{\Delta I_x}{I_o} \cdot \frac{V_y}{R_2} \qquad (49)$$

The net differential output current will be converted to a single-ended output voltage by A1 and R3, R4, R5, R6, just as for the two-quadrant multiplier

$$E_O = \Delta I_c R_3 \qquad (50)$$

(50) can be reduced to

$$E_O = \frac{2R_3}{R_1 R_2 I_O} V_x \cdot V_y \qquad (51)$$

$$-I_o R_2 < V_y < I_o R_2 \qquad (52)$$

$$-I_o R_1 < V_x < I_o R_1 \qquad (53)$$

The scale factor of the multiplier is set by $R_3/R_1 R_2 I_o$, which has the required dimension of V^{-1}.

Performance of the 4-Quadrant Transconductance Multiplier

The overall performance of the variable-transconductance multiplier is excellent, making it the most popular type of electronic analog multiplier. The reasons for the success of the Gilbert linearized multiplier are threefold:

1. Good accuracy: Overall error of less than ±1% of full scale (100mV in 10V) is easily achieved. Errors are proportional to input levels and tend towards zero as the inputs go to zero (except for dc offsets, which can be adjusted to zero). In fact, the "nonlinear" errors can be bounded by a simple linear equation

$$\epsilon(X,Y) = \frac{\epsilon_x}{100} V_x + \frac{\epsilon_y}{100} V_y \qquad (54)$$

where

ϵ_x = specified % nonlinearity on X input

ϵ_y = specified % nonlinearity on Y input

2. Wide bandwidth: up to 10MHz for voltage-output multipliers, over 100MHz with current output. Bandwidth is independent of signal level or input path (X or Y), for bandwidths < 10MHz.

3. Relative simplicity and low cost: The variable-transconductance multiplier can be constructed using "discrete" components, or it can be made in "monolithic" form. In either case, the inherent simplicity and consistency of performance of the circuit make it less expensive than any other four-quadrant multiplier. We will discuss each of these factors in more detail, to relate them to the practical circuit and limitations of the components.

Factors Affecting Accuracy of the Transconductance
Multiplier

So far, in our discussion of the variable-transconductance multiplier, we have assumed that the transistors obey the ideal junction equation, are perfectly matched, and have infinite current gain. We also assumed that all currents in symmetrical paths are equal, except for differences caused by injection of signals. In a practical circuit, the transistors and resistors are not "ideal" and are never (well, hardly ever) perfectly matched. These mismatches and departures from "ideal" behavior give rise to linear errors (input and output offsets, scale-factor errors) and nonlinear errors (2nd and 3rd harmonic distortions).

The linear errors can be theoretically adjusted to zero, and practically adjusted to negligible levels, as discussed in the introduction to this chapter. Four trim points are indicated in Figure 11:

1. X Offset: used to adjust linear Y feedthrough to zero

2. Y Offset: Adjusts linear X feedthrough to zero

3. Output Offset

4. Scale Factor, or "gain"

Nonlinear Errors

The primary source of nonlinearity in the variable-transconductance multiplier is current unbalance or offset-voltage mismatch between the two differential pairs, Q1A-B and Q3A-B. A $500\mu V$ mismatch between the offsets of these pairs will cause 1% (of full-scale) nonlinearity and feedthrough on the X input. This nonlinearity will be proportional to V_x^2, as illustrated in Figure 12. Fortunately, since it is possible to "match-up" pairs in a discrete circuit, or lay out "identical" transistors in an integrated circuit, for an average offset mismatch less than $500\mu V$, the X nonlinearity is usually less than 1%.

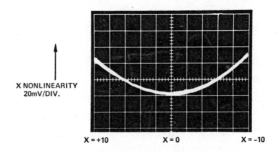

X NONLINEARITY
20mV/DIV.

X = +10   X = 0   X = -10

Figure 12. Parabolic X nonlinearity

Because the second-order X nonlinearity is relatively independent of the Y input signal amplitude, the X^2 nonlinearity can be significantly reduced by cross-coupling a fraction of the X input signal into the Y input, as will shortly be described.

Another source of potential X^2 nonlinearity is unbalance of the currents through the diode-connected transistors, Q2A, Q2B, when $V_x = 0$. The currents can be equalized by using closely-matched resistors in the X current sources.

The X input can exhibit considerable third-order (S-shape) nonlinearity under some conditions, as Figure 13 illustrates. The cubic distortion is caused by an ohmic component of emitter resistance in the differential pairs Q1A-B, Q3A-B. The ohmic (or constant) resistance decreases the transconductance from the theoretical value of qI_c/kT and thus causes nonlinearity. Since high-speed multipliers are operated at high current, these ohmic nonlinearities will be seen to force a speed-vs.-accuracy tradeoff.

The Y input of the transconductance multiplier has relatively low nonlinearity, typically ±0.1% to ±0.2%. The offset-voltage mismatch in the differential pairs, Q1A-B and Q3A-B, and the initial Y input currently unbalance have negligible effect on the Y nonlinearity and feedthrough, so it is consistently low.

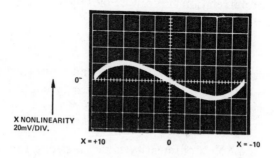

Figure 13. Cubic X nonlinearity

The X and Y input-voltages to differential-current converters can introduce nonlinearity if the emitter resistors are not large compared to kT/qI_c (26Ω at $I_c = 1mA$, $T = 300°K$).

Dynamics of the Transconductance Multiplier

The transconductance multiplier has wide bandwidth and fast transient response, since it is basically a current-mode circuit. Current-output bandwidths of 100MHz and greater can be obtained by operating the multiplier transistors at emitter currents of 10mA or more. However, circuits designed for the best dc accuracy operate at much lower currents: 10μA to 1mA, with bandwidths of 1 to 10MHz. The bandwidth limitation is primarily due to the output amplifier, which converts the difference of the collector currents to an output voltage.

The bandwidth of the 4-quadrant variable-transconductance multiplier is the same for the X or Y input, and is independent of signal level, except for the slew-rate limit of the output amplifier.

Linearizing the Transconductance Multiplier

The 4-quadrant variable-transconductance multiplier circuit, Figure 11, has predominantly second-order nonlinearity and feedthrough on the X input, for the reasons discussed above. The nonlinearity on the Y input is usually negligible compared to the "X" distortion. If all of the first-order errors − linear feedthrough, output offset, scale-factor error − are adjusted to zero, then the multiplier input-output relationship can be closely approximated by

$$E_o = K V_x V_y \pm \delta V_x^2 \, f(V_y) \qquad (55)$$

If the nonlinear term, $\delta V_x^2 \, f(V_y)$ is independent of (or not strongly influenced by) V_y, then the δV_x^2 nonlinearity could be cancelled by adding or subtracting a portion of the X input signal to the Y input, as shown in Figure 14.

Fortunately, the δX^2 nonlinearity is not a strong function of the Y input (i.e., $f(Y)$ is nearly constant), so the cancellation scheme works reasonably well in practice. Usually, the X^2 component of feedthrough (Y = 0) can be reduced to less than 0.1% of full scale (60dB null suppression), and the X nonlinearity ($V_y = 10V$) can be reduced by a factor of 2, with a corresponding reduction of overall error.

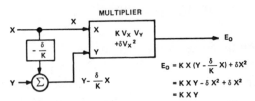

Figure 14. Improving linearity by cancelling second-harmonic distortion

A method of applying the X linearization to a multiplier is shown in Figure 15. This approach relies on fairly low source resistances, 100Ω or less, and the availability of both + and − (differential) Y inputs. On many multipliers (e.g., all Analog Devices multipliers of this type), the Y_o trim terminal can be used as the −Y input, for the linearization circuit (but not always with the same sensitivity).

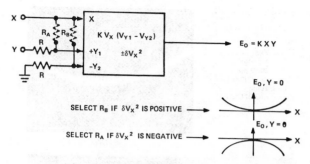

Figure 15. Applying linearization to a multiplier

Figures 16 to 18 show the results of applying the linearization to a multiplier. Note especially the reduction in both low-frequency and high-frequency feedthrough.

The cross-coupling linearization technique could be applied to the Y input, but the Y nonlinearity is generally already so low that "diminishing returns" sets in.

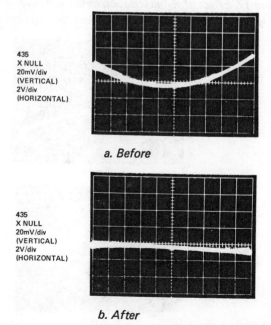

a. Before

b. After

Figure 16. Effect of X-linearization of a transconductance multiplier $X = \pm 10V$, $Y = 0$, 20mV/div. vertical scale

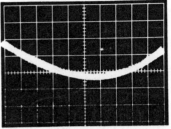

X$_{NON-LINEARITY}$
Y = +10V
20mV/div
(VERTICAL)
2V/div
(HORIZONTAL)

a. Before

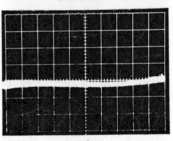

X$_{NON-LINEARITY}$
Y = +10V
20mV/div
(VERTICAL)
2V/div
(HORIZONTAL)

b. After

Figure 17. Effect of X-linearization of the same multiplier as Figure 16. X = ±10V, Y = +10V, 20mV/div. vertical scale

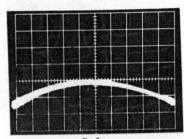

X$_{NON-LINEARITY}$
Y = -10V
20mV/div
(VERTICAL)
2V/div
(HORIZONTAL)

a. Before

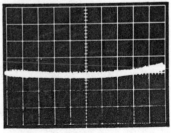

X$_{NON-LINEARITY}$
Y = -10V
20mV/div
(VERTICAL)
2V/div
(HORIZONTAL)

b. After

Figure 18. Effect of X-linearization of the same multiplier as Figure 16. X = ±10V, Y = −10V, 20mV/div. vertical scale

The cross-coupling linearization technique can be applied to any multiplier that has second-order nonlinearity on one or both inputs. The amount of reduction of overall error will depend on the degree to which the nonlinearity on one input is independent of the signal level at the other input.

In general, complete cancellation of the second-order nonlinearity on one input will occur for only one value of the other input. For example, the second-order component of X *feedthrough* (X = ±F.S., Y = 0) may be completely cancelled, but the second-order X *nonlinearity* will be partially cancelled, or may even increase under some circumstances.

LOG-ANTILOG MULTIPLIERS

The log-antilog multiplier is an electrical analog of the C and D scales on a slide rule, since it forms the product of two or more variables by addition of their logarithms

$$X \cdot Y = \epsilon^{(\ln X + \ln Y)} \tag{56}$$

The accuracy and temperature-stability of log-antilog multipliers is excellent, approaching the performance of the more-complex pulse-modulation multipliers. Errors of less than 0.25% of full-scale, with drifts of 0.01%/°C are readily achieved. Although operation of the basic log-antilog multiplier is restricted to one quadrant (typically the first quadrant), it can be offset to operate in four quadrants, as will be explained later. (The offsetting technique, or absolute-value-sign-magnitude technique mentioned earlier, can be applied to any 1-quadrant multiplier.)

Circuit Description

The log-antilog multiplier circuit is closely-related to the transconductance multiplier circuit, in that it relies on the logarithmic properties of silicon-junction transistors.

The basic building block of the log-antilog multiplier is the Paterson diode, or "transdiode" log amplifier, described in detail in Chapter 3-1. This circuit makes the best use of the log properties of the transistor (especially at low currents) and is also the easiest to combine into more-complex circuits, such as an analog multiplier.

The operation of the basic transdiode log amplifier, Figure 19, will be reviewed here for convenience.

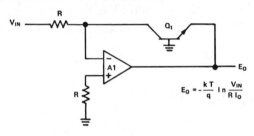

Figure 19. Basic transdiode log amplifier

If we assume that operational amplifier A1 has zero offset current and voltage, then the collector current of Q1 will be V_{in}/R. The output of A1 drives the emitter of Q1, so that the emitter-base voltage of Q1 is

$$E_o = V_{EB} = -\frac{kT}{q} \ln \frac{V_{IN}/R}{\alpha_N I_{ES}} \tag{57}$$

$$\alpha_N \cong 1$$

I_{ES} = emitter saturation current, $\sim 10^{-14}$ A

Let $\qquad \alpha_N I_{ES} = I_o$

The output of A1 is therefore proportional to the logarithm of the input voltage, and also variable with temperature, both through kT/q and through I_o. The temperature-dependence will be cancelled when the log amplifier is used in a multiplier circuit.

A schematic of a two-input log-antilog multiplier can be seen in Figure 20. The two inputs, V_x, V_y, drive two independent transdiode log amps, A1-Q1A and A2-Q2A. The base of Q2A is at ground potential, while the base of Q1A is tied to the emitter of Q2A. Therefore, the voltage at the emitter of Q1A will be proportional to the sum of the logs of V_x and V_y, as follows

$$V_{EB_{2A}} = - \frac{kT}{q} \ln \frac{V_x}{R_x I_{o_{2A}}} \tag{58}$$

$$V_{EB_{1A}} = - \frac{kT}{q} \ln \frac{V_y}{R_y I_{o_{1A}}} \tag{59}$$

$$V_3 = V_{1_A} + V_{2_A} \tag{60}$$

$$-V_3 = \frac{kT}{q} \left(\ln \frac{V_x}{R_x I_{o_{2A}}} + \ln \frac{V_y}{R_y I_{o_{1A}}} \right) \tag{61}$$

$$V_3 = - \frac{kT}{q} \ln \frac{V_x \cdot V_y}{R_x R_y I_{o_{2A}} I_{o_{1A}}} \tag{62}$$

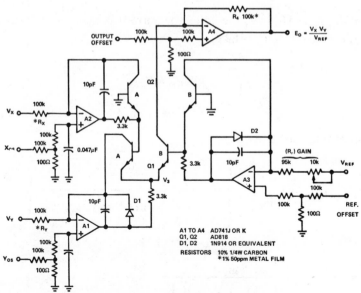

Figure 20. Log-antilog multiplier

The next step is to take the antilog of V_3, in a way that will cancel the temperature-dependence. Note that V_3 appears across the base-emitter circuits of the "B" sides of Q1 and Q2 in series.

$$V_3 = V_{EB1A} + V_{EB2A} = V_{EB1B} + V_{EB2B} \qquad (63)$$

Assuming a constant reference input, V_{REF}

$$V_{EB2B} = -\frac{kT}{q} \ln \frac{V_{REF}}{R_r I_{o_{2B}}} \qquad (64)$$

Solving (63) for V_{EB1B}

$$V_{EB1B} = V_{EB1A} + V_{EB2A} - V_{EB2B} \qquad (65)$$

$$V_{EB1B} = \frac{kT}{q} \ln \frac{V_x V_y R_r I_{o_{2B}}}{V_{REF} R_x R_y I_{o_{2A}} I_{o_{1A}}} \qquad (66)$$

For $V_{EB1B} > 100\text{mV}$, the collector current is exponentially related to the base-emitter voltage,

$$I_{CIB} = I_{o_{1B}} \epsilon^{qV_{EB1B}/kT} \qquad (67)$$

Combining (66) and (67),

$$I_{C1B} = I_{o_{1B}} \exp\left\{ \frac{q}{kT} \frac{kT}{q} \ln \frac{V_x V_y R_r I_{o_{2B}}}{V_{REF} R_x R_y I_{o_{2A}} I_{o_{1A}}} \right\} \qquad (68)$$

$$I_{C1B} = \frac{I_{o_{1B}} I_{o_{2B}} V_x V_y R_r}{I_{o_{1A}} I_{o_{2A}} V_{REF} R_x R_y} \qquad (69)$$

If transistors Q1 and Q2 are monolithic duals, the I_0 terms cancel

$$\frac{I_{0_{1B}}}{I_{0_{1A}}} = \frac{I_{0_{2B}}}{I_{0_{2A}}} = 1 \qquad (70)$$

The output amplifier A4 and feedback resistor R4 will convert I_{C1B} to a voltage

$$E_0 = R_4 \cdot I_{C1B} \qquad (71)$$

$$E_0 = \left\{\frac{R_4 R_r}{R_x R_y}\right\} \frac{V_x V_y}{V_{REF}} \qquad (72)$$

$$V_x, V_y \geq 0, V_{REF} > 0$$

Thus, the circuit of Figure 20 will multiply and divide with a scale factor that is independent of temperature (to the degree that the resistances track, which can be excellent). The output-input transfer function is also independent of the transistor current gains (β).

Performance of the Log-Antilog Multiplier

The actual performance of a practical log-antilog multiplier closely approaches the ideal as given. The static accuracy error and temperature drift are very low. The primary sources of static errors in the log-antilog multiplier are:

1. Transistor log conformity errors: For X or Y inputs near full-scale, the current in the log transistors, Q1A, Q2A, is about $100\mu A$. At this current level, the effects of ohmic emitter resistance become noticeable and will result in about 0.1% nonlinearity. Limiting the full-scale current to $100\mu A$ prevents greater nonlinearity.

2. Input current and offset voltage of the operational amplifiers, A1-A4 introduce "offset" errors at the X, Y, and reference inputs and signal outputs. Of the order of about 5mV, these offsets can be easily trimmed to less than 0.1mV by offsetting the reference (i.e. "+") inputs of amplifiers A1-A4.

3. Resistance tolerance: this causes an error in the scale factor, which can be adjusted by the "gain" pot.

4. Offset voltages in transistor pairs Q1A-B, Q2A-B cause scale-factor error of 4% per millivolt of offset. Gain-trim removes this error.

The temperature stability of the log-antilog multiplier is excellent. The scale-factor drift will be about $0.01\%/°C$ with 50ppm resistors for R_x, R_y, R_r, and R_4. The input and output offset drift is determined by the op amps, and so will be about $20\mu V/°C$ for $V_{REF} = 10V$. For lower values of V_{REF}, the input offset drift will be multiplied by $10/V_{REF}$ at the output.

As is true for other log circuits, the bandwidth of the log-antilog multiplier is proportional to the magnitudes of the inputs. This effect is due to decreased loop gain, with a corresponding increase in loop time constant, at reduced currents. Typically, the multiplier will have 100kHz bandwidth for 10V inputs, decreasing to 1kHz at 0.1V.

The total error of the log-antilog multiplier will be less than $\pm 10mV$ (out of 10V) when the input and output offsets and scale factor have been adjusted. The error will decrease with decreasing inputs and will typically be less than 0.1% of output, plus a fixed output offset, over the 0 to +10V output range.

Offsetting a 1-Quadrant Multiplier for
Operation in 4 Quadrants

Any 1-quadrant multiplier may be made to operate in 4 quadrants, by properly offsetting the inputs and output. The multiplier itself remains a 1-quadrant device, operating about a bias point centred within its usual unipolar range. The offsetting scheme can be developed by considering the effect of an offset on the X and Y inputs.

$$E_o = K_1(V_x + X_{os})(V_y + Y_{os}) \tag{73}$$

$$E_o = K_1(V_xV_y + X_{os}V_y + V_xY_{os} + X_{os}Y_{os}) \tag{74}$$

The effect of the input offsets is to introduce an output offset $X_{os}Y_{os}$ and two linear feedthrough terms, $X_{os}V_y$ and V_xY_{os}. If $X_{os} > |V_x|_{max}$ and $Y_{os} > |V_y|_{max}$, then V_x and V_y can be either positive or negative, and E_o in (74) will still be positive. If the undesired terms — those other than $K_1V_xV_y$ — are subtracted from (74), then E_o can be positive or negative: the desired result.

$$E_o = K_1V_xV_y + K_1(X_{os}V_y + V_xY_{os} + X_{os}Y_{os}) - K_0 - K_2V_x - K_3V_y$$

$$\tag{75}$$

If $K_0 = K_1X_{os}Y_{os}$, $K_2 = K_1Y_{os}$, and $K_3 = K_1X_{os}$, then

$$E_o = K_1V_xV_y \tag{76}$$

for V_x and V_y of any polarity.

The offsetting and input coupling are shown in the block diagram, Figure 21. This offsetting scheme can be used with the log-antilog multiplier shown in Figure 20. It adds considerable complexity to the initial adjustment of the multiplier, and the reference voltage (V_{REF}) must be constant, or the "feedthrough" and offset will not stay cancelled. In addition, the 4-quadrant log-antilog multiplier will be slower for negative X and Y inputs (less current in the log transistors) than for positive inputs, making the scheme less effective for waveforms symmetrical about zero.

In spite of these shortcomings, the offset multiplier can be adjusted for errors of 0.1% of full scale, with nonlinearities of the order of 0.05%.

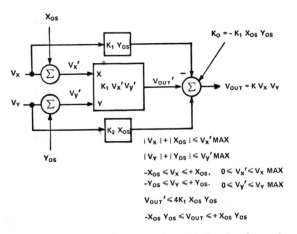

$$|V_X| + |X_{os}| \leqslant V_X' \text{MAX}$$

$$|V_Y| + |Y_{os}| \leqslant V_Y' \text{MAX}$$

$$-X_{os} \leqslant V_X \leqslant + X_{os}, \quad 0 \leqslant V_X' \leqslant V_X \text{ MAX}$$

$$-Y_{os} \leqslant V_Y \leqslant + Y_{os}. \quad 0 \leqslant V_Y' \leqslant V_Y \text{ MAX}$$

$$V_{OUT}' \leqslant 4K_1 X_{os} Y_{os}$$

$$-X_{os} Y_{os} \leqslant V_{OUT} \leqslant + X_{os} Y_{os}$$

Figure 21. Offsetting 1-quadrant multiplier for 4-quadrant operation. Scale factor change in multiplier produces output offset and feedthrough shift at summed output.

PULSE-MODULATION MULTIPLIERS

The pulse-modulation multiplier operates on the principle that the area under a rectangular pulse is proportional to the product of the pulse amplitude and pulse duration (Figure 22).

Figure 22. Basic principle of pulse modulation

It then follows that the average magnitude of a train of rectangular pulses is proportional to the product of the pulse amplitude and ratio of *on* time to period (duty cycle). (Figure 23).

A multiplier may be constructed using this technique. One input is used to control the amplitude of the pulse, the other the duty cycle. The resulting pulse train is low-pass filtered, yielding the average value, which is proportional to the product of the two inputs. A block diagram of a simple, two-quadrant pulse-modulation multiplier is shown in Figure 24.

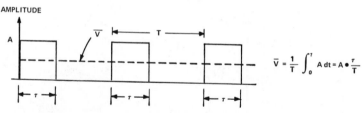

$$\overline{V} = \frac{1}{T} \int_0^\tau A\, dt = A \bullet \frac{\tau}{T}$$

Figure 23. Average value of train of square pulses is proportional to product of amplitude and duty cycle

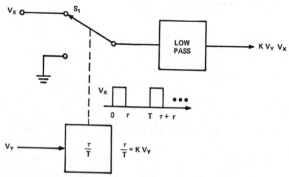

Figure 24. Two-quadrant pulse-modulated multiplier block diagram

The Y input controls the duty cycle of a pulse train, which in turn drives a switch S1. The switch alternates between the input and ground, dwelling at the input for a time proportional to the duty cycle. The output of the averaging filter will be proportional to the product $V_x V_y$. The X input can be either positive or negative, but the Y input is limited to positive values, since the duty cycle, τ/T, cannot be "negative."

The pulse-modulation technique can be extended to four-quadrant operation by using a "balanced" switching and duty-cycle generator, so that a zero Y input results in a 50% duty cycle, as the block diagram in Figure 25 illustrates.

Performance of the Pulse Modulation Multiplier

Pulse-width-pulse-height modulation is inherently the most accurate method of performing analog multiplication. Errors of less than 0.1% of full scale and nonlinearities of 0.02% can be readily

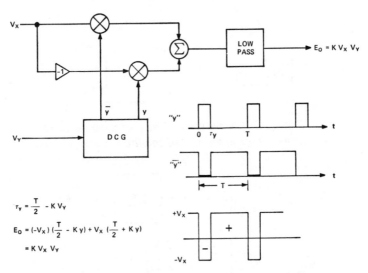

Figure 25. Four-quadrant pulse-modulation multiplier

achieved. The high accuracy is the result of using the nonlinear elements (FET's or transistors) as switches, rather than relying on the exact relationship between their "gain" and input voltage.

Though the pulse-modulation multiplier is ideally 100% accurate, there are several sources of error that limit the accuracy of practical multipliers. Most of the limitations arise from the non-ideal behavior of real switches and duty-cycle generators. However, there is one limitation inherent in the modulation technique itself: the signal frequency must be much less than the averaging frequency to allow sufficient averaging time. Analog averaging will always leave a finite (but usually negligible) ripple component on the output. Generally, the carrier/frequency should be at least 10 to 100 times the signal frequency.

The carrier frequency is in turn limited by component-determined errors:

1. Capacitance between the switch-control terminal and the signal path, e.g., gate-to-channel capacitance of a FET. This capacitance couples a charge into the signal path each time the switch is turned on or off, resulting in an offset voltage. The offset may change with signal level, resulting in a non-

linearity. The "dumped charge" effect can be minimized by using low-capacitance switches or lower carrier frequency to reduce the average charge (current) coupled into the signal path.

2. On-off resistance of switches: FET or CMOS switches (even reed relays) have measurable *on* and finite *off* resistance. As long as the ratio of *off* to *on* is high (> 10,000), errors from this source will be small. If the ratio is low, then some of the input signal will leak into the output when the switch is *off*, increasing the feedthrough.

3. Linearity of the duty-cycle generator: The variable duty-cycle pulse generator is potentially the most significant source of nonlinearity. The controlling input, i.e., V_y, must determine the ratio of *on* to *off* time precisely, over a fairly wide range, especially in a 4-quadrant multiplier. As the duty cycle is reduced, any fixed timing errors, e.g., delays, become a more-significant portion of the *on* time, as shown in Figure 26, introducing nonlinearity.

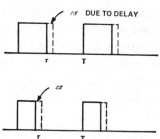

Figure 26. Nonlinear error produced by fixed delay asymmetry in duty-cycle-generator-plus-switch. $\Delta\tau$ is the same whether τ is large or small, a deviation from proportionality.

The nonlinearity of the duty-cycle generator can be reduced to an arbitrarily-low level by using a closed-loop circuit, illustrated in the block diagram of Figure 27.

The input voltage, V_y, is compared to the average value of a chopped reference voltage. The output of the comparator controls the *on* to *off* time of the chopper, so that the average value of the voltage out of the chopper will equal V_y in the steady-state.

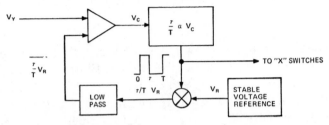

Figure 27. Closed-loop duty-cycle generator

The relationship between duty cycle, τ/T, and comparator output voltage is not important, as long as it is single-valued. The linearity of the overall system is determined by the threshold accuracy of the comparator and the averaging time. Nonlinearities of less than 0.01% can be achieved by this approach.

MULTIPLIER SPECIFICATIONS

Perhaps the best way of gaining an understanding of multiplier specifications and their dependence on multiplier circuit design is to review the specifications as set forth in a multiplier data sheet. The accompanying comparative table lists specifications for modular multipliers using the three techniques discussed in this chapter: transconductance (432, 429), pulse-modulation (427), and log-antilog (434).

The 432 is a low-cost 4-quadrant transconductance multiplier, with 1%-2% error, good bandwidth, and small size. It is comparable to the integrated-circuit AD533 with external trims, drawing a little more power than the internally-trimmed AD532.

The 429 is a fast (10MHz), 4-quadrant transconductance multiplier with low error (0.5%) and low nonlinearity. It is a no-compromise discrete design that uses monolithic dual transistors in the multiplier section and a fast discrete-component output amplifier.

The 427 is a high-accuracy (0.25% error) pulse-modulation 4-quadrant multiplier. The use of a high-frequency carrier (3MHz) allows a signal bandwidth of 100kHz, 100 to 1000 times greater than the usual bandwidth for pulse-modulation multipliers.

The 434 is a 1-quadrant log-antilog multiplier that combines high accuracy (0.25% error) and versatility, since both multiplication and division can be performed simultaneously.

SPECIFICATIONS (pp. 246–247)

The first four lines of the comparative-specification table[*] summarize the salient features of the multiplier, to guide the reader immediately to the ones most likely to fill the needs of his application. The "Aids for the Designer," Chapter 4-4, provides considerable detail on multiplier selection, so it will not be covered here.

MULTIPLICATION CHARACTERISTICS

This block of specifications deals with overall static errors from all sources that are covered in detail in the succeeding Specification blocks (Offset, Scale Factor, Nonlinearity, Feed-through).

Output Function: Defines the ideal functional relationship between the two input voltages, V_x, V_y, the output voltage E_o, and the scale constant, V_r. All errors are defined as deviations from this transfer function and are specified as percentages of full scale, 10V. A typical transfer function is

$$E_o = \frac{V_x V_y}{10V} \tag{77}$$

For $V_x = V_y = 10V$,

$$E_o = \frac{10 \times 10}{10} = 10V \tag{78}$$

---

[*]This table is an abbreviated example involving a few contrasting modular multiplier types, with information valid as of Summer 1973. For further information on these or the many more types available within each class, as well as the many IC types, it is suggested that the reader consult the most recent edition of the Analog Devices *Product Guide* or supplements.

MULTIPLIERS/DIVIDERS (Discrete)
SPECIFICATION SUMMARY (Typical @ 25°C and ±15VDC unless otherwise specified)

MULTIPLICATION TECHNIQUE	TRANSCONDUCTANCE	
Model[1]	Economy	Accurate Wideband
	432J (432K)	429A (429B)
Price *1-9	$29 ($45)	$109 ($139)
Price 10-24	$27 ($43)	$104 ($129)
Full Scale Accuracy[2]	2% (1%)	1% (0.5%)
Divides and Square Roots	YES	YES
Multiplication Characteristics		
Output Function	XY/10	XY/10
Error, Internal Trim (±)	2%(1%) max	1%(0.5%) max
Error, External Trim (±)	1.0% (0.6%)	0.7% (0.3%)
Accuracy vs. Temperature (±)	0.06%/°C (0.04%/°C)	0.05%/°C (0.04%/°C max)
Accuracy vs. Supply (±)	0.1%/%	0.03%/%
Warm up Time to Specifications	1 min	1 sec
Output Offset (±)		
Initial	20mV (25mV max)	20mV (10mV) max
Average vs. Temperature 0°C to +70°C	2mV/°C (1mV/°C)	2mV/°C (1mV/°C max)
Average vs. Supply	10mV/%	1mV/%
Scale Factor (±)		
Initial Error	1%(0.5%)	0.5% (0.25%)
Non Linearity (±)		
X Input (X = 20V p-p, Y = ±10VDC)	0.8% (0.6% max)	0.5% (0.2%) max
Y Input (Y = 20V p-p, X = ±10VDC)	0.4% (0.3% max)	0.3% (0.2%) max
Feedthrough		
X = 0, Y = 20V p-p 50Hz	80mV (50mV) p-p max	25mV (10mV) p-p max
with external trim	30mV p-p	8mV (5mV) p-p
Y = 0, X = 20V p-p 50Hz	120mV (100mV) p-p max	50mV (15mV) p-p max
with external trim	N/A	35mV (10mV) p-p
Feedthrough vs. Temperature, each input	1mV p-p/°C	2mV p-p/°C
Bandwidth		
−3dB Small Signal	1MHz	10MHz
Full Power Response	700kHz	2MHz min
Slew Rate	45V/μsec	120V/μsec min
Small Signal Amplitude Error (±)	1% @ 40kHz	1% at 300kHz min
Small Signal Vector Error (±)	1% @ 10kHz	1% at 50kHz min
Settling Time for ±10V Step	1μsec to 2%	0.5μsec to 1%
Overload Recovery	3μsec	0.15μsec
Output Noise		
5Hz to 10kHz	600μV rms	500μV rms
5Hz to 5MHz	3mV rms	2.5mV rms
Output Characteristics		
Voltage at Rated Load (min)	±10V	±11V
Current (min)	±5mA	±11mA
Load Capacitance Limit	0.001μF	0.01μF
Input Resistance		
X/Y/Z Input	10MΩ/10kΩ/36kΩ	10kΩ/11kΩ/27kΩ
Input Bias Current		
X/Y/Z Input	2μA each	+100nA/+100nA/±20nA
Maximum Input Voltage		
For Rated Accuracy	±10.1V	±10.5V
Safe Level	±Vs	±16V
Power Supply (Vs)		
Rated Performance	±15V	±14.7 to ±15.3V
Operating	±12 to ±18V	±14 to ±16V
Quiescent Current	±4.5mA	±12mA
Temperature Range		
Rated Performance	0°C to +70°C	−25°C to +85°C
Operating	−25°C to +85°C	−25°C to +85°C
Storage	−55°C to +125°C	−55°C to +125°C
Package Outline	QC-2	FA-4
Case Dimensions	1.1" X 1.1" X 0.4"	1.5" X 1.5" X 0.6"
	28 X 28 X 10.2mm	38.1 X 38.1 X 15.2mm

*Summer, 1973. Price is listed here as a measure of relative cost, not primarily as a commercial inducement. Those interested further should consult recent Product Guides or

PULSE TYPES	LOG-ANTILOG
High Accuracy	
427J　(427K)	434A　(434B)
$159 ($210)	75(87)
$143 ($189)	69(77)
0.25% (0.2%)	0.5% (0.25%)
YES	YES
XY/10	YZ/X
0.25% (0.2%) max	0.5% (0.25%) max
0.15% (0.1% max)	0.3% (0.1%)
0.02%/°C max	0.02% (0.02%/°C max)
0.02%/%	0.02%/%
1 min	1 min
5 mV	2mV (2mV max)
0.2mV/°C (0.2mV/°C max)	1mV/°C (1mV/°C max)
1mV/%	1mV/%
0.1% (0.05%)	0.2% (0.1%)
0.08% (0.04%) max	0.2% (0.1%)[2]
0.08% (0.04%) max	0.2% (0.1%)[2]
20mV p-p max	+2 mV Peak Max
4mV p-p	–
20mV p-p max	+2mV Peak Max
5mV p-p	–
0.2mV p-p/°C	–
100kHz	100kHz[3]
30kHz	30kHz
2V/μsec	2V/μsec
0.1% at 4kHz	
1% at 700Hz	
20μsec to 0.1%	40μsec to 0.1%
10μsec	20μsec
50μV rms	300μV rms
1mV rms	1mV rms
±10.2V	+11V
±7mA	+5mA
0.01μF	0.01μF
10kΩ/10kΩ/33kΩ	100kΩ/90kΩ/100kΩ
±3μA/±3μA/±10μA	10nA/100nA/10nA
±10.5V	+10.5V
±16V	±16V
±14.8 to ±15.3V	±14.4 to ±15.6
±14.8 to ±16V	±10V to ± 18V
±16mA	±10mA
0°C to +70°C	-25°C to +85°C
-25°C to +85°C	-55°C to +125°C
-55°C to +125°C	-55°C to +125°C
D-2	
1.6" X 3.0" X 0.6"	1.5 X 1.5 X 0.6
40.6 X 76.2 X 15.2mm	38.1 X 38.1 X 15.2mm

NOTES:

[1]Parentheses indicate specification for the high performance (K version) model of each multiplier when it differs from the J or A version. For example, order Model 427J for 0.25% accuracy, Model 427K for 0.2% accuracy.

[2]434 is a one quadrant device: specs are for inputs between 0 and +10V only

[3]Bandwidth depends on level of input. Specs given for 10V

price lists, or the nearest Sales office, since prices are subject to change. See also Table 2 in Chapter 3-3.

A scale factor $(1/V_r)$ of $1/10/V$ is almost universal at present, but others, such as $1/V$, $1/5/V$, or $1/100/V$ have been used. The scale factor can also be adjustable (or even variable over a wide range, as shown for the 434 log-antilog multiplier). Where the scale factor is adjustable or variable, the multiplier specifications are usually given for a $1/10/V$ scale factor, and deviations from these limits are elaborated as a function of scale factor.

Actual error (V) and percentage error (of 10V F.S.) are related as follows:

$$\epsilon = V_{measured} - \frac{V_x V_y}{10V} \tag{79}$$

$$\% \text{ Error} = 100 \; \frac{\epsilon}{10V} = 10 \cdot \epsilon \tag{80}$$

Error, Internal Trim: the maximum difference between the multiplier's actual and ideal output values for any pair of dc input voltages within the multiplier input range at 25°C without the intervention of any external adjustments. The error is expressed as a percentage of full scale (80); thus, 1% error is $0.01 \cdot 10V = 100mV$.

The maximum error almost always occurs for full-scale inputs ($\pm10V$), as discussed in detail under nonlinearity. The error includes offset, feedthrough, nonlinearity, and scale-factor errors. This specification characterizes the "accuracy" of the multiplier.

As a practical matter, the measurement is made at the "end-points" of the four quadrants, $(V_x, V_y) = (+10V, +10V)$, $(-10V, +10V)$, $(-10V, -10V)$, $(+10V, -10V)$.

The maximum error for the 432J is $\pm2\%$, which implies that full-scale output may be between $\pm9.8V$ and $10.2V$; the 427K has one-tenth the maximum error of the 432, i.e., $\pm0.2\%$ or $\pm20mV$ (untrimmed).

Error, External Trim: the error remaining after the X and Y feedthroughs and output offset have been nulled out using external potentiometers or voltage dividers. This is a measure of the irreducible component of error, approximately equal to the nonlinearity (see also *scale factor* and *nonlinearity*).

Accuracy vs. Temperature (Error vs. Temperature): The rate at which the error, as defined above, changes with temperature. It is expressed as a percentage of full scale (10V) per degree centigrade. This coefficient includes the effects of output offset drift, feedthrough drift, and scale-factor drift, and so can be used to predict the maximum error expected over a temperature range as follows (e.g., $T_H > 25°C$)

$$\text{Error(V)} = \frac{1}{10} \left\{ |\% \text{ error}|_{25°C} + \left| \frac{\Delta(\% \text{ error})}{\Delta T} \right| (T_H - 25°C) \right\} \quad (81)$$

For example, the 429B has an error of 0.5% maximum at 25°C, and an error drift of ±0.04%/°C(max). To calculate the maximum error at 70°C:

$$\epsilon_{70°C} = \frac{1}{10} \left\{ 0.5 + 0.04 (70 - 25) \right\} \quad (82)$$

$$\epsilon_{70°C} = 0.1(0.5 + 1.8) = 0.23V = \pm230mV \quad (83)$$

The error calculated in this fashion represents error at or near full-scale output, where error due to scale-factor drift predominates. If both inputs are less than 1/3 of full scale (1/10 full-scale *output*) the drift is considerably less, since output offset drift predominates.

Accuracy vs. Supply (Error vs. Supply): the sensitivity of the multiplier output voltage to changes in power-supply voltage, expressed as %(full-scale)/%(supply-voltage change). It includes the effects of scale factor, feedthrough, and offset vs. supply at dc.

Example: for the 432, this error is specified at $\pm 0.1\%/\%\Delta V_s$.

$$0.1\% \text{ (full scale)} = 10\text{mV}$$

$$1\%\Delta V_s = 150\text{mV}$$

Therefore, the output of the 432J will change

$$\frac{10\text{mV}}{150\text{mV}} = \pm 0.067\text{V/V} = 67\text{mV/V} \tag{84}$$

Another way of looking at power-supply rejection is to recognize that the multiplier's internal reference circuit attenuates changes in the power-supply voltage. This ranges from a power-supply rejection ratio (PSRR) of 15:1 (PSR $\cong$ 23dB) for the 432 to 75:1 ($\cong$ 38dB) for the 427K. In general, the more accurate the multiplier, the less sensitive it is to supply changes.

Warmup Time to Specifications: The time elapsed after the dc power is applied to the multiplier, before the errors are expected to be within the specified limits. While this does not include the time required for the multiplier to stabilize *completely*, it does indicate how long it will be before changes in output due to warmup will be small compared to the specified error.

In general, most modular multipliers are operating at their rated specifications within a few milliseconds after turn-on, since their internal temperature rise is only a few °C. Also great care is taken in the design and packaging to minimize the temperature coefficients and internal thermal gradients.

OUTPUT OFFSET

Initial Output Offset: the output voltage for $V_x = V_y = 0V$. This specification gives the maximum offset at 25°C with no external adjustment. In all cases, this offset can be adjusted to zero with an external potentiometer or voltage divider. Offset is the principal error when the output is less than 1V.

Initial offset has a non-zero value due to shifts in encapsulation and tolerances of internal trims. The higher the accuracy rating of the multiplier, the less the initial offset.

Average Offset vs. Temperature: the dependence of the output offset on temperature. Unlike operational amplifiers, the average offset vs. temperature in multipliers is independent of the initial offset.

Example — 429B: Offset = ±10mV *max* untrimmed
Offset vs. temperature = ±1mV/°C *max*

To calculate maximum offset at 70°C,

$$E_{os70°C} = \left|E_{os25°C}\right| + \frac{\Delta E_{os}}{\Delta T} \; (70°C - 25°C) \tag{85}$$

$$E_{os70°C} = 10mV + 1(45) = ±55mV \; max \tag{86}$$

Average Offset vs. Supply: the sensitivity of output offset to changes of supply voltage, expressed as millivolts per % change of supply voltage at dc. Like the total error vs. supply, this quantity can be expressed in volts/volt, inversely as power-supply rejection ratio (PSRR) for offset, and in log (dB) form: PSR = $20\log_{10}$PSRR. For example, the 429 and the 427 have offset sensitivity of 1mV/1%ΔV_s, or 1mV/150mV. The offset PSRR is thus 150, and the PSR is about 43dB.

SCALE FACTOR

Scale Factor — Static, or low-frequency — The difference between the average scale factor and the ideal scale factor of 1/10/V. Errors due to this factor are expressed in % of output signal; that is, a 0.5% scale-factor error will cause a 50mV error at E_o = 10V, and a 5mV error at E_o = 1V. The scale-factor error includes only the average linear gain error (i.e., the error in the slope of a "best straight line" through the range of output for one input constant, the other swinging through its range). The nonlinear component is discussed under *nonlinearity*.

The scale-factor error can be adjusted to zero at any one point. However, the nonlinearity makes it impossible to adjust the scale-factor error to zero over the entire X-Y operating range. The *average* scale-factor error can be adjusted for minimum error over limited regions (e.g., one or two quadrants), or for the best compromise over all values of input.

NONLINEARITY

Nonlinearity: the irreducible component of error. The specification represents the peak difference between the multiplier output and the theoretical output with the average scale-factor error adjusted to zero under the specified test conditions. Schemes for testing nonlinearity will be found in Figures 40, 46, and 47 at the end of this chapter. Since the output and input waveforms should have the same shape (one input constant), the test circuit displays the difference between the multiplier output voltage and an input that swings over the entire range, while the other input is held constant. The average scale-factor error (slope) is adjusted out.

Typical nonlinearity curves for the 432 and the 427, measured in this way, can be seen in Figures 28 a-d, for one polarity of constant voltage. For each of these curves, corresponding curves exist (not necessarily of the same shape) for the opposite polarity. Note that the curves are smooth and without discontinuities at the origin. The parabolic shape of the 432's X nonlinearity indicates that the X input has primarily second-harmonic distortion (proportional to X^2). The S-shaped 427 nonlinearity curves indicate predominantly cubic distortion.

FEEDTHROUGH

Feedthrough. Ideally, the output of the multiplier should be zero if either input is zero, independently of the signal applied to the other input. Actually, a certain fraction of the non-zero input will

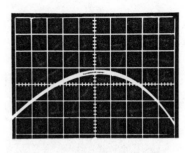

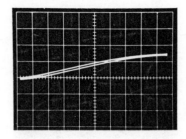

a. Model 432 X-input nonlinearity
for ±10V input signal,
Y = 10V, vertical scale: 20mV/div.

b. Model 432 Y-input nonlinearity
for ±10V input signal,
X = 10V, vertical scale: 20mV/div.

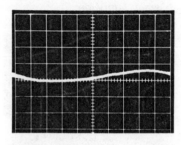

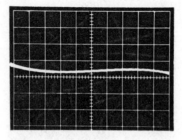

c. Model 427 X-input nonlinearity
for ±10V input signal,
Y = 10V, vertical scale: 10mV/div.

d. Model 427 Y-input nonlinearity
for ±10V input signal,
X = 10V, vertical scale: 10mV/div.

Figure 28. Typical nonlinearity curves

"feed through" and appear at the output. The feedthrough signal is composed of two components, one linear, the other nonlinear. The linear component is the product of the voltage on the varying input and the effective offset voltage of the "zero" input. This can be trimmed to zero by introducing an equal and opposite offset at the trim input (X_o, Y_o).

The nonlinear component, which cannot be reduced by zero by an offset adjustment, is due to the nonlinearity of the multiplier circuit. Graphically, it is the intersection of the nonlinearity sur-

face with the XZ and YZ planes (Figures 3 and 4, this chapter).
Figures 29 (a-d) show typical X and Y feedthrough waveforms for

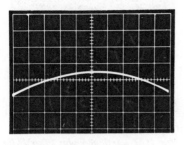

a. Model 432 X Feedthrough,
X = ±10V, Y = 0, vertical
scale: 50mV/div.

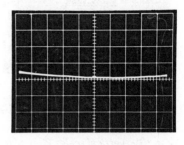

b. Model 432 Y Feedthrough,
Y = ±10V, X = 0, vertical
scale: 50mV/div.

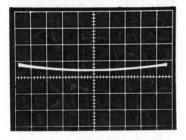

c. Model 427 X Feedthrough,
X = ±10V, Y = 0, vertical
scale: 10mV/div.

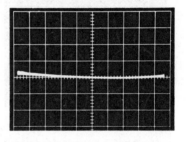

d. Model 427 Y Feedthrough,
Y = ±10V, X = 0, vertical
scale: 10mV/div.

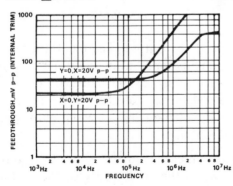

e. Model 429
Feedthrough vs. frequency

Figure 29. Typical feedthrough curves

the 432 and the 427. Note that the 432 has pronounced parabolic X feedthrough, which greatly resembles the nonlinearity at the extremes.

Feedthrough vs. Frequency. Feedthrough increases with frequency, due to capacitive coupling between the inputs and the output stage. Figure 29e is a plot of both X and Y feedthrough vs. frequency for the 429.

BANDWIDTH (High-Frequency Dynamic Parameters)

Bandwidth, –3dB Small-Signal: The output frequency at which the scale-factor of the multiplier has decreased to 0.7 times its dc value. "Small" signal usually means an output of less than 5% of full scale, e.g., 1Vp-p for a ±10V(FS) multiplier. Bandwidth is usually measured with a full-scale dc voltage on one input, a 1Vp-p sine wave on the other. It can be seen on the chart that the two transconductance multiplier types have wider bandwidths that either the pulse-modulated or log-antilog types.

The term "output frequency" is significant. For example, the low-frequency output of a multiplier connected as a squarer (X = Y), for sine-wave input, is a double-frequency sine wave with an amplitude of one-half the square of the input amplitude, biased by a like amount. The output amplitude will be down 3dB for an input of lower frequency than for the dc X sine case, because of the frequency doubling. On the other hand, the "dc" component of the output can remain unaffected up to considerably higher frequencies.

Full-Power Response: the maximum frequency at which the multiplier output can product full-scale voltage at rated current, without noticeable distortion. This is measured by applying 10Vdc to one input and a 20Vp-p sine wave to the other (and vice versa). Again, the transconductance multipliers are much faster than the pulse-modulation or logarithmic types.

Slew(ing) Rate: the maximum rate of change of output voltage for large signals. It is measured with one input at 10V, the other a step swing of 10 or 20V. A typical 429 step response, showing the slewing rate, is shown in Figure 30. The approximate relationship between slewing rate and full-power bandwidth is

$$S \simeq A \, 2 \, \pi \, f_P$$

where

S = Slewing rate, in volts/microsecond

A = Peak sine-wave amplitude, in volts

f_P = Measured frequency for full output, in MHz

For example, the 429 has f_P = 2MHz(*min*). For A = 10V,

$$S \simeq 10 \times 2\pi \times 2 = 126V/\mu s \tag{88}$$

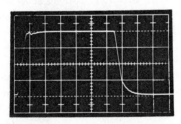

Figure 30. Step response, showing slew rate of Model 429.
Voltage scale: 5V/div; time scale: 200ns/div.
V_x = *20V p-p 400kHz square wave*
V_y = *+10.0V*

Small-Signal Amplitude Error. This is the frequency at which the amplitude response, or scale factor, is down by 1% (0.1%, for high-accuracy types), measured with a "small" signal, e.g., 10% of full-scale. If, for a given type, this frequency turns out to be 1/3 or less of the full-power frequency, then signals as large as full scale fit the definition of "small signals." For example, $f_{-1\%}$ for both the 429 and the 432 is less than 1/6 of f_P; therefore $f_{-1\%}$ applies to any signal in the whole ±10V range.

The 1% error bandwidth is related to small-signal bandwidth and the rolloff rate (depending on the number of poles in the transfer function). For responses governed by a single pole, the –1% error bandwidth occurs at about 1/7 of the –3dB bandwidth

$$|A| = 0.99 = \left| \frac{1}{1 + j\omega/\omega_0} \right| = \frac{1}{\sqrt{1 + (\omega/\omega_0)^2}} \tag{89}$$

whence $(\omega/\omega_0) \cong 1/7$

It is interesting to note that the output bandwidth or speed of transconductance and pulse-modulation multipliers is essentially independent of signal level (except for slewing rates), or of choice of input (X or Y). And, of particular interest, the measured bandwidth is independent of any dc bias level added to the measuring "small" signal.

Vector Error: the frequency f_V at which the instantaneous, or vector difference between an input signal and the output, of the same frequency, becomes equal to 1%. For a single-pole rolloff (first-order lag), it is the frequency at which the phase shift becomes 0.01% of a radian, or 0.57° − 1/100 of the -3dB frequency. Vector error is due primarily to phase shift, since the attenuation of magnitude at f_V is only 0.05% (Figure 31).

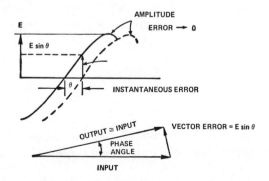

Figure 31. Vector error (at unity gain), a measure of instantaneous error

Settling Time for ±10V Step: the time required for the output voltage to approach within a specified percentage of its *final* value, in response to a 10V (full scale) input step. The time is measured from the instant the step is applied, until the output has entered the specified error band (for the last time), and therefore includes transport delay, slewing time, and linear-settling time.

Overload Recovery: the time required for the output of the multiplier to return to within its linear region, after a 50% over-voltage (where permissible, 15V for 10V-scaled devices) has been removed from the input.

OUTPUT NOISE

Output Noise: The rms value of the noise at the output of the multiplier, in a 5Hz to 10kHz bandwidth, measured with both inputs at zero. Noise is not appreciably affected by input voltage, so the specified value can be applied to any input level in the operating region. Peak-to-peak noise is usually taken to be about 6.6 times the rms level for Gaussian noise (see Figure 14, Chapter 2-3); multiplier noise can usually be safely assumed to be Gaussian.

Wideband Noise: For low-bandwidth multipliers, this includes any out-of-band effects, such as carrier leakage to the output of pulse-height-pulsewidth types (like the 427). The variable-transconductance types have fairly constant noise spectral density over their bandwidth, with no out-of-band components.

OUTPUT CHARACTERISTICS

Output Voltage at Rated Load: the minimum output voltage range at dc with the multiplier supplying the specified load current.

Output Current: The minimum current available from the multiplier output at full-scale output voltage.

Load Capacitance: The maximum value of capacitance that can be connected to the output, with no oscillations resulting, in the *multiply* mode.

INPUT RESISTANCE

The resistance between the input terminal and power common. This may be an actual resistor, or the effective input resistance of an input-amplifier circuit. Typically, input resistance is in the $10k\Omega$ to $100k\Omega$ range, a reasonable level, since multipliers are usually driven from low-impedance sources, such as closed-loop op-amp outputs.

INPUT BIAS CURRENT

The current flowing into or out of the input terminal with zero volts at the input. This is due to the bias current of the internal circuit, e.g., base current of input transistors.

MAXIMUM INPUT VOLTAGE

For Rated Accuracy: the maximum voltage that, when applied to either or both inputs, will produce an output voltage within the specified error limits. Usually, it provides a slight overrange capability. The multiplier will work with higher inputs, as long as the product of the inputs is within the output voltage range.

Safe Level: The Maximum Voltage that Will Not Damage the Input Circuit. The notation $\pm V_s$ means that the input can be no greater than the supply; if the supply is zero (or disconnected), the input must be zero (applicable especially to IC multipliers and the 432). For the other types, it is stated as an absolute-maximum voltage; i.e., the 429 will be safe with $\pm 16V$ in and zero or rated supply voltage.

POWER SUPPLY

V_s *for Rated Performance*: the power-supply voltage at which all *min/max* error specifications are guaranteed; normally $\pm 15V$, $\pm 2\%$.

Operating: the range of power-supply voltages over which the multiplier will operate normally, but with increased error, as calculated from the power-supply-rejection coefficients. Over this range, the multiplier will accept ±10V inputs and provide ±10V output. For some multipliers, graphs of input and output voltage swing vs. V_s are provided.

Quiescent Current: the current drawn from the $\pm V_s$ supplies, with the inputs and outputs at zero volts. Under full-output conditions, this current will increase by an amount approximately equal to the load current, since most multipliers have Class AB output stages.

TEMPERATURE RANGE

Rated Performance: the range over which the temperature coefficients apply, and other parameters remain within min/max limits.

Operating: the temperature range over which the multiplier will operate, with generally slight degradation of specified temperature coefficients.

Storage: the maximum temperature extremes that the multiplier can withstand, without power applied.

PACKAGE OUTLINE

This refers to a standard Analog Devices drawing, showing the pin configuration and mechanical dimensions; since multipliers vary widely in size and pinout, it is a good idea to check this out.

Case Dimensions (self-explanatory). The higher-accuracy, pulse-modulation types (427) are at present larger than the transconductance or log types (429, 434). The smallest modular case is the transconductance IC type, such as the 432. IC's are available in TO-116 hermetic 14-pin dual in-line and TO-100 10-pin metal-can packages.

CHECKLIST OF MULTIPLIER PARAMETERS

A. Static or Low-Frequency Errors (Accuracy)

1. Output Offset Voltage
2. X and Y Feedthrough
3. X and Y Nonlinearity
4. Total Error
5. Changes of Above Parameters with Temperature or Power-Supply Voltage

B. Dynamic Performance

1. –3dB Small-Signal Bandwidth
2. Phase Shift vs. Frequency
3. Full-Output Bandwidth
4. Slewing Rate
5. Rise Time
6. Settling Time
7. Frequency for 1% Vector Error
8. Frequency for 1% Amplitude Error
9. Nonlinearity vs. Frequency
10. Feedthrough vs. Frequency
11. Differential Phase Shift
12. Overload Recovery Time

C. Input and Output Characteristics

1. Input Resistance
2. Input Current
3. Output Voltage
4. Output Current
5. Output Resistance
6. Input and Output Voltage Limits vs. Power-Supply Voltage
7. Quiescent Current

TESTING

TEST EQUIPMENT

Depending on the parameters to be measured, and the number of multipliers to be tested, the equipment used to test multiplier characteristics can range from a self-contained multiplier test set to a simple arrangement of ordinary laboratory instruments. Some of the most-useful test equipment is listed below:

1. Digital Voltmeter — essential for measuring dc offsets, and input and output voltages for determining "accuracy" of multipliers. 4½-digit resolution, with $<\pm0.02\%$ error is adequate for most measurements. 1Vdc and 10Vdc ranges will be the most used.

2. Precision dc Voltage Reference — to supply input voltage for "accuracy" measurements, stable reference for non-linearity tests. Should be capable of supplying both plus and minus 10.000V at 1mA *simultaneously*, adjustable in 100mV steps down to zero volts.

3. Function Generator — provides low-frequency sinusoidal input signal for crossplot tests, and square waves or pulses for dynamic tests. Generator output voltage should be adjustable from zero to 20Vp-p into 1kΩ over frequency range of 1Hz to 1MHz (5 or 10MHz is desirable for testing the faster multipliers).

4. Variable dual 15-volt Power Supply, 50mA output current, with adjustable current limit. A variable supply is useful for measuring multiplier input and output voltage limits as a function of supply voltage.

5. Oscilloscope — for crossplots and dynamic tests. Calibrated, dc-coupled vertical and horizontal inputs are required for crossplots. Vertical deflection factors of 5mV/cm (for testing "high-accuracy" multipliers) to 5V/cm are most useful.

Horizontal deflection factors of 0.5V/cm to 5V/cm are adequate. Bandwidth of 100kHz on both axes is sufficient for "static" error measurements.

A wideband oscilloscope — at least 10MHz bandwidth — is essential for dynamic tests

6. Precision adder/subtractor — for measuring nonlinearity. This can be constructed according to the schematic Figure 32.

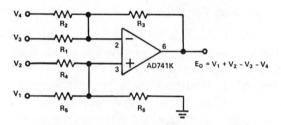

$E_O = V_1 + V_2 - V_3 - V_4$

R_1 TO R_6: 10kΩ PRECISION RESISTORS, TOLERANCE ±0.1%,
TEMPCO ≤ 50ppm; ratio-match R_1 AND R_2 TO R_3 AS CLOSELY
AS POSSIBLE, DO THE SAME FOR R_4 AND R_5 TO R_6.

Figure 32. High-precision adder subtractor

TEST CIRCUITS

The crossplot is one of the most powerful and useful techniques for performing sensitive adjustments and measuring multiplier errors, for example feedthrough and nonlinearity, by plotting such quantities as a function of the input variable. This is most easily done by displaying the error on the vertical axis of the oscilloscope and using the multiplier input signal to drive the horizontal input.

A crossplot test setup for measuring X feedthrough is shown in Figure 33. In this case, the X input of the multiplier (an X-Y plotter could be used in place of the oscilloscope, to obtain a large-scale permanent record) is driven by a 20Vp-p 10Hz sine wave, and the Y input is grounded. The output of the multiplier is connected to the vertical channel of an oscilloscope with sensitivity of 20mV/cm, direct-coupled. The sinusoidal X drive signal is connected to the horizontal input of the oscilloscope, sensitivity 2V/cm direct-coupled (±10V F.S.). The oscilloscope trace is centered on the screen (zero input and zero feedthrough is at the Origin).

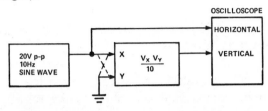

Figure 33. Cross plot test setup connected to measure X feedthrough. For Y feedthrough exchange X and Y inputs.

A photograph of the X feedthrough of a transconductance multiplier produced with the aid of this test setup is shown in Figure 34. The symmetrical parabolic shape indicates that the nonlinear component of X feedthrough is proportional to X^2. The peak value of X feedthrough is 50mV, occurring at X = +10V and −10V. Figure 35 illustrates the effect of an additive linear component to the X feedthrough, produced by Y_{os}. The parabola is no longer symmetrical; the +10V end is higher than the -10V

TABLE: MULTIPLIER TEST MATRIX

Test	V_x	V_y	E_o	Read Error On
Offset	0	0	$0 \pm E_{os}$	DVM
X Feedthrough	20V p-p	0	⌢	Scope
Y Feedthrough	0	20V p-p	⌢	Scope
X Nonlinearity	20V p-p	+10V	⌢	Scope
X Nonlinearity	20V p-p	−10V	⌢	Scope
Y Nonlinearity	+10V	20V p-p	⌢	Scope
Y Nonlinearity	−10V	20V p-p	⌢	Scope
Full-Scale Errors				
I	+10V	+10	$+10V \pm \epsilon$	DVM
II	−10V	+10V	$-10V \pm \epsilon$	DVM
III	−10V	−10V	$+10V \pm \epsilon$	DVM
IV	+10V	−10V	$-10V \pm \epsilon$	DVM

end, with a difference of 40mV. This indicates that there is 20mVp-p of "linear" X feedthrough (that can be cancelled out by a Y_{os} adjustment).

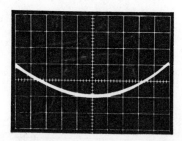

Figure 34. Measurement of X feedthrough showing nonlinear (parabolic) component only, X = ±10V. Vertical scale: 20mV/div.

The Y feedthrough can be cross-plotted by interchanging the X and Y inputs on the test setup.

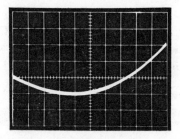

Figure 35. Measurement of X feedthrough with Y_{os} not optimized, X = ±10V. Vertical scale: 20mV/div. Additive linear term = 40mV p–p; Y_{os} = 20mV.

The crossplot technique can be extended to the measurement of nonlinearity, as illustrated in Figure 36.

Figure 37.   DC Accuracy (Error), V_{OUT}, I_{OUT}, Z_{OUT}

Figure 38.   Offset

Figure 39.   Low-Frequency Feedthrough, Crossplot

Figure 40.   Nonlinearity, Crossplot

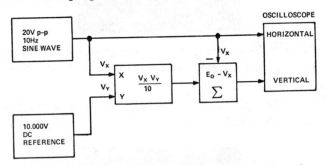

Figure 36. Cross plot setup for measuring X nonlinearity at Y = 10V. For Y = –10V, summing block should compute $E_o + V_x$. Interchange X and Y to measure Y nonlinearity.

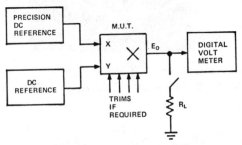

Figure 37. Test setup for measuring DC accuracy, output voltage and current range at rated load, output resistance.

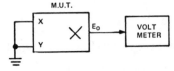

Figure 38. Output offset measurement

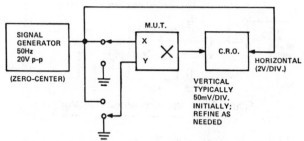

Figure 39. Low-frequency feedthrough crossplot

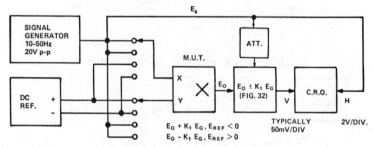

Figure 40. Nonlinearity crossplot

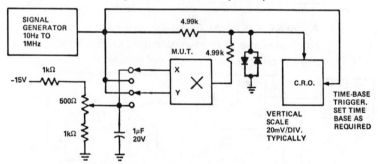

Figure 41. Vector (instantaneous) error, settling time

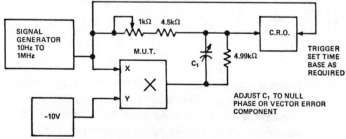

Figure 42. 1% error bandwidth, X nonlinearity vs. frequency (3rd & 4th quadrants)

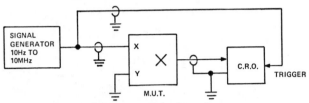

Figure 43. X feedthrough vs. frequency

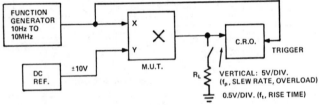

Figure 44. Phase shift, differential phase shift

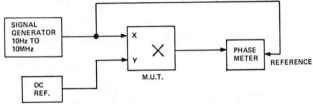

*Figure 45. Full-power frequency, slewing rate,
overload recovery; small-signal amplitude response
and rise time; output current and voltage*

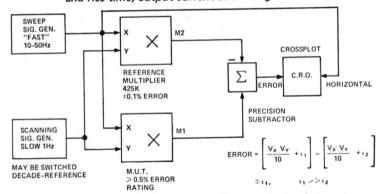

*Figure 46. Sophisticated X nonlinearity test, using accu-
rate multiplier as reference. The X input is swept at a
"reasonable" frequency, and the Y input signal swings
slowly over its range. The inputs are reversed to check Y
nonlinearity. If Y is swept continuously, the signal
envelope is the worst-case error magnitude.*

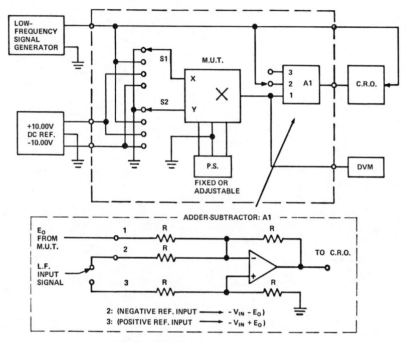

Figure 47. Multipurpose test box

Note: These test circuits are principally designed for testing 4-quadrant devices. Testing of single-quadrant devices is in some respects similar; however, here are a few differences:

1. Tests using a negative voltage at one input and passive summation of input and output, taking advantage of the negative gain of the device in the 2nd and 3rd and the 3rd and 4th quadrants, cannot be used. Either precise subtraction or CRO differential inputs (if sufficiently accurate) should be used.

2. Single-quadrant devices generally require half-scale biasing of the input signal generator output; peak-to-peak swing is 10V (for 0 to 10V devices).

3. For a complete response picture, logarithmic devices may require several sets of small-signal tests, employing small signals biased at intervals, e.g., 9V ±1V, 0.9V ±0.1V, etc.

III

Dividers (Ratio Circuits)

Chapter 3

An analog "divider" circuit produces an output voltage or current proportional to the ratio of two input voltages or currents. For convenience and clarity, in this chapter, it is to be assumed that the inputs and outputs are voltages (unless noted otherwise).

$$E_o = K\frac{V_z}{V_x} = V_r\,\frac{V_z}{V_x} \tag{1}$$

The denominator is denoted V_x, the numerator V_z, and the output E_o. The dimensional scale factor K (or V_r), is usually 10 volts. If the ratio of the inputs is unity, the output is equal to K. The input/output relationship for an ideal analog divider is summarized in Figures 1, 2, and 3.

K = SCALE FACTOR (VOLTS) = 10V (MOST COMMON)

K = 1V IS USEFUL FOR $V_z > V_x$

Figure 1. Block diagram of divider

The operating region of the variables (quadrants of operation) is defined by the polarity and magnitude ranges of the numerator and denominator inputs, and of the output. Figure 2 depicts the operating region of the inputs of a 2-quadrant divider with normal polarity relationships (bipolar numerator, positive denominator).

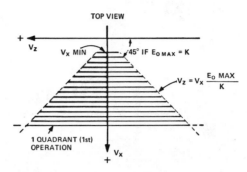

Figure 2. Top view of 2-quadrant divider-function surface, showing constant-denominator elements.

If the numerator and denominator are both restricted to a single polarity (usually positive), the divider is said to operate in a single quadrant, indicated by shading. Generally, the denominator is restricted to a single polarity, since the transition from one polarity to another would require the denominator to pass through zero, which would call for infinite output (unless the numerator were simultaneously zero).*

Besides excluding the vicinity of $V_X = 0$, the operating region of a practical analog divider does not cover an entire quadrant or half-plane, because the maximum allowable numerator magnitude depends on the denominator magnitude and either the output range or the scale factor.

$$V_{z_{max}} = \frac{E_{o_{max}}}{V_r} \cdot V_x \qquad (2)$$

In the case, where $E_{o_{max}} = K \; (= 10V)$, the input region is bounded by the 45° line, $V_z \leqslant V_x$. For small values of V_x, the operating region is further limited by the minimum value of denominator $V_{x_{min}}$ that will allow reasonable performance.

The input-output relationship of an ideal divider can be visualized by imagining the three-dimensional surface relating the three

---

*It is possible to construct a 4-quadrant divider that accepts bipolar numerator and denominator (except for a "forbidden zone" in the vicinity of zero denominator) and provides an output with proper polarity relationships, but it has few useful applications.

variables. Figure 2 is a top view of the surface, showing the constant-x elements; for each value of V_x, the output is linear in V_z. Figure 3a is a view of the surface in perspective. It is a developed surface generated by two sets of straight-line elements, (1) constant V_x and (2) constant E_o. Figure 3b is a top view showing the contours of constant E_o. E_o is seen to be equal to zero along the V_x axis, equal to K at the intersection of the $V_x = V_y$ plane and the $E_o = K$ plane, and linearly proportional to V_z

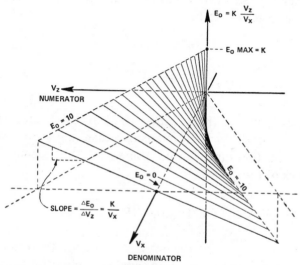

a. Two-quadrant divider input-output surface, showing constant-denominator elements

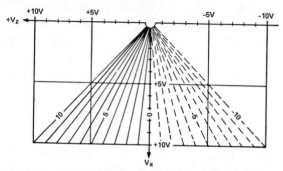

b. Contours of equal E_o, 1V contour interval, for $E_o = 10V_z/V_x$, $V_x > 0$

Figure 3. Two quadrant divider — input/output surface

where it intersects any plane perpendicular to the V_x axis. E_o is inversely proportional to V_x where the surface intersects any plane perpendicular to the V_z axis. The surface approaches verticality as V_x approaches zero, tending towards $+ \infty$ for positive V_z and towards $- \infty$ for negative V_z; however, it is truncated considerably earlier by the intersection of the $E_o = K$ plane and the $V_x = V_z$ plane.

Theoretically, the output will approach the value $\pm K$ as $\pm V_z$ and V_x approach zero together

$$\lim_{|V_z| = V_x \to 0} K \cdot \frac{V_z}{V_x} = K \qquad (3)$$

In contrast with the theory, the output of real dividers is generally undefined for denominators smaller than some minimum value, typically in the range from 10mV to 1V, depending on the properties of the device.

ERRORS OF ANALOG DIVIDERS

Division has long been the most difficult of the four arithmetic functions to implement with analog computing devices. This difficulty stems primarily from the nature of division: the magnitude of a ratio becomes quite large, approaching infinity, for a denominator that approaches zero (and a non-zero numerator). Thus, an ideal divider must have potentially "infinite" gain and infinite dynamic range. For a real divider, both of these factors are limited by the magnification of drift and noise at low values of V_x.

In other words, the "gain" of a divider for the numerator is inversely dependent on the value of the denominator (Figure 4). On the other hand, if the ratio of numerator to denominator remains constant as their magnitudes vary, the quotient is constant (Figure 5).

The output of a practical analog divider will differ from the theoretical ratio of its inputs by an amount that is, in general,

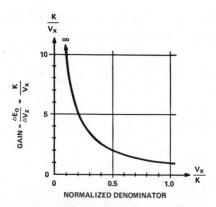

Figure 4. Divider gain as a function of denominator voltage

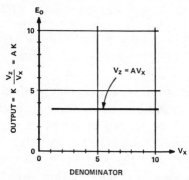

Figure 5. Divider output for constant ratio (A) of numerator to denominator

dependent on the magnitudes of the inputs. The overall error is the net effect of several factors, of which the most important are:

Type of Error	Approximate Range of Magnitude
1. Numerator offset, Z_{os}	1% to 0.001% of $V_{z_{max}}$
2. Denominator offset, X_{os}	1% to 0.001% of $V_{x_{max}}$
3. Output offset, E_{os}	1% to 0.01% of $E_{o_{max}}$
4. Scale-factor error, ΔK	1% to 0.05% of K
5. Nonlinearity, $f(V_z, V_x)$	5% to 0.05% of V_z, V_x

The effects of these errors can be seen more plainly when they are introduced into the "ideal" divider equation

$$E_o = (K + \Delta K) \frac{V_z + Z_{os}}{V_x + X_{os}} + E_{os} + f(V_z, V_x) \qquad (4)$$

The equation can be rewritten to sort out the effects of the combined errors on the output

$$E_o = \underbrace{(K + \Delta K)}_{\substack{\text{scale-} \\ \text{factor} \\ \text{error}}} \underbrace{\frac{V_z}{V_x + X_{os}}}_{\substack{\text{ratio} \\ \text{error}}} + \underbrace{\underbrace{\frac{(K + \Delta K)Z_{os}}{V_x + X_{os}}}_{\substack{\text{input offset} \\ \text{referred to} \\ \text{output}}} + \underbrace{E_{os}}_{\substack{\text{output} \\ \text{stage} \\ \text{offset}}}}_{\text{total output offset}} + \underbrace{f(V_z, V_x)}_{\substack{\text{non-} \\ \text{linearity}}} \quad (5)$$

Considering these terms separately,

1. The scale-factor error, ΔK, is independent of the level of V_z or V_x. However, as will be shown, there is an additional error due to X_{os}, which may be viewed either as an additional X-linearity error or as a variable scale factor on V_z, depending on the interpretation of V_x and the portion of the range that is used. If $X_{os} = 0$, the term simply represents the ideal division, with a ΔK error. As X_{os} becomes more significant in relation to V_x, if affects the slope of the E_o/V_z relationship. If X_{os} is negative, (and V_z non-zero), as V_x approaches the positive value, $-X_{os}$, the ratio tends to "blow up" (Figure 6). If V_z is precisely zero, the output

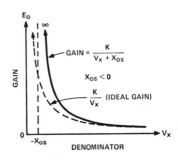

Figure 6. Divider gain error at output as a function of denominator X ($X_{os} < 0$)

will be limited to $K + \Delta K$, small comfort, since K is usually full scale, and an infinitesimal deviation of V_z from zero will drive the output into limits. If X_{os} is positive, the gain will not become infinite within the range of V_x, but large linearity errors will result for small values of V_x (Figure 7). In particular, at $V_x = X_{os}$, the gain will have halved.

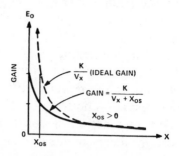

Figure 7. Divider gain error at output as a function of denominator X ($X_{os} > 0$)

2. The "input offset" error, referred to the output,

$$(K + \Delta K) \, \frac{Z_{os}}{V_x + X_{os}} \; \cong \; K \, \frac{Z_{os}}{V_x + X_{os}}$$

The numerator offset, Z_{os}, is subjected to a gain $K/(V_x + X_{os})$. If X_{os} is zero and Z_{os} is non-zero, this term tends to "blow up" as V_x approaches zero; in any event Z_{os} will be magnified for all $V_x < K$. The value of X_{os} serves to modify the "blowup point" (asymptote): if X_{os} is negative, the offset error will become "infinite" when V_x has the positive value $-X_{os}$. If X_{os} is positive, the offset error will become high, but not infinite (small comfort again!). If $Z_{os} = 0$, then the input offset error, referred to the output, will be zero (except at $V_x + X_{ox} = 0$, when it becomes equal to K).

3. The output-stage offset is independent of V_z and V_x, and, since it undergoes no magnification, it is generally a negligible source of error, compared to the output errors produced by the input offsets. Its contribution is most salient in such applications

as linearizing, where the dynamic range of the denominator is not usually large; for such applications, E_{os} should be trimmed to zero, and the effect of its temperature coefficient should be considered.

4. The output nonlinearity $f(V_z, V_x)$ is identifiable in terms of the nonlinearity of the straight-line elements (Figure 3), with all other errors tweaked to zero. V_z (numerator) nonlinearity is the departure of E_o from proportionality to V_z at constant V_x. Figure 8 is a typical plot of numerator nonlinearity, measured by subtracting KV_z/V_x from the output. Nonlinearity as a function of denominator (actually as a function of *ratio*) is defined in terms of deviation of the measured ratio from the theoretical ratio as V_x and $\pm V_y$ are varied together in a constant ratio (the radial "spiral-staircase" straight-line elements in Figure 3b).

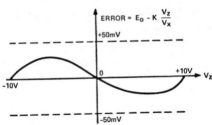

Figure 8. Nonlinearity as a function of numerator voltage (V_z). Denominator = constant

Besides these easily-determined deviations, stated in terms of the familiar concept of *linearity*, it is also possible (but less practical) to consider the *fidelity to hyperbolic form*, comparing the output, as a function of V_x (V_z held constant), with the ideal output. Errors due to denominator offset, numerator offset, and scale-factor error are excluded. Generally, the nonlinearity takes the form of limited gain at small values of denominator voltage, as shown in Figure 9. Gain, K/V_x, is plotted vertically against V_x horizontally, for constant V_z. As V_x is reduced, the gain increases hyperbolically, until, at small values of V_x, a peak is reached, then the gain decreases to zero at $V_x = 0$. This kind of gross "denominator nonlinearity" is fairly common in analog dividers, since a zero X-input may correspond to shutting off a current or voltage in the circuit, reducing the gain to zero.

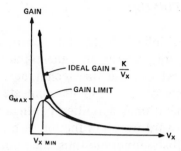

Figure 9. Gain limit at low values of denominator voltage

Divider Errors — Summary

Based on all of the above considerations, a good analog divider is identified by the following properties:

1. Fidelity to the Ratio Function: "Gain" (K/V_x) must vary inversely with the denominator over a wide range of denominator values.

2. Numerator and denominator input errors, such as offsets, noise, and drift, must be much less than the smallest input signals.

3. If requirements (1) and (2) are met, the output of the divider should be constant for constant ratios of numerator and denominator, independent of their magnitudes. For example, $10/10 = 0.01/0.01 = 1$, and $1/10 = 0.001/0.01 = 0.1$.

DIVIDER CIRCUITS

This section deals with three of the most common divider circuits.

1. Inverted multiplier
2. Direct variable-transconductance divider
3. Log-antilog divider

The design of these circuits is largely based on the multiplier circuits discussed in Chapter 3-2; they are similar in principle, in circuitry, and in physical appearance; in fact, some are identical. While other techniques for division exist, the three listed above are the most popular, and a detailed discussion of their design and performance should provide adequate insight into analog-dividers-in-general to suit any practical purpose.

INVERTED MULTIPLIERS

The "inverted multiplier" is by far the most commonly-used analog divider circuit. Nearly all general-purpose 2-input multipliers can (and most do) use this technique to achieve division. The circuit consists of a multiplier connected as the feedback element of an operational amplifier configuration, as shown in Figure 10. The forward transfer function of this circuit will be the inverse of the feedback function; thus, the multiplication function is inverted to form a divider.

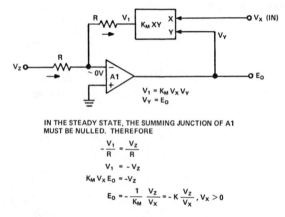

IN THE STEADY STATE, THE SUMMING JUNCTION OF A1
MUST BE NULLED. THEREFORE

$$-\frac{V_1}{R} = \frac{V_Z}{R}$$

$$V_1 = -V_Z$$

$$K_M V_X E_O = -V_Z$$

$$E_O = -\frac{1}{K_M}\frac{V_Z}{V_X} = -K\frac{V_Z}{V_X}, V_X > 0$$

Figure 10. "Inverted-multiplier" divider circuit

In more specific terms, the multiplier, like a voltage-controlled potentiometer, controls the loop gain of the negative feedback circuit around the op amp. As the X input voltage to the multiplier is decreased, the gain from the Y input to the multiplier output is reduced proportionally, reducing the negative feedback (and the loop gain). Since the multiplier output must balance the Z input, the multiplier Y input must be increased proportionally. Since the multiplier Y input is supplied by the output of the circuit, the Z input is magnified in the same ratio that X is decreased.

If the X (or denominator) input is reduced to zero, then the feedback is zero, and the gain between the Z input and the amplifier output will be the open-loop gain of the op amp. If the op amp and the multiplier are "ideal," then the forward gain will be infinite for zero denominator. Of course, real op amps do not have infinite gain, and real multipliers always have finite feedthrough, so real dividers have finite gain for zero denominator (however, noise and offset errors tend to render the finite-gain question academic, at least for low frequencies, where op amps generally have plenty of open-loop gain).

Any multiplier circuit, whatever its operating principle, can (in concept) be made to divide in this manner. But practical problems, such as stabilizing the closed loop, incompatible forms of inputs, slow response, high cost, etc., tend to narrow the field. Since it is the case (as we have shown in Chapter 3-2) that the variable-transconductance multiplier offers an excellent overall combination of cost, speed, accuracy, and (in I.C. form) size, it is reasonable to assume that inverting such multipliers to form dividers would be popular as a means of division.

Performance of Practical Inverted-Multiplier Dividers

Any multiplier can, in concept, be converted into a divider by adding an operational amplifier and two resistors, as shown in Figure 10. However, most general-purpose, 4-quadrant multipliers include an output amplifier and associated resistors, and call for *external* closure of the output-amplifier loop. They can readily be inverted to 2-quadrant dividers by reconnecting the multiplier inputs and the feedback resistor, as shown in Figure 11.

The performance of the inverted-multiplier divider circuit depends primarily on the performance of the multiplier, since (except for wideband devices) the op amp usually has insignificant errors compared to the multiplier. Depending on the desired divider performance, variable-transconductance multipliers, with errors ranging upwards from 1%, or the more-accurate pulse-modulation types, with errors in the vicinity of 0.1%, might be used. Characteristics and circuitry of these multipliers are discussed in Chapter 3-2.

A 1% multiplier usually has appreciable offsets and nonlinearity; as a divider, the dynamic range of its denominator is limited to about 10:1, i.e., a 10✕ magnification of error and drift. The lower

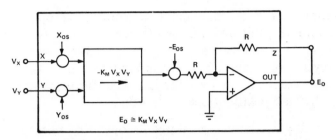

a. Multiplier connection, showing effective location of offset errors

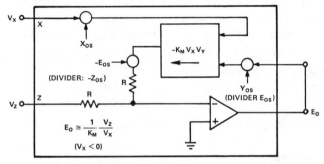

b. Divider connection, showing equivalent divider offset errors

Figure 11. Relationship of multiplier and divider errors in "inverted-multiplier" divider

errors and drift of the 0.1% multiplier will increase the useful denominator dynamic range to 100:1, but the errors and drift are still significant beyond a 10:1 range (Figure 12).

Specified multiplier performance can be used to predict divider errors. Their relationship is outlined in Table 1. Equation (5) can be rewritten in terms of the multiplier parameters to describe divider performance based on multiplier specs:

$$E_o = \frac{1}{K_m - \Delta K} \cdot \frac{V_z}{V_x + X_{os}} + \frac{1}{K_m} \cdot \frac{E_{os}}{V_x + X_{os}} + Y_{os} + f(E_o, V_x) \qquad (6)$$

where K_m is the multiplier scale constant = $1/V_r$

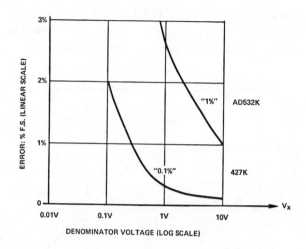

*Figure 12. Total error vs. denominator for 1% and 0.1%
"inverted multiplier" dividers*

TABLE 1. RELATIONSHIP OF DIVIDER PARAMETERS TO MULTIPLIER PARAMETERS

Multiplier Parameter	Corresponding Divider Parameter	Principal Output Component of Divider
A. "Linear" Effects		
1. Output offset, E_{os}	Numerator offset, Z_{os}	KE_{os}/V_x
2. X-Input offset, X_{os}	Denominator offset, X_{os}	$KV_z/(V_x + X_{os})$
3. Y-Input offset, Y_{os}	Output offset, E_{os}	Y_{os}
4. Scale factor, $K_m = 1/V_r$	Scale factor, $K = V_r$	KV_z/V_x
B. Nonlinear Effects (other errors minimized)		
5. X Nonlinearity	Nonlinearity of constant ratio, $V_y = E_o$	$(K_m V_x V_y - E_o)/K_m V_x$ (V_y const) $X_{NL}/K_m V_x$
6. Y Nonlinearity	Numerator nonlinearity	$(K_m V_x V_y - E_o)/K_m V_x$ (V_x const) $Y_{NL}/K_m V_x$
C. Dynamic Error (incremental)		
7. Bandwidth (-3dB frequency)	Bandwidth (−3dB frequency)	$f_{-3dB} V_x/K$

Figure 13 is a graph of incremental frequency response for two representative dividers at unity gain ($K/V_x = 1$) and higher gains within their respective practical ranges of division.

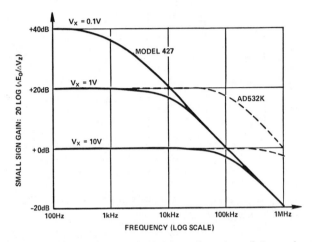

Figure 13. Small-signal response as a function of denominator voltage for two multiplier-dividers connected in the "divide" mode

Since "inverted-multiplier" dividers have such limited performance, and present many pitfalls to the unwary designer, their use (involving low-cost general-purpose IC multipliers, at any rate) should be limited to such applications as linearizing, that involve a small dynamic range of denominator variation, and where lowest device cost is essential. The following guidelines are offered to make the best of a "bad thing."

1. Avoid using an "inverted-multiplier" unless low cost is essential and adequate pains can be taken to be sure of best results. Be prepared to consider as an alternative a specialized internally-trimmed 1- or 2-quadrant divider that has guaranteed-maximum error specifications over the expected denominator and ambient-temperature ranges.

2. If a general-purpose multiplier/divider is used as a divider, some thought should be given to performance, primarily in the areas of bandwidth, accuracy, and drift errors vs. denominator. For many applications, a faster or more-accurate multiplier will be

found necessary than may have been expected at first. If possible, external trims should be provided for input and output offsets, and the trim procedure outlined in *Testing and Adjusting Dividers* (later in this chapter) should be followed. The following specific points should be considered in applying Guideline 2.

A. Allow for increased error, at room temperature, as the denominator magnitude decreases. Always use preamplification for small signals to scale the denominator to 10V full scale; use the smallest dynamic range of denominator that the application will tolerate. Use a multiplier/divider having higher accuracy than is required at full-scale denominator: As a rule-of-thumb, a multiplier used for division will have about 3X larger total error (than multiplication) at $V_x = 1V$, if external trimming is used. Without external trimming, it will have 10X greater total error. For a graphical comparison of divider errors, among a number of devices, see Figure 24.

Example: A system requires a 2-quadrant divider with less than 1% (of 10V full-scale) error and a 1V to 10V denominator range. Choose a multiplier with approximately (1/3)% error (e.g., one might consider 427J−0.25% or 426L−0.5%) if trimming is permissible (one objection to trimming might be that it would complicate field replacement, since each replacement unit would have to be "tweaked-in"). If trimming is excluded, a multiplier divider with (1/10)% error would be required; stated another way, a 0.1% multiplier/divider will have a ±1% = 0.1V error in the *divide* mode with a 1V denominator.

B. Allow for increased offset and scale-factor drift, and increased noise, at decreased denominator magnitudes. As discussed earlier, the noise and offset drift errors are inversely proportional to denominator magnitude. Furthermore, the scale factor will also drift, because of the effect of denominator drift on the magnitude of the apparent scale factor.

Example: System requirements dictate a maximum offset of ±20mV for a 10:1 denominator range and ±5°C temperature range. Allowable multiplier offset drift is:

$$\frac{20mV}{5°C} \times \frac{1V}{10V} = 0.4mV/°C$$

Figures 26 and 29 provide information about the drift of specific devices with temperature, as a function of denominator voltage.

C. Allow for decreased bandwidth as denominator decreases. The small-signal bandwidth of all "inverted-multiplier" dividers is directly dependent on denominator magnitude. Since most of these dividers have -6dB/octave gain rolloff, the bandwidth is linearly related to the denominator level:

$$f_c = f_{c_{max}} \frac{V_x}{K}$$

where f_c is the -3dB bandwidth, and K is the divider scale factor, usually 10V.

Bandwidth vs. denominator data for several dividers are plotted in Figures 27 and 28.

D. Provide adjustments for numerator, denominator, and output offsets, if possible. A scale-factor trim may be used, but it is often unnecessary, because scale factor can be adjusted elsewhere in the system, as a gain adjustment on either input variable or at the destination of the output. For best accuracy, trim the divider according to the procedure outlined in *Testing and Adjusting Dividers*. If all three offset trims cannot be used, choose at least one of the following:

1. Numerator offset trim (Z_{os} or E_{os}) controls the shift in output offset with varying denominator. This trim is essential when the denominator range is greater than 3:1.

2. Output offset (Y_{os} if the feedback is via the Y-input of the multiplier) controls the fixed portion of the total output offset. The Y_{os} adjustment can be used to minimize the total output offset for a denominator range of 3:1 or less.

3. Denominator offset (X_{os} if the X input is the denominator) controls the change in apparent scale factor as the denominator is varied. The X_{os} trim can be eliminated if the range of V_x is expected to be limited to about 3:1 (from full-scale input).

2-QUADRANT VARIABLE-TRANSCONDUCTANCE DIVIDER

The direct variable-transconductance divider is based on the linearized variable-transconductance multiplier circuit explained in Chapter 3.2. The basic input/output transfer equation for the multiplier circuit is

$$E_0 = \frac{V_x V_y}{K_m I_{REF}} \tag{7}$$

If I_{REF} is used as an input, then the circuit becomes a 3-input simultaneous multiplier-divider. The schematic of the 2-quadrant variable-transconductance divider appears in Figure 14, which is

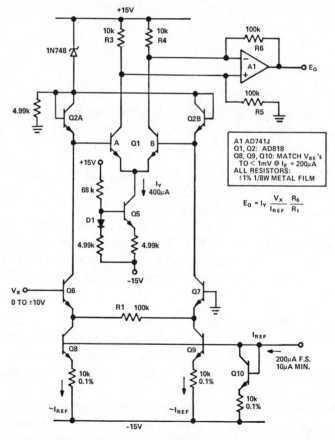

Figure 14. Practical two-quadrant variable-transconductance divider

quite similar to Figure 10, Chapter 3-2. The current I_{REF} determines the standby current through the diode-connected transistors (Q2). As I_{REF} increases, the dynamic resistance of the diodes decreases

$$r_E = \frac{kT}{qI_{REF}} \tag{8}$$

The X input voltage produces a division of the total current between the two diodes (and hence a greater voltage drop across one than across the other). The difference current is

$$\Delta I = \frac{V_x}{R_1} \tag{9}$$

Clearly, if an increase in I_{REF} produces lower resistance in the diodes, the incremental voltage drops across each caused by ΔI will decrease. Conversely, as I_{REF} decreases, the change of voltage across the diodes, as a function of V_x, increases.

The difference in diode voltage is amplified at fixed gain (assuming constant Y input) by the differential pair Q1 A-B. Thus, the overall gain is inversely dependent on I_{REF} (as Eq. 7 indicates), and the ideal division equation will be followed over a fairly wide range of I_{REF}, typically 20:1. The dynamic range of response to I_{REF} is limited primarily by the β's of the diodes and differential transistors, and by the increase in emitter resistance of the X-input amplifier as I_{REF} is reduced.

This version of the variable-transconductance divider has very wide bandwidth; up to 5MHz can be achieved, and the bandwidth is not strongly dependent on the denominator magnitude. For instance, the AD531 integrated-circuit multiplier-divider uses just this scheme for division — it has a nearly constant bandwidth of 750kHz for a 20:1 range of denominator. Discrete versions of this divider can have 5MHz bandwidth over a 10:1 range of denominator. The accuracy of this circuit can be reasonably good, with errors of about 0.5% at $I_{REF} = 200\mu A$ and 2% at $I_{REF} = 10\mu A$.

IMPROVED 2-QUADRANT VARIABLE-TRANSCONDUCTANCE DIVIDER

The accuracy and dynamic range of the 2-quadrant transconductance divider described above can be greatly improved by a few refinements. The modified circuit can divide accurately over a 1000:1 (10mV to 10V) denominator range and can easily achieve less than ±0.5% error over a 100:1 range without requiring external trimming. In addition, the numerator nonlinearity is extremely low, ±0.05%, and independent of denominator magnitude.

The variable-transconductance circuit can also be considered as a log circuit and analyzed in terms of the logarithmic behavior of its elements, since the slope of the natural logarithm of a number is inversely dependent on the magnitude of the number (r_E in Eq. 8 is such a slope).

$$\frac{d(\ln x)}{dx} = \frac{1}{x} \tag{10}$$

$$\int \frac{1}{x}\,dx = \ln x + C \tag{11}$$

In the variable-transconductance divider, the denominator controls the magnitude of x (I_{REF}) and therefore the "gain" for the numerator, or V_x (in Fig. 14) signal.

The improved 2-quadrant divider uses a differential log-antilog function to directly synthesize the division function. A simplified schematic of the divider circuit, similar to that of the Model 436 Divider, is shown in Figure 15.

The denominator voltage, V_x is applied to two symmetrically-arranged transdiode log circuits (see Chapter 3-1), Q1A-A1 & Q1B-A2, through R1 and R2. The numerator voltage, V_z, is applied to Q1A-A1 directly through R3, and inverted ($-V_z$) through R4 to Q1B-A2.

The numerator, V_z, and denominator, V_x, voltages are converted to currents that are summed at the inputs of A1 and A2. Since R3

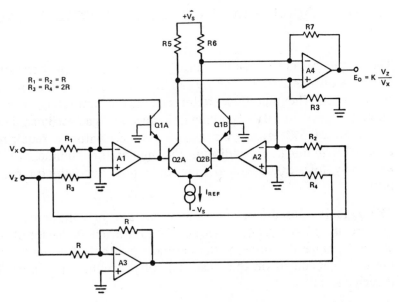

Figure 15. Two-quadrant variable-transconductance divider

and R4 are 2X R1 and R2, the currents in Q1A and Q1B are proportional to $V_x + \frac{1}{2}V_z$ and $V_x - \frac{1}{2}V_z$. The output voltages of A1 and A2 are therefore proportional to the logarithms of the sums and differences, since

$$V_{EB} = -\frac{kT}{q} \ln \frac{I_c}{a_N I_{ES}} \qquad (V_{cb} = 0) \qquad (12)$$

The voltages at the emitters of Q1A and Q1B are applied to a differential antilog circuit, Q2A-Q2B, which operates at a constant sum (reference current I_{REF}). The form of the currents to be differenced in A4 is

$$I_c = a_N I_{ES} (\epsilon^{qV_{BE}/kT} - 1) \qquad (13)$$

assuming that $V_{CB} = 0$.

The difference of the collector currents of Q2A-Q2B is converted to an output voltage $2 \Delta I_c R_7$, by the collector-loading and summing resistors R5 to R8, and amplifier A4. By an analysis similar to that for the Gilbert transconductance multiplier in Chapter 3-2, it is fairly easy to show that

$$\Delta I_c = \frac{I_{REF}}{2} \cdot \frac{I_z}{I_x} \qquad (14)$$

where

$$I_z = \frac{V_z}{2R}, \quad I_x = \frac{V_x}{R} \qquad (15)$$

$$E_o = \frac{R_7 I_{REF}}{2} \cdot \frac{V_z}{V_x}, \quad |I_z| < |I_x| \qquad (16)$$

The ratio relationship between V_z and V_x is precise to the extent that the transistors obey the ideal junction equations, and the effects of the op-amp input offset currents (A1, A2) can be ignored. Practically speaking, the limitations of the amplifiers are more significant than those of the transistors, since (it has been shown that) the transistors follow the ideal current-voltage relationship from at least 10pA to 100µA (7 decades, or a dynamic range of 10^7), while low-cost bipolar-input amplifiers have input offset currents of 0.5nA (AD308) to 5nA (AD201A). The dynamic range for 1% error, due to op-amp input current, is then 0.01 X 100µA/0.5nA, or 2000:1.

The symmetrical arrangement of the circuit is essential to its operation in two quadrants with low distortion. In fact, all currents in symmetrical paths must be perfectly balanced, or second-harmonic distortion will be introduced into the numerator and denominator. Interestingly, the oscilloscope photographs showing numerator nonlinearity, Figures 16 & 17, show only third-order (S-shaped) distortion, due to emitter resistance in Q2A&B. Matching of the X and Z input resistors, and the balanced circuit configuration combine to eliminate second-order distortion in the input log amplifiers.

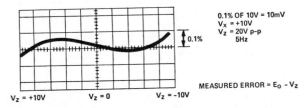

Figure 16. Nonlinearity of two-quadrant divider as a function of numerator input. Denominator is constant at +10V

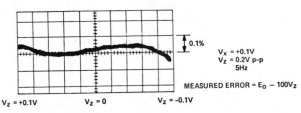

Figure 17. Same as Figure 16, but denominator is 0.1V; numerator swing is ±0.1V for full-scale ±10V output swing

The bandwidth of this variable-transconductance divider is not strongly dependent on the denominator magnitude, as Figure 18 shows. The reason for this is that the output section, Q2A&B and A4, operate at an essentially constant high ($200\mu A$) current level, while the log amps, A1 & A2, operate in a quasi-current mode, with very low (~0.3V) output swings and high loop gain.

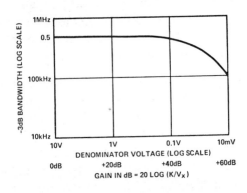

Figure 18. Small-signal bandwidth (-3dB) of the two-quadrant variable-transconductance divider, as a function of denominator voltage.

To summarize, the wide dynamic range, low errors, and wide bandwidth of the variable-transconductance divider are the result of four features of the circuit:

1. Numerator and denominator inputs to summing junctions of operational amplifiers ensure lowest-possible input offset, noise, and drift. The input errors are still magnified by K/denominator at the output, but the input drifts are less than $10\mu V/^\circ C$, resulting in an output offset drift of $1mV/^\circ C$ for $V_x = K/100$.

2. Fidelity to the Division function: The log-antilog synthesis of the linear ratio is theoretically exact for all finite denominators, $0 < x < \infty$. The only limitation in dynamic range lies in deviations from ideal performance of the hardware; the transistors operate over 7 decades, the op amps over 3 decades (limited by offsets and drift). Errors less than 0.5% of full scale can be achieved over a 100:1 denominator range. Figure 19 shows the effect of denominator voltage on total error.

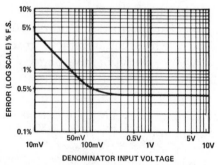

Figure 19. Total error of the two-quadrant variable-transconductance divider, as a function of denominator voltage

3. Low numerator distortion (0.05%): The symmetry of the circuit (and available components) permit low distortion, independent of the denominator (Figs. 16 and 17).

4. Wide bandwidth (500kHz), low output-stage drift: Operation of the output section at a constant high current level ($200\mu A$) produces wide bandwidth and low output-stage offset drift. The small internal voltage swings and high loop gain of the input log amplifiers reduce the dependence of bandwidth on the denominator, as Figure 18 demonstrates.

Besides all its performance advantages, the concept and circuitry of the variable-transconductance divider are relatively simple, compared to almost any other 2-quadrant-divider approach.

LOG-ANTILOG DIVIDER

The log-antilog multiplier circuit discussed earlier (Chapter 3-2, Figure 20) also makes an excellent divider. In fact, it is probably the most accurate one-quadrant divider circuit available.

If the V_{REF} input of the log-antilog multiplier circuit is used as a denominator and relabeled V_x, and the X input is relabeled V_z, the circuit becomes Figure 20 (this chapter), with the transfer function

$$E_o = \frac{V_y V_z}{V_x} \qquad (17)$$

An additional advantage of the log-antilog circuit is that it is a three-input circuit that performs multiplication and division simultaneously and with equal accuracy, greatly increasing its usefulness for a wide variety of applications, such as implicit solutions of all types, including square roots, rms, and vector equations (see Chapters 2-3, 2-5, and 3-6). For square-rooting, with wide dynamic range, one simply connects the output to the denominator input; then $E_o = 10V_z/E_o = \sqrt{10V_z}$.

In effect, the circuit will do the work of two independent one-quadrant multiplier/dividers, thus simplifying the implementation of equations requiring multiplication and division.

Circuit Description

The operation of the log-antilog multiplier-divider circuit, Figure 20, has been described in detail in Chapter 3-2; only a brief summary will be given here.

The three input variables, X, Y, and Z (V_x or I_x, V_y or I_y, V_z or I_z), are applied to three independent transdiode log amplifiers, A1-Q1A, A2-Q2A, and A3-Q2B. The outputs of the log amplifiers,

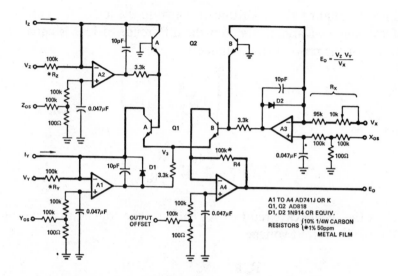

Figure 20. Log-antilog divider-multiplier. Heavy lines trace signal paths.

equal to the emitter-base voltages of the transistors, are proportional to the logarithms of the input variables. For example, for V_z

$$-V_A = \frac{kT}{q} \ln \frac{V_z}{R_z I_{ES}} \qquad (18)$$

The sum of the base-emitter voltages around the loop from the base of Q2A to the base of Q2B is

$$0 = V_{BE2A} + V_{BE1A} - V_{BE1B} - V_{BE2B} \qquad (19)$$

Substituting the log relationships between input currents and base-emitter voltages, and cancelling matching constants, as discussed in Chapter 3-2,

$$\ln \frac{V_z}{R_z I_o} + \ln \frac{V_y}{R_y I_o} - \ln \frac{I_{c1B}}{I_o} - \ln \frac{V_x}{R_x I_o} = 0 \qquad (20)$$

Since the sum of the logarithms is equal to the log of the product of the summed arguments, and the difference of logs is equal to the log of the ratio of the arguments,

$$\ln \frac{I_{c1B}}{I_o} = \ln \left[\frac{V_z V_y}{I_o V_x} \cdot \frac{R_x}{R_z R_y} \right] \tag{21}$$

and

$$I_{c1B} = \frac{V_z V_y}{V_x} \cdot \frac{R_x}{R_z R_y} \tag{22}$$

R4 in the feedback circuit of A4 converts the collector current of Q1B to the output voltage

$$E_o = \frac{R_4 R_x}{R_z R_y} \cdot \frac{V_z V_y}{V_x} = K \frac{V_z V_y}{V_x} \tag{23}$$

Note that the output is independent of temperature, and that the scale factor is determined by only four resistors, which can easily be matched, both for initial value and for temperature coefficient.

Performance of the Log-Antilog Divider

The log-antilog circuit is capable of high accuracy, wide-dynamic-range division and multiplication for three reasons:

1. Very low errors at the signal inputs: The input amplifiers A1, A2, A3 can have offsets less than $100\mu V$ and input currents of 5nA or less, with offset voltage drifts of $10\mu V/°C$ or less. This results in an input error of 0.1% or less for inputs as low as 100mV.

2. Use of summation of the logarithms of currents allows a wide dynamic range; the V_{BE} of a log transistor changes only about 60mV per decade at room temperature, thus many decades may be accommodated without danger of saturation or other problems inherent in dealing with wide-dynamic-range signals directly. In fact, the dynamic range is limited primarily by the offset current of the input amplifiers A1 to A3 (0.1nA typically),

that sets a lower limit on the input voltage or current in low-cost general-purpose devices using bipolar transistors. The high end of the dynamic range is limited to 1mA or less (usually $100\mu A$) by emitter and base resistances in the log transistors.

3. Low nonlinearity due to excellent log-conformity of monolithic dual transistors for currents between 1nA and $100\mu A$: Overall nonlinearity of 0.05% can be achieved in a properly designed circuit.

As an indication of the accuracy and dynamic range that can be achieved in practice, Figure 21 shows the error (as a function of denominator) for the Analog Devices 434 log-antilog divider. The 434B typically has less than 0.2% error over 2 decades of denominator, and 1% or less error over 3 decades — without any external adjustment. If the numerator and denominator input offsets are adjusted externally, the error can be held to 0.2% over three decades: 1000-to-1.

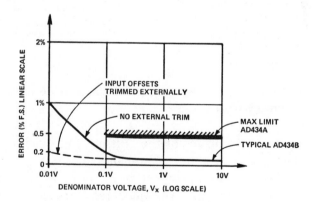

Figure 21. Log-antilog divider: total error as a function of denominator voltage

In common with other logarithmic circuits, the bandwidth of the log-antilog divider is dependent on input signal magnitude. For instance, if the bandwidth is 100kHz for a 10V numerator input, it will be about 10kHz for a 1V numerator input. This is a direct result of the change in loop gain with signal level in transdiode log amplifiers.

Using a 1-Quadrant Analog Divider in 2 Quadrants

A one-quadrant divider, such as the log-antilog circuit discussed here, will accept only single-polarity numerator and denominator inputs. In many cases, it is essential to have a two-quadrant divider that will accept a bipolar numerator (e.g., a sine-wave centered around zero volts) and a unipolar denominator.

Any one-quadrant divider will work as a two-quadrant divider if a fraction of the denominator input is used to bias or offset the numerator input, as outlined in Figure 22. The divider circuit itself remains unipolar, with an output offset of ½ of full scale. The offset can be removed by subtraction (Figure 22) or by ac coupling (Figure 23) to yield an output centered about zero.

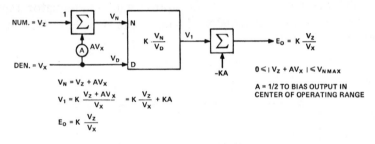

$$V_N = V_Z + AV_X$$

$$V_1 = K \frac{V_Z + AV_X}{V_X} = K \frac{V_Z}{V_X} + KA$$

$$E_O = K \frac{V_Z}{V_X}$$

$$0 \leqslant |\, V_Z + AV_X \,| \leqslant V_{NMAX}$$

A = 1/2 TO BIAS OUTPUT IN
CENTER OF OPERATING RANGE

Figure 22. Offsetting a 1-quadrant divider for operation in two quadrants

In general, the performance of the "offset" two-quadrant divider will not be as good as the performance of the single-quadrant divider, but it will be much better in terms of accuracy and dynamic range than the two-quadrant "inverted-multiplier" dividers.

If the log-antilog divider is offset for 2-quadrant operation, the performance will be reduced in two areas:

1. Bandwidth: The bandwidth will depend on both numerator and denominator levels. As the numerator swings towards its negative extremity, the magnitude of the input to the circuit tends towards zero. This reduces the bandwidth and can cause distortion on negative half cycles.

2. Output offset variation with denominator level: Since the zero-output point of the offset divider is really the half-scale point of the basic circuit, the nonlinearity of the circuit will cause the offset to shift with denominator (a common problem with one-quadrant circuits that are offset to provide a semblance of 2-quadrant performance).

Measurement of the performance of the 434A log-antilog divider in the circuit of Figure 23 typically yields the following results:

Denominator Voltage (V_x)	Output Error Variation from value @ $V_x = 10V$	-3dB Bandwidth (Numerator or Denominator)
+10	0	55kHz
+1	+20mV	5.5kHz
+0.1	+25mV	550Hz

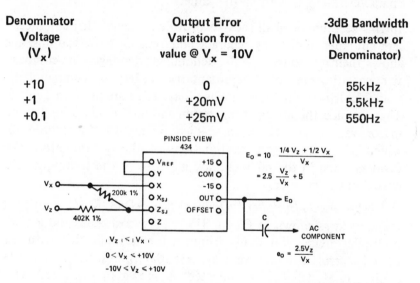

Figure 23. Offsetting AD434 for two-quadrant operation

A one-quadrant divider may be used in two quadrants if it is preceded by an absolute-value circuit having polarity sensing, and if the output is applied to a sign-magnitude circuit (e.g., Chapter 3-5, Figure 15) to restore the polarity. While eliminating the offset problem, this circuit (using a log-antilog divider) will still have bandwidth difficulties, because the signal will slow down each time the numerator crosses through zero.

DIVIDER SPECIFICATIONS

Analog dividers have not, in the past, been as thoroughly specified as analog multipliers. There are at least two reasons for this state of affairs:

1. Until recently, division existed principally as a form of application of a multiplier that could be connected as a divider. Most suppliers of modular and IC multiplier-dividers have placed primary emphasis on the underlying device, the multiplier.

2. Until recently, most analog dividers have had poor to fair performance, for the reasons discussed earlier in this chapter. Since their usefulness was somewhat limited, many manufacturers concluded that there was little point to attempting complete characterization so early in the game.

In this section, we shall seek to provide a format that embraces a large number of the key accuracy and dynamic specifications, and shows their relationship to the denominator voltage. The comparative specifications of a representative variety of commercially-available divider circuits in use as of late 1973 are listed in Table 2. They include the errors at maximum and minimum useful denominator values, and are accompanied by graphs (Figures 24-30) which show the errors as functions of the denominator. The dividers listed in the table employ all three of the techniques discussed in this chapter:

1. *"Inverted-Multiplier" Divider* The AD532K IC multiplier (connected as shown in Figure 31) is a low-cost transconductance multiplier/divider, internally trimmed for multiplication with less than 1% error; external trims are essential for best performance as a divider. Model 427J (connected as shown in Figure 32) is a high-performance pulse-modulation multiplier/divider. Though one of the best "inverted-multiplier" dividers available, it still requires external trim adjustments to make best use of its excellent multiplying characteristics in division.

2. *Direct Variable-Transconductance Divider (2-Quadrant)* The AD531K (connected as shown in Figure 33) uses the basic transconductance cell (Figure 14). Its error is less than 1%, and it has a bandwidth somewhat less than 1MHz. It is the best IC divider available at this writing. Model 436 (connected as shown in Figure 34) uses the improved transconductance circuit of Figure 15 to achieve errors less than 0.5% *over a 100:1 dynamic range*, and 500kHz bandwidth. It requires no external trimming, but outperforms all other 2-quadrant dividers.

3. *Log-Antilog Divider* Model 434B (connected as shown in Figure 35) has the highest accuracy of any 1-quadrant divider available at this writing. It operates over a 100:1 range of denominator with a maximum error of 0.25%, without external trimming.

INTERPRETING THE SPECIFICATIONS

Transfer Function is the ideal relationship between the divider inputs and output, including the scale factor, which is either fixed or set at a nominal 10V. The AD532K can accept a differential denominator input (and the AD531K can accept a differential "X" numerator input), but it will be assumed that $X_1 = V_x$ and $X_2 = 0$ when considering the performance characteristics listed here. The 434B and AD531K can multiply and divide simultaneously, because of the three input variables. The specifications listed here will be considered for constant Y input equivalent to a 10V scale factor. The AD531K is assumed to be connected in the circuit of Figure 33, with a voltage input, V_D.

Quadrants of Operation defines whether the divider will accept a bipolar numerator (*two-quadrant*) or a unipolar numerator (*one-quadrant*). The denominator is limited to one polarity for virtually all analog dividers — including these. The denominator polarity differs from type to type; its polarity and range are given in the "Denominator, X" specification.

Total Error (*accuracy*) @ *Maximum Denominator* specifies the error, or difference between the actual and the theoretical output of the divider at full-scale denominator and (usually) full-scale output. As the denominator is decreased in magnitude from its full-scale value, the error increases (AD532K, 427J) or stays about the same (AD531K, 434, 436), as Figure 24 shows. As noted by the groupings of devices, the AD531K, AD532K, and 427J require 3 external adjustments to meet the typical figure for total error in Table 2 (the adjustment procedure is described in the *testing* section). This requirement is typical for all general-purpose multiplier/dividers. In contrast, the two specialized types are internally trimmed, and need no external trimming to meet the *maximum* error specifications listed in Table 2.

TABLE 2. SPECIFICATIONS OF DIVIDER CIRCUITS[1,2]

Parameter	WITH EXTERNAL TRIM		
	AD531K	AD532K	427J
Transfer Function	$\dfrac{10(X_1 - X_2)}{V_D}$	$\dfrac{10Z}{X_1 - X_2}$	$10\dfrac{Z}{X}$
Configuration (Figure)	33	31	32
Quadrants of Operation	2	2	2
Total Error (%), Max. Denominator (V) (Figure 24)	1, +10	1, −10	0.2, −10
Total Error (%), Min. Denominator (V)	3, 0.5	3, −1.0	2, −1.0
Small-signal Bandwidth (−3dB, kHz), Max. Denominator (V) (Figure 27)	1000, +10	1000, −10	100, −10
Bandwidth (−3dB, kHz), Min. Denominator (V)	1000, +0.5	100, −1.0	1, −0.1
Output Offset Drift (mV/°C), Max. Denominator (V) (Figure 28)	1, +10	1, −10	0.25, −10
Output Offset Drift (mV/°C), Min. Denominator (V)	2, +0.5	10, −0.5	25, −0.1
INPUT CHARACTERISTICS			
Numerator, Z			
Voltage Range (V)	±10	±10	±10
Maximum safe voltage	±V_s	±V_s	±V_s
Input Resistance (kΩ)	10MΩ("X")	36	33
Input Current (μA)	8*	10	3
Input Offset Voltage (μV)	N.S.†	N.S.†	N.S.†
Offset Voltage Drift (μV/°C)	N.S.†	N.S.†	N.S.†
Denominator, X			
Voltage Range to Meet Spec[3] (V)	N.S.†	N.S.†	N.S.†
Voltage Range to Meet Spec, External Trim (V)	+0.5 to +10	−1 to −10	−0.1 to −10
Maximum Safe Voltage	±V_s	±V_s	±V_s
Input Resistance (kΩ)	30	10MΩ	10
Input Current (μA)	0.5	3	3
Offset Voltage (μV)	N.S.	N.S.	N.S.
Offset Voltage Drift (μV/°C)	N.S.	N.S.	N.S.
OUTPUT CHARACTERISTICS			
Voltage Range (V, minimum)	±10	±10	±10
Current Range (mA, minimum)	±5	±5	±5
Resistance (Ω)	1.0	1.0	0.1
Capacitive Load (pF)	1000	1000	1000
POWER SUPPLY			
Specified Performance (V)	±15	±15	±15 ±1%
Operating (±V)	12−18	12−18	12−18
Quiescent Current (±mA)	4(AD531 only)	4	15
PHYSICAL SIZE (mm)	I.C.§	I.C.§	41 × 76 × 15
PRICE ($ U.S., 1-9)	45	36	159

NOTES

[1] All specifications typical at 25°C, ±15V supply, unless noted otherwise, circuit connected in divider configuration of Figures 31−35

[2] All errors specified in % are % of 10V Full Scale (1% = 0.1V)

[3] External trim not required to meet specs, except as noted

WITHOUT EXTERNAL TRIM[3]

	436	434B
	$10\dfrac{Z}{X}$	$Y\dfrac{Z}{X}$, $10\dfrac{Z}{X}$
	34	35
	2	1
	0.5, +10*	0.25, +10*
	0.5, +0.1*	0.25, +0.1*
	500, +10	100, +10
	300, +0.1	1, +0.1
	0.3, +10	0.1, +10
	1, +0.1	1, +0.1
	±10	0 to +10
	±V_s	±V_s
	10	100
	±0.1	±0.01
	±100*	±100*
	15	15
	+0.1 to +10	+0.1 to +10
	5mV to +10	5mV to +10
	±V_s	±V_s
	25	100
	0.1	0.01
	±100*	±100*
	15	15
	±10	±10
	±5	±5
	0.1	0.1
	1000	1000
	±15 ±3%	±15 ±3%
	12—18	12—18
	10	10
	38 × 38 × 15	38 × 38 × 15
	80 (Approx.)	87

*Maximum specification

†Not specified

§IC's may be available in choice of packages and chip form; consult product data sheets

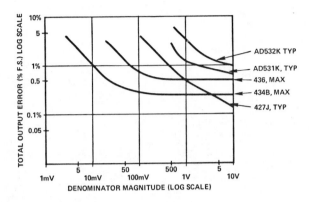

Figure 24. Total error vs. denominator at constant ratio = 10V full-scale output, 1% = 100mV

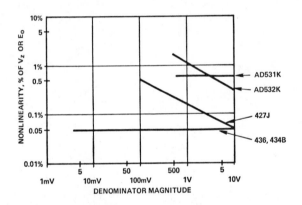

Figure 25. Nonlinearity (at constant denominator) vs. denominator, % of numerator (log scales)

The total error includes the errors from all sources, and so represents a worst-case condition. There are six major sources of error included:

1. Numerator nonlinearity (Figure 25)
2. Denominator nonlinearity
3. Scale-factor error
4. Numerator offset, referred to the output
5. Denominator offset, referred to the output
6. Output amplifier offset

Errors (3) through (6) can be theoretically adjusted to zero at a fixed temperature and denominator level, while the input non-linearities are inherent and cannot be removed (except perhaps for some reduction of second-order error by the "cross-feed" technique discussed in 3-2, applied to divider-connected transconductance multipliers). Rather than specify each of these six errors individually, it is usually easier to measure the overall effect of the errors, which is by definition the total error.

Total Error at Minimum Denominator specifies the error, as just defined, at the minimum useful denominator magnitude. This represents the worst-case operating conditions for a divider, since the errors are at a maximum. The "inverted-multiplier" dividers, 427J and 532K, show the greatest increase of error at small denominator voltages, as might have been expected. The log-antilog and variable-transconductance dividers show no increase in the specified error, even with the denominator voltage at 1/100 of full scale, as one might have predicted from the analysis of these circuits.

The increase of error of the 532K and 427J is caused principally by the magnification of the numerator and denominator offsets by $10/V_x$, and by the denominator nonlinearity, which causes an apparent scale-factor error. The AD531K maintains error at the 1% level over a 10:1 range, then the error increases sharply due to decreasing current in the X input circuit, causing the gain to go to zero (Figure 9). The 436 and the 434B have very small ($100\mu V$) input offsets and very low nonlinearity; they have relatively-little difficulty with small denominator voltage.

Total-Error Drift (with Temperature) vs. Denominator. The change in total error per °C change in ambient temperature increases as the denominator decreases, as shown in Figure 26. This effect is more noticeable with the inverted-multiplier types, 427J and AD532K. The 434B and 436 show a slight increase in drift at the low end (100mV denominator).

"Total-error drift" is the sum of the total output-offset drift and scale-factor drift. The numerator and denominator nonlinearities are relatively insensitive to temperature.

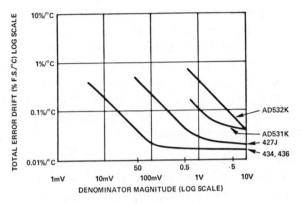

Figure 26. Total error drift vs. denominator voltage (% full-scale/°C)

<u>*Bandwidth (Small-Signal, -3dB) at Maximum Denominator:*</u> the frequency at which the numerator or denominator input-to-output "gain" is reduced to 70% (-3dB) of its dc value. The input must be "small," that is, less than 10% of either full-scale or denominator magnitude, to avoid excessive distortion (which would invalidate sine-wave analysis). The small-signal bandwidth of the 427, AD532K, and 434, are directly dependent on denominator magnitude, as the graph (Figure 27) shows.

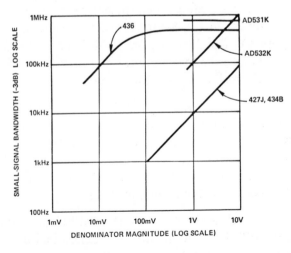

Figure 27. Bandwidth (-3dB) vs. denominator (K = 10V)

Interpreting denominator voltage as "1/gain," (Figure 28), the 436 is seen to be unique, because its bandwidth is essentially independent of denominator; it achieves a 30MHz gain-bandwidth product at 100mV denominator! Similarly, the AD531K has about 750kHz numerator bandwidth over a 20:1 denominator range; it is thus the fastest in terms of absolute bandwidth (over a limited range of denominator). (Gain = $10V/V_x$ = 10/0.1 = 100; if bandwidth = 300kHz, Gain-bandwidth = 100 × 0.3 × 10^6 = 30MHz.)

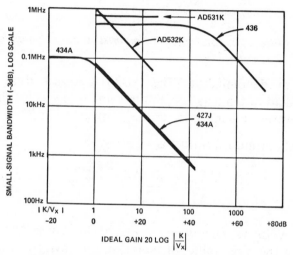

Figure 28. Bandwidth (–3dB) vs. gain (K/V_x)

<u>*Output Offset Drift vs. Denominator:*</u> The rate at which the total output offset changes with temperature, as a function of the denominator. The two components of this drift are numerator-input offset drift, multiplied by gain (K/V_x), and output-stage offset drift. As Figure 29 shows, the 427, 434, and 436 have drifts of about 0.2mV/°C at V_x = 10V, while the AD532K drifts at about 1mV/°C. The drift of the 427 and AD532K increase as the denominator is reduced, because the numerator offset drift is greater than the output-stage drift; hence the sum is dominated by the 1/X relationship. Since the numerator offset drift of the 434 and 436 is small (about 10μV/°C, compared to the output-stage drift of 300μV/°C – the total offset drift at G = 1 or V_x = 10V), the total output drift increases by only a factor of 2 for a 30:1

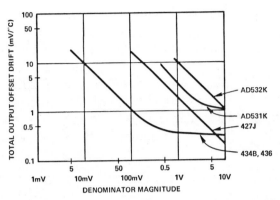

Figure 29. Total output offset drift vs. denominator voltage (log scales)

change in denominator. The AD531K offset drift is almost independent of temperature, but it is not as low as for the 434 and 436; however, it is much lower than for any other IC.

The total output offset drift is

$$\frac{\Delta V_{os}}{\Delta T} = \frac{K}{V_x} \frac{\Delta Z_{os}}{\Delta T} + \frac{\Delta Y_{os}}{\Delta T}$$

Output Noise vs. Denominator Figure 30 shows the relationship between the rms value of noise at the divider's output, in a constant bandwidth of 5Hz to 10kHz, and denominator magnitude. The total noise at the output is essentially

$$V_N = \sqrt{\left[E_{Nz} \frac{K}{V_x}\right]^2 + \left[E_{Nx} \frac{K}{V_x}\right]^2 + E_{No}^2}$$

where

E_{Nz} = equivalent numerator input noise: Vrms, 5Hz to 10kHz

E_{Nx} = equivalent denominator input noise: Vrms, 5Hz to 10kHz

E_{No} = equivalent output-stage noise: Vrms, 5Hz to 10kHz

K = scale factor = 10V

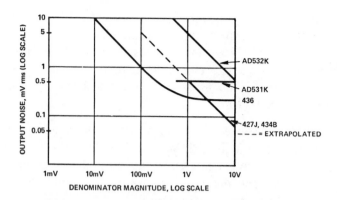

Figure 30. Output noise vs. denominator (5Hz to 10kHz bandwidth)

The 436 has the best overall noise performance, while the 427 and 434 are quieter for denominator voltage in the range 2V to 10V. As a general-purpose multiplier/divider, the AD532K is reasonably quiet; the AD531K output noise is almost independent of denominator level. If "peak-to-peak" noise is defined as 6X rms, 1mV rms = 6mV peak-to-peak.

Input Characteristics

These specifications define the input voltage range and errors at the numerator and denominator inputs.

Numerator (Z) Voltage Range is the maximum span of numerator voltage for which the error specifications apply. The numerator voltage is limited to a magnitude less than or equal to the denominator voltage for the dividers described here* (see Figures 2 and 3). The AD532K, AD531K, 427J and 436 will accept positive and negative numerator voltages (i.e., they are 2-quadrant devices), but the 434 is limited to positive numerators (1 quadrant).

Numerator: Maximum Safe Voltage is the maximum that can be applied to the divider input continuously without causing damage.

---

*But K may be scaled for <10V for the AD531 and the 434; and the numerator may exceed the denominator by the factor $E_{o_{max}}/K$.

Numerator: Input Resistance is the effective input resistance between the numerator input terminal(s) and power-supply common. The input resistance ranges from $10k\Omega$ on the 436 to $10M\Omega$ for the AD531K. The resistance of the signal source should be 0.1% or less of the numerator input resistance to minimize variation of scale factor.

Numerator Input Current is the bias current flowing into or out of the numerator terminal, with $V_z = 0$. Usually, this is a negligible source of error, even for the AD532K, but it should be considered in the choice of the resistance-to-ground for applications where the inputs are to be capacitance-coupled.

Z-Input Offset Voltage: The numerator offset is not specified for general-purpose multiplier/dividers (427J, AD531K, and AD532K). The $100\mu V$ Z-offset of the 436 and 434B will cause an error of only 0.1% at 100mV input; the error can be further reduced by externally trimming.

Z-Input Offset-Voltage Drift characterizes the sensitivity of the Z-input offset to temperature. It is not usually specified (at this writing), but it is nevertheless a useful number, since the numerator offset drift is the primary source of output offset drift for values of denominator voltage substantially less than full-scale.

$$\text{Output offset drift} = \frac{K}{V_x} \frac{\Delta Z_{os}}{\Delta T} + \frac{\Delta E_{os}}{\Delta T}$$

Denominator, X, Voltage Range to Meet Specifications (Without External Trim) is the maximum span of denominator voltage for which the specified accuracy is maintained, with no external adjustment of offsets or scale factor. Since the AD531K, AD532K, and 427J require trimming for reasonable performance as dividers, no limit is specified for them. The 436 and 434B will operate over a range of denominator exceeding 100:1 with low error.

Denominator Range to Meet Specifications With External Trim: Trimming the offsets expands the dynamic range of the 436 and 434 to 2000:1, and provides a reasonable dynamic range for the 427, AD531K, and AD532K. Note that the 427 and AD532K require negative denominators, while the 434B and 436 require positive denominators. The AD532K will accept positive denominator voltage if the input terminals (X_1 and X_2) are interchanged.

Maximum Safe Denominator Voltage is the maximum voltage that can be applied to the divider input continuously without causing damage.

Denominator Input Resistance: The AD532K has a very high input resistance (3MΩ), while the other three range between 25 and 100kΩ. All five types should be driven from sources with low resistance to minimize source-loading errors. The 3μA typical input current of the AD532K partially offsets the advantage of its high input resistance for source resistances of 10kΩ or more.

Denominator Input Current is the current flowing into or out of the denominator terminal with $V_x = 0V$.

Denominator Offset Voltage is a constant offset voltage effectively in series with the denominator input. The low offset of the 436 and 434 causes minimal change in apparent scale factor over a 100:1 denominator range:

$$\text{Error @ } V_x = 10V; \ \frac{10^{-4}}{10} \cdot 100\% = 0.001\%$$

$$\text{Error @ } V_x = 0.1V; \ \frac{10^{-4}}{0.1} \cdot 100\% = 0.1\%$$

Denominator Offset-Voltage Drift characterizes the sensitivity of the denominator offset drift to temperature. The specified $\Delta X_{os}/\Delta T$ of 15μV/$^{\circ}$C for the 436 and 434B can cause a change in error of 0.7% over a 100:1 denominator range, and 0° to 70°C temperature range.

Output Characteristics

This group of specifications describe the output terminal properties of the divider.

Voltage and Current Range: The two-quadrant dividers, AD531K, AD532K, 427J, and 436 will supply a minimum of ±10V at ±5mA at their outputs. The one-quadrant 434 swings a minimum of 0 to +10V at 5mA. Types requiring an external loop closure around the output amplifier can employ current boosters "inside the loop" to beef up the output for output current beyond 5mA.

Output Resistance: All five dividers have low-impedance outputs, resulting in minimal change in output voltage as load current is changed. However, since output impedance is affected by loop gain, the general-purpose multipliers used as "inverted-multiplier" dividers will suffer an increase of output impedance as denominator decreases.

Capacitive Load is the minimum amount of capacitance that can be connected directly between the divider output terminal and ground without causing the device to oscillate.

Power Supply

Specified Performance: The power-supply voltage and tolerance required for the divider to meet its accuracy specifications. The design-center power-supply voltage for all five types is ±15V.

Power Supply Operating Range is the range of power-supply voltage that the divider will accept and still operate as a divider. If external trims are used, the circuit can usually be adjusted to meet the same accuracy specifications as for ±15V supply.

Quiescent Current is the current drawn from the power supplies at zero output voltage and current. The quiescent current-drain will be augmented by the load current, since the output amplifiers have "Class B" output stages.

Physical Size: The IC's (AD531K, AD532K) are by far the smallest, and the pulse-modulation multiplier/divider, 427J, is the largest. IC types are available in hermetically-sealed ceramic 14-pin dual in-line packages, and the AD532 is also available in the hermetic TO-100 10-pin metal can; IC types may also be made available in chip form for applications requiring hybrid construction.

*Price** The IC's are the least expensive, but they offer the lowest performance (except for speed of the AD531K), and trims are required for reasonable performance. Surprisingly, the most-expensive multiplier/divider, 427J, does not offer the highest performance as a divider. The high-performance 436 and 434B outperform the 427, at lower cost, but are not as inexpensive as the IC types. If one considers the fact that the 436 and 434 have excellent performance without external adjustments, and even

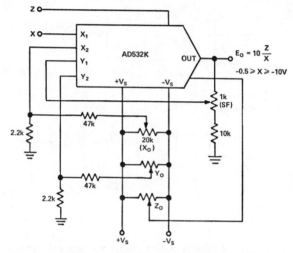

Figure 31. AD532K divider circuit (inverted multiplier)

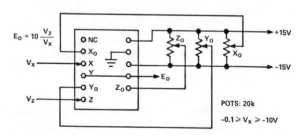

Figure 32. 427J divider pin connections (inverted multiplier), pinside view

*Summer, 1973. Price is listed as a measure of relative cost, not primarily as a commercial inducement. Those interested further should consult recent Product Guides or price lists, or the nearest Sales office, since (a) prices are subject to change, and (b) IC prices are more so.

better performance when "tweaked-up," and that the 434 multiplies and divides simultaneously, it may turn out that the 434 and 436 are less expensive to use than IC's in many applications.

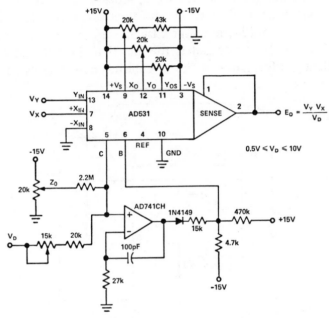

Figure 33. AD531K divider circuit — direct variable — transconductance (circuit for voltage input to denominator)

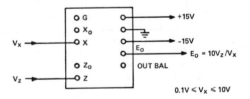

Figure 34. 436 divider connections, pinside view (high performance 2-quadrant variable transconductance)

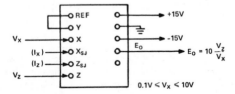

Figure 35. 434B divider pin connections (log-antilog), pinside view

TESTING AND ADJUSTING ANALOG DIVIDERS

Good test and adjustment procedures are exceptionally important for analog dividers, since their performance spans a wide range and is often critically dependent on adjustment. For example, a general-purpose "2%" multiplier-divider (e. g. AD532J) can have a worst-case error of 2 volts, or 20%, when used as a self-contained divider with a 10:1 denominator range. In contrast, a specialized high-accuracy divider (e.g. 434B) can have a worst-case error of only 25mV under the same conditions — almost *100 times* less error than the general-purpose device! The accuracy of the "2%" divider can be improved by at least a factor of 3 by the adjustment procedure described in this section.

The tests for divider performance fall into the same three general categories as for multipliers

1. Static accuracy or error
   a. Total error
   b. Output offset
   c. Numerator and Denominator offsets
   d. Numerator and Denominator nonlinearity
   e. Scale-factor error
   f. Dependence of errors on denominator
   g. Dependence of errors on temperature and power-supply voltage

2. Dynamic errors
   a. Small-signal bandwidth
   b. Large-signal bandwidth
   c. Slewing rate
   d. Settling time
   e. Dependence of the above on denominator

3. Terminal or interface parameters
   a. Input resistance and voltage range
   b. Output voltage and current
   c. Power-supply voltage and current

In this section, the principal emphasis will be placed on tests for static and dynamic errors, since methods for measuring the terminal parameters are straightforward.

TEST EQUIPMENT FOR DIVIDER TESTING

Precision dc Reference: Accuracy to within 0.01% of setting, plus ±100μV, from ±10V to 0V in steps of 100mV or less.

Precision Decade Voltage Divider: 0.01% ratiometric "accuracy" with output buffer and inverter (see Figure 47). This is *essential* for accuracy measurements on wide-dynamic-range dividers, and quite useful for testing ordinary dividers.

Digital Voltmeter with at least 4½-digit resolution, 0.02% accuracy error. The 1V and 10V ranges will be the most-frequently used.

Sine-Wave or Function Generator: Frequency range 1Hz to 5MHz. Adjustable dc offset is handy for testing 1-quadrant divider and applying ac signals to denominator. Output amplitude range of 2mVp-p to 20Vp-p is the most useful. A calibrated output attenuator will facilitate measurements on wide-dynamic-range dividers.

Dual Power Supply with adjustable output voltage (±15V nominal) and adjustable current limit to prevent costly "accidents."

Oscilloscope with calibrated vertical *and horizontal* voltage inputs for cross-plot tests. Desirable sensitivity ranges are: Vertical, 10mV/cm to 5V/cm; Horizontal, 100mV/cm to 5V/cm. Bandwidth of 300kHz is adequate for crossplots. At least 5MHz vertical-axis bandwidth is required for dynamic tests.

Divider Test Socket, with all connections made and trimpots included, facilitates the test and adjustment of both modular and IC devices.

TEST AND ADJUSTMENT OF STATIC ERRORS

Cross-plotting the divider error against the denominator input is the most revealing, and also the most efficient method of testing the accuracy of an analog divider. This approach minimizes the ambiguity and lack-of-feel inherent in a dc point-by-point test or adjustment procedure.

The crossplot test setup, Figure 36, can be used for testing and adjusting any 1- or 2-quadrant divider. The most-important part of the setup is the low-frequency (10Hz or less) offset sine-wave generator that provides the sweep voltage for the divider and the horizontal axis of the oscilloscope. A unity-gain inverter, A1, provides a 0 to +10V sweep for dividers requiring positive inputs.

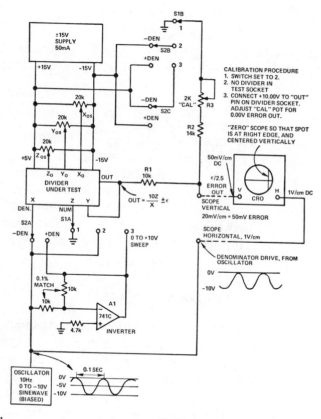

S1 Switch Position	Function	TEST CONDITIONS		Ideal Output	ADJUST
		Numerator, Z	Denominator, X		
1	Numerator & Output Offset	0V	0 to −10V (or +10)	0V	Z_o = NUM, Y_o = OUTPUT OFFSET
2	Denominator Offset, $V_z = V_x$	0 to -10V	0 to −10V (or +10)	+10V (−10V)	X_o = DEN OFFSET
3	Denominator Offset, $V_z = -V_x$	0 to +10V	0 to −10V (or +10)	−10V (+10V)	X_- = DEN OFFSET

Figure 36. Divider error crossplot circuit

Since the output of the divider should ideally be constant (no slope or curvature) with the sweep-test conditions used in this setup, it is only necessary to subtract an ideal constant from the actual divider output to determine the error, as a deviation from zero on the display. The example given below is for devices requiring negative denominator voltage:

Test	Inputs	Ideal Output	Subtract
1. Numerator offset	$V_Z = 0$, $V_X = 0$ to $-10V$.	0V	0V
2. Output offset	$V_Z = 0$, $V_X = 0$ to $-10V$	0V	0V
3. Denominator offset	$-|V_Z| = V_X = 0$ to $-10V$	+10V	+10V
4. Denominator offset (1st quadrant of 2-quadrant dividers)	$V_Z = -V_X$, $V_X = 0$ to $-10V$	$-10V$	$-10V$
5. Scale factor	$-|V_X| = V_X = 0$ to $-10V$	+10V	+10V

The subtraction is accomplished via a simple resistive divider R1 and R2, R3, referenced to the power supply. A more precise reference voltage, and a more sophisticated subtractor could be used, but the simple approach is adequate for testing the majority of general-purpose multiplier-dividers. If a good supply is used, and the R1-R2, R3 network carefully adjusted, the test circuit will be accurate enough for testing the specialized dividers, such as the 434.

Using the Divider Crossplot Tester

The easiest way to understand the operation of the crossplot tester is to follow through the adjustment of a typical two-quadrant divider, such as that illustrated in Figures 37 through 42. The procedure applies to any type of divider circuit, whether "inverted multiplier," log-antilog, transconductance, etc., and whether in modular or IC form. The sources of the errors, (for example, numerator offset) for the various divider types have been discussed earlier in this chapter. At any rate, these errors can be treated in a "black-box" fashion, as the procedure illustrates.

Proper adjustment of the "offset" errors dramatically reduces the divider's errors at room temperature, as Figures 37-38 and 40-41 illustrate. However, with changes in temperature, the errors can increase to values comparable to the untrimmed values, particularly for divider-connected general-purpose "multiplier/dividers." The temperature-drift effects can be easily monitored with the sweep-test circuit if the divider itself is placed in a temperature-test chamber (along with any adjustment circuitry that is to be subjected to ambient temperature variations). The circuit is adjusted for minimum error at 25°C, then the change in error over the temperature range of interest (say, 0° to 70°C) is observed.

(CONDITIONS: $V_Z = 0$, $V_X = 0$ to $-10V$, $\epsilon = V_{OUT}$
TEST SET FUNCTION SWITCH SET TO 1.)

The scope trace will be sharply curved up or down at the right edge of the screen, as the denominator, V_X, approaches zero. Notice that the total output offset is 400mV or 4% of Full Scale at a denominator of -2 volts.

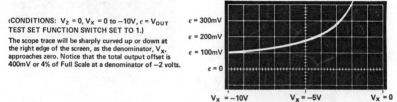

Figure 37. Total output offset vs. denominator test

CONDITIONS: $V_Z = 0$, $V_X = 0$ to $-10V$, $\epsilon = V_{out}$
TEST SET FUNCTION 1.

Adjust the Z_O potentiometer to flatten the scope trace - - note the dramatic reduction in offset as $V_X \rightarrow 0$. The error trace will not necessarily be centered on the "0" line - - this will be adjusted in the next step.

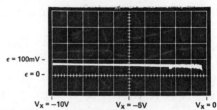

Figure 38. Numerator offset, Z_O, adjustment test

CONDITIONS $V_Z = 0$, $V_X = 0$ to $-10V$
TEST SET FUNCTION 1.

Adjust the Y_O potentiometer to align trace with center 0 error line. Note the increase in noise as $V_X \rightarrow 0$. This indicates the increased input-to-output gain of the divider, which approaches infinity as the denominator approaches zero. The total output offset is dramatically reduced from the initial value in Figure 37.

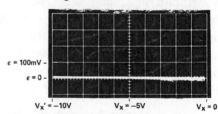

Figure 39. Output offset, Y_O, adjustment test

(CONDITIONS: $V_Z = V_X = 0$ to $-10V$
TEST SET FUNCTION 2.)

The scope trace will curve up or down as $V_X \rightarrow 0$. The increase in error is due primarily to denominator offset, since numerator offset was adjusted in Figure 38.

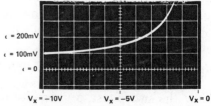

Figure 40. Total output error vs. denominator test

(CONDITIONS: $V_z = V_x = 0$ to $-10V$
TEST SET FUNCTION 2.)

Adjust X_O potentiometer to flatten trace,
which will usually not be aligned with "0"
error line, due to scale factor error.

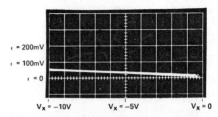

Figure 41. Denominator offset, X_O, adjustment test

(CONDITIONS: $V_z = 0$ to $+10V$, $V_x = 0$
to $10V$, TEST SET FUNCTION 3.)

Readjust X_O potentiometer if necessary to
flatten trace; switch back to conditons of
Figure 41 and check for best compromise in
flatness as $V_x - 0$. If a scale factor trim is
provided, adjust it for minimum difference
between error trace and zero error line for
input conditions of Figures 41 and 42.

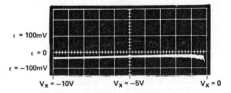

*Figure 42. Denominator offset, X_O, and scale factor
adjustment test*

Measuring Denominator Frequency Response

Figure 43 shows a test setup for measuring the frequency response of the divider's denominator input as a function of denominator level. The denominator input of all commonly-available analog dividers is restricted to a single polarity: plus or minus (or, with differential inputs, either), so the signal generator must be offset to maintain V_{min} and V_{max} within the denominator input limits. The difference (i.e., p-p amplitude) may be maintained at a considerably smaller level (typically 10% of the offset) to minimize waveform distortion: if the ac test signal spans an appreciable fraction of the dc offset, the output will be noticeably distorted, since it is proportional to $K/(V_{DC} + V_{AC})$.

A square-wave can be used to determine denominator rise time, slewing rate, and settling time. In general, the time response will be slower for signals going toward zero denominator, and faster for signals approaching full-scale denominator.

Measuring dc Accuracy- and Temperature-Drift

While the sweep test, or crossplot technique, for measuring divider accuracy, is the fastest and the best way to get an overall picture

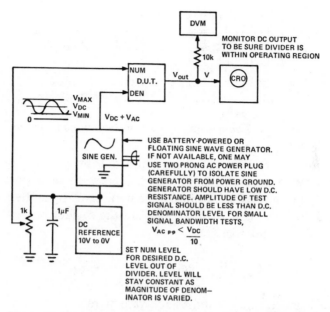

Figure 43. Test setup — denominator frequency response vs. denominator magnitude

of error as a function of denominator, it is difficult to measure the absolute error precisely (as necessary for calibration) with this technique. If the crossplot circuit is carefully calibrated, it can be used for absolute measurements. However, it tests the divider at a constant ratio of numerator to denominator, and so does not readily show numerator (constant-denominator) nonlinearity.

Furthermore, since the numerator and denominator are swept over their ranges, and spend a relatively-short time at low levels, errors due to thermal effects may not be visible. The sweep can be slowed, or a low-frequency square-wave can be used, but they are not as revealing as a dc test.

Measuring Numerator Nonlinearity

Numerator nonlinearity is measured by comparing the divider output with the numerator input at constant denominator voltage, with typical results as shown in Figure 44. Since the gain or attenuation through the divider is inversely dependent on the denominator level, the input or output must be linearly amplified

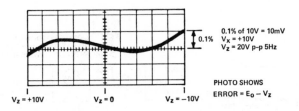

0.1% of 10V = 10mV
$V_x = +10V$
$V_z = 20V$ p-p 5Hz

PHOTO SHOWS
ERROR = $E_0 - V_z$

Figure 44. Two-quadrant divider, numerator nonlinearity

or attenuated to facilitate the input/output comparison. The easiest way to accomplish this for dividers with gains potentially greater than unity is to attenuate the numerator signal by a factor γ, equal to X/K, and then to compare the divider output to the input of the attenuator:

$$E_o - E_o\,(\text{ideal}) = K\,\frac{\gamma V_z{}'}{V_x} - V_z{}'$$

If a precision attenuator and an accurate dc reference are used, the denominator nonlinearity can be estimated from this test: the difference between the divider output and the numerator input will have an average slope equal to the "gain" error, or departure of the gain from the ideal value K/X. The scope photo, Figure 44, shows the nonlinearity of a 436 divider, measured with the test setup of Figure 45.

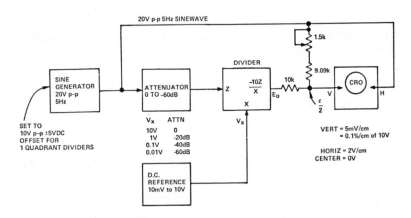

Figure 45. Test setup for numerator nonlinearity

DC Measurements of Accuracy and Temperature Drift

The dc error of the divider can be easily measured with the test setup of Figure 46. A precise voltage divider (e.g., Figure 47), is the most important piece of test equipment, since it provides a numerator voltage that is a precise fraction of the denominator voltage, independently of the absolute calibration of the dc voltage reference.

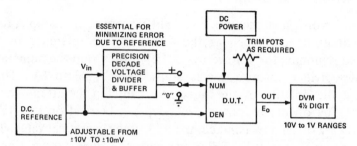

Figure 46. Divider DC accuracy test setup

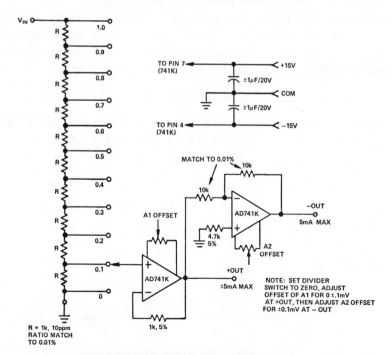

Figure 47. Decade divider and buffer

Since the divider (ideally) takes the ratio of the numerator and the denominator inputs, small errors (<1% of setting) in the calibration of the absolute magnitude of the dc reference will have negligible effect on the accuracy of measurement if the numerator is a precisely-known fraction of the denominator. It is convenient (and desirable) to have available a dc reference with good resolution (e.g., 1mV steps); it is especially useful for testing the high-performance log-antilog dividers.

The dc error-test procedure is straightforward. For example, for a 2-quadrant divider (such as the 427), set the reference to the desired denominator voltage (e.g., -10V), then step the numerator voltage divider through 1-volt steps from -10V to 0V. Then, reverse the polarity of the numerator and step the attenuator to +10V. The divider output at each step can be read on the DVM, and then written down and compared to the theoretical value. The same procedure can be followed, with the divider operating at different ambient temperatures, to determine the temperature coefficient of divider error. Table 3 is a sample temperature-test form that includes the minimum number of measurements necessary to determine the offset and overall accuracy drift of a 2-quadrant divider over the 0° to 70°C operating range.

Measuring Numerator Frequency Response

This test is straightforward for a 2-quadrant divider, and is not very involved for a 1-quadrant divider. The test setup, Figure 48,

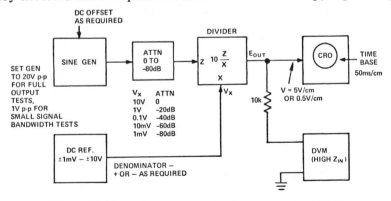

Figure 48. Test setup — numerator frequency response vs. denominator magnitude

TABLE 3. TEST CHART FOR ACCURACY AND OFFSET DRIFT VS. TEMPERATURE FOR A TWO QUADRANT DIVIDER

Denominator V_x	Numerator V_z	Theoretical E_o	E_o Measured		
			0°C	25°C	+70°C
−10.000V	−10.000V	+10.000			
−10.000V	+10.000V	−10.000V			
−10.000V	0.000V	0.000V (Offset)			
− 1.000V	− 1.000V	+10.000V			
− 1.000V	+ 1.000V	−10.000V			
− 1.000V	0.000V	0.000V			

Absolute Error = $V_{MEAS.} - V_{THEO.}$ = δ

% of Full Scale Error = $\dfrac{\delta}{10V} \cdot 100$ = %FS = 10δ

%/°C DRIFT = $\dfrac{\delta\% \, FS}{T_2 - T_1}$

Measurements should be made at smallest expected denominator. 1V was picked for this example since it is the minimum useful denominator for most general—purpose two quadrant "inverted multiplier" dividers.

illustrates the basic scheme. A signal generator with a continuously-variable output-amplitude control may be used instead of the constant-amplitude source and calibrated attenuator. However, it is important to remember that the signal on the numerator input must have peak amplitude less than the magnitude of the denominator for dividers with a 10V scale factor (constant K or variable V_y). One-quadrant dividers require a dc offset for the numerator proportional to that of the denominator; so the numerator offset will have to be adjusted at each denominator level.

The bandwidth test procedure is

1. Set the dc reference to the desired denominator voltage.

2. Set the oscilloscope vertical sensitivity to 5V/cm for large-signal tests; use dc coupling so that the total output voltage is

displayed — this will make saturation and offset effects more noticeable. Set the oscilloscope sensitivity to 0.5V/cm or less for small-signal tests. AC coupling will be required for 1-quadrant dividers, since their output will be "offset" due to the dc bias required on the numerator input.

3. Set the amplitude and offset of the signal generator so that the divider output, as displayed on the scope, is within the required limits, e.g., 1Vp-p, at a frequency at least a decade below the expected -3dB frequency.

(The numerator dc offset can be most easily adjusted by connecting a DVM to the divider output, as indicated in Figure 48. The offset is then adjusted so that the dc output is at the desired level within the operating range of the divider — for example, +5V for full-output frequency tests on a 434 one-quadrant divider.)

4. Sweep the generator up in frequency until the divider's output amplitude falls off — or peaks — to the 3dB ($\mp30\%$) point, or there is noticeable distortion at the output. Repeat this procedure at several denominator voltages within the expected operating range.

A square-wave input can be used to determine slewing rate, rise time, and settling time in the same manner as for a multiplier or an operational amplifier.

Nonlinear IC's

Chapter 4

"The Analog Art shows no signs of yielding to the Dodo's fate. The emergence and maturation of monolithic processing finesse has perhaps lagged a bit behind the growth of the Binary Business. But whereas digital precision is forever bounded by bits, there is no limit excepting Universal Hiss to the ultimate accuracy and functional variety of simple analog circuits."

Barrie Gilbert, January, 1973

Although a great majority of configurations and performance classes of the devices to which this book is devoted are principally available in the form of modular discrete–circuit assemblies, a sea–change is occurring.

Having thoroughly saturated the linear–function area of analog circuits with a plethora of general– (and special–) purpose operational amplifiers, integrated–circuit technology has now turned to the development of nonlinear-function circuits.

An increasing number of non-linear functions, previously available only in module form, are becoming available as integrated circuits. Except for comparators (which combine op-amp and digital characteristics), the greatest progress made by analog IC's to date has been in circuits that perform the basic functions of multiplication, division, squaring, rooting, and multiplying D/A conversion.

SIMILAR BUT DIFFERENT

For the most part, the similarities between the IC and the module approaches are straightforward: they both perform similar circuit operations and can be generally applied to solve the same system-design problems. The differences fall into two categories: generic differences between IC's and modules, such as cost and size; and technological differences, which affect circuit design and performance, bringing both advantages and limitations.

The most salient difference is cost. Not only are costs of comparable IC devices competitive at the time they are introduced, but the trend is inexorably down, especially as usage increases to large quantities — a powerful incentive to the system designer to use them for new applications. Not long ago, modular multipliers with <2% overall error were sold in the $25 to $30 price range. Now, equivalent IC multipliers (AD533JH) sell for less than $6 in hundred lots! The next few years will see "garden-variety" IC multipliers selling at prices not much higher than those of today's general-purpose IC op amps.

Other generic differences are small size and high reliability. The small IC packages (both the hermetically-sealed ceramic dual in-line package and the hermetically-sealed TO-100 can) offer significant savings of space in all applications, some of which would not even be possible using the modular discrete-component package. (Such devices as the AD532, needing no external trims or other components, offer the ultimate in small size). The increased potential reliability of monolithic assembly techniques, relative to their equivalent modular functions, has been amply documented.

Because discrete-circuit modules are not limited in either circuit complexity, physical configuration, or package size, they continue to offer performance advantages over their IC counterparts, but the gap has already narrowed to negligible proportions for general-purpose moderate-bandwidth 1-2% devices. Where state-of-the-art accuracy, speed, stability, and/or ease-of-use must be obtained, modules are still (at this writing) the first — and perhaps the only — choice.

Modules using discrete circuitry can employ pulse-modulation circuits to obtain errors less than 0.1%, while *present* IC devices, using variable transconductance, are limited to ±0.5%, with external trim. ("Cross-feed" trims can reduce this by a factor of 2, as explained in Chapter 3-2.) The pulse-modulation technique, inherently more accurate, is difficult to employ in integrated-circuit form, because of the stringent demands it imposes on circuit components produced on a monolithic chip. However, new approaches to IC multiplier designs hold considerable promise for the future.

Discrete designs can also take advantage of monolithic matching of selected elements as building blocks, providing combinations that are difficult to come by with reasonable yield on a single monolithic chip, at today's state-of-the-art. On the other hand, automatic laser trimming of IC multipliers, on the chip, can help make IC multipliers more competitive in price/performance/convenience.

In the area of dynamic performance, IC transconductance cells are inherently wideband (as incomplete current-to-current devices). Complete monolithic multipliers are generally limited to a few MHz of bandwidth because of limitations on the speed of lateral PNP transistors used in the level-shifting output amplifier. Incomplete current-output devices (so-called multipliers, but actually just current cells) require moderate numbers of external components, including a level-shifting amplifier, all of which tends to reduce speed, as well as to contribute additional errors that must be added to the errors of the basic monolithic device. The economic choice between basic transconductance cells, with considerable external circuitry, and modules or complete IC's with guaranteed performance, tends to favor the latter, except for specific non-precision high-speed application areas where the nature of the current cell is not a barrier to its use.

The IC multipliers discussed in this book are complete self-contained operational blocks, requiring only a power supply, trims (in some cases), and normal circuit-implementation techniques to obtain working circuits.

Finally, there are differences in testing philosophy that affect the performance of IC multipliers. The cost of production-testing module-type multipliers manually is not a large part of the cost of these devices. On the other hand, in order to realize the potential cost savings of IC multipliers, automated test procedures are an absolute necessity. The principal drawback of automated high-speed testing is that there is insufficient warmup time to allow all parameters to reach their final values. Consequently, both predictive techniques and conservative safety margins must be used by the manufacturer (i.e., Analog Devices) to make sure that the user achieves the guaranteed specifications in a normal, fully-warmed-up circuit application. The bonus for the user is that quite often the devices he purchases from a conscientious IC manufacturer may be considerably better than the specifications would indicate.

DESIGNING IC MULTIPLIERS

It is only within the past five years that the basic circuit configurations were devised which made accurate monolithic multipliers possible. It took about two of these years for the industry to provide the means of fabricating devices having errors of a magnitude that would be competitive with discrete-component circuit modules.

The key technological factors that had to be awaited were:

1. Near-ideal NPN transistors exhibiting high current gains, close matching, and very close conformity to ideal logarithmic junction characteristics over wide current and temperature ranges.

2. Linear, stable, close-tolerance monolithic resistors.

3. An important factor that helped monolithic multipliers attain performance standards suitable for general-purpose applications is the better understanding of subtle sources of errors arising from the chip layout, such as thermal effects, non-negligible resistance of the aluminum metallization, non-matching of apparently identical transistors, etc.

Both the user and the designer of monolithic nonlinear circuits are faced with the need for compromises. Those problematic factors that concern the designer are:

1. The complete system, comprising input amplifier(s), the main functional core, the output amplifier, and all auxiliary circuits, such as reference-voltage and bias supplies, must be achieved with whatever devices can be concentrated on a chip no larger than, say, 80 mils (2mm) square.

2. All the power generated by the circuit has to be dissipated by a relatively small package, in contrast to the substantial size and mass of a discrete-circuit module. So the compromise between high load-driving capability, on the one hand, and negligible warmup and long-term drifts, is especially severe. In addition, the close thermal coupling between the power-dissipating output stage and the temperature-sensitive logarithmic devices calls for very careful layout to minimize thermal unbalances between critical transistor-pairs.

3. Some very useful components are just not available. Selected devices, such as computer-matched transistors, calibrated and guaranteed reference diodes, and fast, high-gain PNP transistors obviously cannot be incorporated into the design. The most serious problem is the bandwidth limitation caused by the use of lateral PNP transistors; most of the other problems can be satisfactorily dealt with.

4. The number of adjustments needed to get the device within its stated accuracy must be minimized. Apart from their inconvenience for the user and their adverse effect on production costs, each external adjustment requires a minimum of one bonding pad and one package pin, both of which are usually at a premium (one bonding pad consumes as much area as a typical NPN transistor, and an extra pin may make a low-cost 10-pin version of a device unobtainable). On-chip resistor trimming is feasible (and is used successfully in production of the AD532), but it does add to production cost. Thus, the considerations involved in minimizing worst-case tolerances take on an added importance in monolithic design.

But IC's pose opportunities as well as problems. The advantages for the user are low cost, small size, and high reliability. For the designer, they include:

1. Low incremental cost for adding useful circuit refinements. The addition of even a single matched transistor pair to a discrete design would require very careful consideration and would be expected to increase both the material- and the labor- cost of the module noticeably. In contrast, the inclusion of additional monolithic transistor pairs would not make

a substantial difference to the cost of either materials or labor; it could even lower the total cost by ensuring that a greater proportion of the chips (i.e., yield) performed properly. For example, quite complex circuits can be used to generate a precise voltage reference to replace the calibrated zener diode used by the module designer. In fact, there is every possibility that such circuits, pioneered by engineers battling with the apparent "limitations" of the monolithic medium, will become the preferred type of voltage reference for future discrete modules.

2. Another advantage of the monolithic approach is, ironically, a result of the very close proximity of components on a chip that also causes the difficulties listed under (2) above. It is the unexcelled matching and temperature-tracking of devices *as processed*, that is, without the need for selection. As the technology has matured, these factors are tending to become a secondary source of inaccuracies, rather than the dominant source that they once were. Thus, trimming and testing costs of the monolithic product will continue to decline, in relation to the available performance.

So much for the generalities. Let us now take a closer look at the way a monolithic circuit designer approaches the operational requirements.

COMPARING CIRCUITS

Suppose we wish to design a 4-quadrant multiplier of medium accuracy (say, 1%) having a small-signal bandwidth of about 1MHz. We might begin by listing some of the techniques available and consider their properties in relation to the feasibility of monolithic fabrication.

Quarter-Square. This method requires two precision squaring devices and several operational amplifiers to generate sums and differences. Once the classical approach to precise high-speed multiplication, it does not have much to commend it for monolithic implementation; in any event, the technique is practically obsolete, even for module designs. In fact, low-cost multipliers are now perhaps the best way of performing the squaring operation!

Pulse Modulation. This method would require very high carrier frequencies to achieve 1MHz signal bandwidth and is rather difficult to adapt to 4-quadrant operation. Also, several connections to external capacitors would probably be necessary to provide various timing and filtering functions.

The Hall Effect. It is possible to make Hall plates* with standard NPN processes, but they are not optimum, and the output product is a signal at the millivolt level ("The Hall Effect is a small effect"). A rather strong magnetic field that responds to one of the variables is a necessity; and the two input channels have greatly differing dynamics.

Antilog-of-sum-log (Chapter 3-2). This method can certainly be used in a 1-quadrant monolithic realization. Some objections, however, are that the full four-quadrant capability calls for a substantial increase in complexity over the 1-quadrant case, speed is a function of signal amplitude, and a relatively-large number of amplifiers are needed.

Current-ratio (*linearized variable-transconductance*). This method is uniquely suited to the medium and is used in practically all the monolithic multipliers on the market today. It is simple, basically temperature-insensitive, and operates on differential current signals to achieve four-quadrant operation directly. Bandwidth is inherently wide; it is limited in practice only by the output amplifier, which uses lateral PNP transistors to achieve level-shifting.

DESIGN CONSIDERATIONS

Let us now sit in the designer's chair and see how the monolithic aspects affect his approach to applying the current-ratio cell. The basic core has already been described (Chapter 3-2), but it is instructive to recast it slightly to emphasize certain points.

Figure 1 shows a typical monolithic version; many others are possible. Notice that the two "linearizing diodes" are actually only a

*The Hall Effect: If current flows through a conductor and a magnetic field is applied at a right angle to it, an orthogonal voltage will be developed across the conductor proportional to the product of the current and the flux density. The proportionality constant ("Hall Coefficient") is a function of the material and the conductor's configuration.

pair of emitters in a common collector-base region. This is one of the ways the designer reduces the chip area; an extra emitter can consume as little as 5% of the area required by a fully-isolated transistor. A similar conservation of space is possible by noticing that pairs of output transistors Q4-Q6 and Q3-Q5 share common-collector regions. Again, if separate transistors are replaced by putting two emitter-base structures in a single collector region, the area increase is only about 30%.

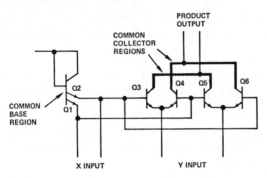

Figure 1. Circuit of monolithic multiplier cell. Compare this with Q1, Q2, Q3 in Figure 11, Chapter 3-2.

These six junctions are placed close together on the chip and must have exactly-matched emitter areas. Often, large emitters are used to reduce the effects of process tolerances. However, there is a need to compromise, because although isothermal matching may improve with emitter size, the larger devices become more widely separated, increasing the probability of increased sensitivity to thermal gradients on the chip; it can be shown that each centigrade degree between certain pairs can introduce 0.07% distortion. Another reason for limiting the emitter area is to maintain a reasonable cutoff frequency.

Area-matching in the multiplier core is of paramount importance, for two reasons. First, it can be shown that a 1mV offset between these junctions can cause a parabolic nonlinearity of about 1% of full scale to be superimposed on the product. Second, offsets introduce large zero-errors, referred to the inputs, because of the large amount of degeneration needed to handle ±10V signals; the ratio is about 200, so a 1mV transistor mismatch becomes 200mV at the X or the Y input (or both — it all depends on where the mismatches arise). Furthermore, this zero-error has a kT/q kind of temperature dependence, amounting to nearly 0.7mV/°C in the example just given. High-quality processing and thoughtful layout have done much to take the sting out of these problems, and new circuit tricks are being added to the designer's repertoire which permit him to completely eliminate all but a trace of residual mismatch.

Another source of errors in this simple cell, which seldom confronts the module designer, arises out of the difficult topology and attendant interconnection problems of the circuit (try to connect everything up on paper without introducing any crossovers; and you must get all input- and output nodes outside the circuit). This is not the place to delve into the interesting ways of dealing with this problem; but as an example of the care that must be exercised, consider the consequences of using a rather circuitous route for one of the aluminum interconnects. At 50mΩ per square, a length difference of 20 mils (1/2mm) in a 0.5 mil (1/80mm) track amounts to a resistance inequality of 2Ω. If this conductor is carrying 1mA, a differential voltage of 2mV is generated. Inside the critical junction loop of Figure 1, such a mistake would ruin any chances of achieving distortion levels of even 1%.

Avoidance of excessive ohmic drops is also a contributing factor to the choice of sub-milliampere operating levels for these transistors. The effect of these drops on linearity explains why VHF multipliers, which need to operate at much higher currents to maintain bandwidth, usually exhibit inferior linearity to instrumentation circuits.

Compensation for errors introduced by finite current-gains (beta) provides another example of the way in which the monolithic designer can take advantage of the medium. In this case, it is that betas of devices all over the chip tend to match up and to track with temperature. It can be shown that the circuit of Figure 1 introduces a scale-factor error which is three times the α error. That is, if Q3 through Q6 had α's of 0.99 (Beta = 100), the overall gain error would be –3%. Although this could be initially trimmed out, any variation of beta, especially at low temperature, would introduce a scale-factor shift. Fortunately, state-of-the-art processes consistently produce transistors having betas of several hundred, so this is not a severe problem. High-accuracy multipliers can take advantage of beta-tracking (and of the current-ratio principle) to employ compensation circuitry which maintains a very stable scale factor, even over the military (–55°C to +125°C) temperature range.

This has been just a brief look into some of the ways the design of monolithic multipliers differs from that of their discrete counterparts. A full treatment of the subject has never been published, and possibly never will, because of its highly-specialized nature. Most of the topics just discussed in relation to multipliers apply just as well to other nonlinear circuits based on logarithmic junctions, particularly those involved in ratio and power-function generation. The comparison with discrete circuitry is equally pertinent.

SPECIFICATIONS AND CHARACTERISTICS

Since both IC and modular nonlinear function circuits share common design principles and are used for the same kinds of applications, and since the operational guidelines for most IC nonlinear devices were set in terms of their modular forebears, it is not surprising that few differences exist in their specifications. In fact, if one bears in mind that (except for the AD532, at this writing) the IC device is specified under externally-trimmed conditions — usually the adjustment of four variable resistors (for those devices that include the output amplifier) — then the only differences are those that exist between any IC and the equivalent modular function (warmup time, power dissipation, size, cost, etc.) Therefore, the definitions, established in Chapter 3-2, of feedthrough, nonlinearity error, gain error, accuracy, etc., apply equally to both IC and modular function-circuits.

Modular circuits, using pulse-width/amplitude modulation principles, multiply with maximum errors below 0.1%; the best of IC's (at this writing — it is necessary to reiterate this *caveat*, because of the rapidly-expanding limits of the IC state-of-the-art), using the transconductance principle, guarantee that multiplying errors will be less than 0.5%. However, the best IC has a typical accuracy (error) vs. temperature specification of 0.01%/°C, due to the excellent temperature tracking inherent in IC construction (a comparable number to that for modules). Modular multipliers offer small-signal bandwidths of 10MHz and minimum slewing rates of 120V/µs; the best performance of a current-output (incomplete) IC multiplier, over a limited range, is 6MHz and 30V/µs, and for voltage-output IC's 1MHz and 45V/µs.

There are on the market several nonlinear IC devices that essentially provide only the basic transconductance multiplying function, with a resulting low-level current output signal that requires the addition of an amplifier and a number of passive components to achieve a usable signal level. This externally-applied circuitry provides its own sources of error (as well as cost), the amount of which depends to a great extent on the choice of active and passive components and the circuit-design virtuosity of the user. In this case, then, the manufacturer cannot provide a guarantee of overall accuracy, but must limit himself (in a manner similar to that of general-purpose, many-degree-of-freedom devices, such as op amps) to specifying individual parameters. It is important to note that Analog Devices has principally chosen to opt for committed circuitry: all AD IC multipliers are complete-on-a-chip circuits, including the output amplifier, and are thus guaranteed for overall performance.

NONLINEAR IC'S vs. IC OP AMPS

Many readers of this book will reach this point with a background of experience in the applications of functional devices, whatever their mode of manufacture (from "bottles" to chips). But there are also many whose experience with analog ("linear") circuits is pretty much limited to IC op amps. It may be helpful for the latter to consider the following comparison of qualitative differences between nonlinear and "linear" analog integrated circuits.

In the vast majority of both nonlinear and linear (i.e., op amp) applications, accuracy of signal reproduction is the specification of greatest concern. Due to the many degrees of freedom of op amp circuits, however, it is rarely that the manufacturer can satisfy this concern with a single specification that guarantees the user his required accuracy. The specifications that contribute to overall accuracy: input bias current, offset current, input voltage offset, voltage-offset drift, CMRR, etc., all add up to different errors, depending on the particular way (of many) in which the op amp is used. Thus, the very-general nature of IC op amps hinders the manufacturer in any attempt to guarantee an overall accuracy specification. He must constrain himself to specifying such parameters as I_b, E_{os}, Gain, CMR, etc., and leave the determination of overall accuracy to the user.

Nonlinear function IC's, on the other hand, perform rather specific tasks. Though their applicability is wide: automatic gain control, true rms, vector summation, absolute value, ratio measurement, etc., in almost all cases they are hooked-up in the same way, and perform the same function, be it multiplying, dividing, squaring, or square-rooting. The multiplicity of degrees-of-freedom, that prevents the IC manufacturer from providing a relevant overall accuracy specification for op amps, does not exist in the case of nonlinear-function IC's. Thus the user is freed from the often-laborious and sometimes confusing requirement that he calculate the circuit's worst-case accuracy error before considering the many parameters that must be traded-off. Except for instances where he is seeking performance levels considerably better than the overall specification, the need to understand, interpret, and calculate the effects of such partial specifications as feedthrough, nonlinearity, and scale-factor error (which are provided nevertheless) is eliminated by their measurement and inclusion in the guaranteed overall accuracy specification.

As noted earlier, though, overall accuracy can be truly guaranteed only when the signal that is the result of the particular nonlinear manipulation is received at the output of the device at a level that requires no further processing or amplification (excluding external trims, the effects of which are included in the specification). The point is that partial-multiplier circuits must be treated as many-degree-of-freedom devices, and the overall performance calculation must include the effects of components added externally, but not included in the manufacturer's specification.

TESTING AND SELECTION

The test circuits given for multipliers in Chapter 3-2 are universally applicable from a technical point of view. However, where multipliers are produced (or consumed) in volume, the integrated-circuit device, with its low cost and high production volume, has made it necessary to turn to computer-controlled automated testing systems, in order to keep the costs commensurate with manufacturing costs by less handling and higher throughput. The automated test system can make more measurements, in a shorter period of time, sort the devices into a number of different categories, and do it without the need for skilled test technicians in routine operation.

A typical bench test of a modular multiplier may involve the setting-up of three pieces of equipment, inserting the module into a socket, manipulating a number of switches (or interconnections), making (say) 10 measurements, and classifying the module (if that particular module has more than one classification) — a procedure which may consume one minute (or more) per device.

Temperature testing requires loading the modules onto boards, which are then put into an environmental chamber and adjusted at room temperature. The chamber ("oven") is then brought to each test temperature, and more measurements are made. Temperature testing thus requires an additional minute per device per temperature level, plus setup time and stabilization time.

On the other hand, the automated test system, after an initial investment in (equipment and) time for writing and debugging the program, can perform about 5 times as many measurements in as little as two-to-three *seconds* per device. When combined with an environmental chamber and data storage, such as magnetic tape, the temperature testing and drift testing (which involves calculating the differences between the same tests at two different temperatures) are also performed in a very short time (about 1 second). In addition to the critical parameters, the high rate of testing allows checking of less-important (but still guaranteed) parameters such as power consumption, bias currents, output swing into a load, and maximum ratings.

The use of automated handling equipment, interfaced to the test system, provides for more-efficient use of the test system by reducing insertion time to less than 1 second.

Automated testing, however, is not a panacea; it has its problems and limitations. One of these is that, with such fast testing, the device does not have enough time to reach its normal operating temperature. This can usually be compensated for by testing to a tighter limit than that which is guaranteed. This obviously necessitates discarding some devices which otherwise might meet all of the specifications (or placing potential premium types in a lower-price category), but the saving in test time far outweighs the yield loss caused by insufficient warmup time.

The major difficulty in testing multipliers is performing the three null adjustments and setting the scale factor ($1/V_r = 0.1/V$), in order to be able to perform the accuracy, linearity, and feed-through measurements. The basic technique used with the automatic system is to vary each null voltage by means of a programmable dc source, in response to a successive-approximation routine written into the software, while the measurement system monitors the output. When the proper output level is reached, the null voltage is stored by a sample-hold, the programmable source is switched to the next null terminal, and the process is repeated.

For gain tests, rather than adjusting the scale-factor by setting an attenuator on the Y input (of the AD530 or AD533), as a user might do, the automatic system adjusts the programmable signal source to the Y input voltage required to give unity transfer from the X input to the output. This value of voltage is then used later to establish the Y input during the various tests that call for the full-scale value of Y.

To avoid an iterative null routine, where the nulls must be repeated, or fine-tuned, the adjustments are performed in the proper order, viz., Y feedthrough ($0 \times Y$, trim X_{os}), X feedthrough ($X \times 0$, trim Y_{os}), E_{os} (0×0, trim V_{os}), and scale factor.* It is accomplished by using a sample-and-difference technique to measure the output-voltage *changes* in response to the input signals, while ignoring the output magnitude (untrimmed offset voltage)

*This process straightforwardly zeroes the output, including the effects of both the Z_{os} and $X_{os}Y_{os}$ terms in equation 13, Chapter 3-2.

during the adjustment of "linear" feedthrough (see Figures 34 and 35, Chapter 3-2).

For example (Figure 2), in nulling the Y feedthrough, the X_o S-H is gated into the *track* condition, essentially connecting the programmable voltage source to the X_o pin of the multiplier. At the same time, the X input is set to zero, and the Y input is alternately switched from +10 to –10V. The output (+Y) is subtracted from the output (–Y). When the difference is zero, the successive-approximation routine is stopped, and the X_o S-H is put into *hold*. The differencing technique eliminates dc offsets and also prevents the untrimmable component of feedthrough, due to multiplier nonlinearity, from affecting the null.[†] The whole null procedure, involving all four adjustments, takes about 1 second to perform.

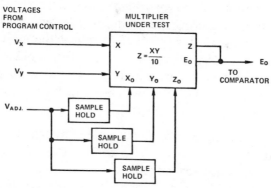

Figure 2. Connections to multiplier in automatic test circuit

For temperature testing, the measured 25°C null voltages are stored on magnetic tape and later played back to establish proper 25°C null conditions for each multiplier, as it is tested at the extremes of its rated temperature range. This technique closely simulates the user's circuit conditions (pots set at 25°C), while allowing the use of automatic handling equipment with resulting throughput of the order of 500 devices per hour. With this test setup, performance over the temperature range can be both guaranteed and verified by testing of 100% of the devices.

Test systems such as this, though few and far-between (at this writing, it is believed to be unique with Analog Devices), account

[†]Feedthrough nonlinearity is quadratic for these devices, as noted in Chapter 3-2.

for the excellent quality/price ratio of complete monolithic multipliers, and for the maintenance of applications-oriented specifications.

DYNAMIC TRIMMING

Until the introduction of the AD532, integrated-circuit multipliers required several external components to allow adjustment to within the specified tolerances. Such adjustments are costly for the user to implement (compared to the basic price of the device), and they introduce additional potential for thermal and accidental errors.

By the combination of the micro-machining capability of a laser with an automatic measuring and positioning system, it has become possible (and economically feasible) to adjust the multiplier offsets and scale factor by changing the values of on-chip thin-film resistors. The result is an integrated-circuit multiplier which can be plugged in and turned on, with no adjustments or external components required.

Figure 3 is a block diagram of an automated dynamic-trim system developed at Analog Devices. The coordinates of the starting- and stopping-point of each resistor to be trimmed are stored on a continuous loop of punched tape. The *device position control* reads the tape and positions the device under the laser beam by applying a number of pulses to digital stepping motors driving an X-Y table. The *power and control logic* module applies power and input voltages to the device in a sequence determined by the logic. It also conditions the output of the device and feeds it to the *measurement and trim control system,* which determines when the output of the device has reached the desired value; it then, via the control logic, turns off the laser beam. The measurement approach used is similar to that described above in relation to the automatic null and test system.

The laser beam, which is coherent light concentrated in a small area to generate a high energy density, oxidizes part of the resistor, causing its resistance to increase permanently.

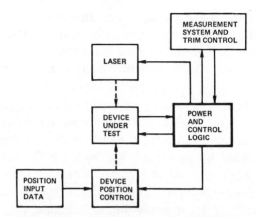

Figure 3. Laser-trimming system configuration

Figure 4 is a simplified schematic of a multiplier cell. In normal operation, a constant voltage is applied at the −X input to adjust the offset of the X-input transistor pair. In dynamic trimming, the X inputs are held at zero volts (so is the −Y input), while the +Y input is switched between a specified ± voltage-pair. The laser then increases the resistance of either R1 or R2, which adjusts the current balance in the stage for minimum linear feedthrough. This is measured by phase-sensitive chopping, and filtering of the device's

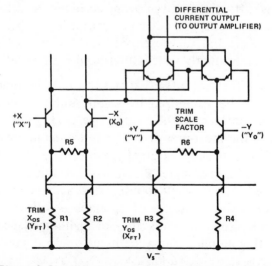

Figure 4. Simplified schematic of multiplier input circuit showing feedthrough and scale-factor trim resistors

output; the laser is turned off when the output of the filter is zero, indicating equal feedthrough at both input levels. The feedthrough for the +X input is adjusted in a similar manner by holding the Y inputs and –X at zero and increasing the resistance of either R3 or R4.

The offset voltage is zeroed by adjusting one of a pair of resistors in the on-chip output amplifier. All inputs are set to zero volts, and the resistance is increased until the output reaches zero volts. The scale factor is set by increasing R_6, which is deliberately made slightly low, while the errors are monitored for maximum input values in all four quadrants. The resistance is increased until the total four-quadrant error is minimized. Because this adjustment affects the offsets, it is necessary to repeat the offset trimming a second time (fine trim) in order to insure the best-possible yield vs. cost.

Once the device has been plugged in and aligned to the X-Y table, the trim procedure is completely automatic.

APPLYING: PRACTICAL USE OF IC'S FOR BEST RESULTS

HINTS

While especial attention is given to the accuracy, feedthrough, and linearity specifications, and their minimization, the best efforts of the manufacturer, and the cost paid by the user, may be wasted if the device is improperly scaled, interfaced-with, or adjusted.

Scaling. The effects of offset drift and cubic nonlinearity and feedthrough can be minimized by scaling inputs and outputs to give full-scale voltages at their respective peak values. Parabolic nonlinearity is the most-prevalent type in transconductance multipliers, and it can usually be greatly-reduced by the methods of Figures 14 and 15 in Chapter 3-2. Low-level input signals should be preamplified to, say, ±10V to insure best accuracy and dynamic range.* Figure 5 shows the use of op amps as followers-with-gain to provide both isolation and scaling of high-impedance low-level signals. Inverting amplifiers may be used to adjust gain in applica-

*Note that if both inputs are scaled to half-scale, instead of full-scale, 3/4 of the dynamic range is wasted. If the output of the circuit must be only 1/4-scale, it is better to use an additional attenuator following the multiplier's full-scale output.

tions where the polarity relationships require inversion, for example, if a divider requires negative denominator input from a positive signal.

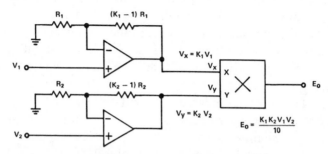

Figure 5. Op amps as followers-with-gain to scale multiplier inputs to ±10V

Buffering. If either signal source is at high impedance, the input impedance of a multiplier may result in significant loading error. This is composed of specified resistance levels, bias current (and its variation with time and temperature), and the added shunt resistance of any resistive voltage dividers introduced for scaling purposes. In such cases, the source should be buffered by a follower-connected op amp, especially if the amplifier can also provide needed amplification.

Trimming. Most multiplier data-sheets provide trim procedures for optimizing performance in various modes: multiplication, division, squaring, rooting, etc. For some modes, certain trim adjustments can be eliminated. Trims may even be incorporated into adjacent portions of the circuit, if appropriate, and the device trim-terminals grounded, to minimize overall error of a *circuit* containing the multiplier.

The use of cross-feedthrough trims has been mentioned early in the chapter, as a means of squeezing the best performance out of a low-cost device. Usually, only an X trim is necessary; Y feedthrough is close to the point of diminishing returns.

Since internally-trimmed devices are almost never exactly "on" the design-center setting, performance can be improved in critical cases by introducing external trims. Trimming may be used preferentially to improve performance in those quadrants where accuracy is most important.

Finally, if the multiplier is a crucial element of a very high-precision circuit, calling for overall error less than 0.1%, it may be feasible to keep a high-accuracy multiplier at constant temperature, map out its error surface, and use several low-cost multipliers in a function-fitting configuration (Chapter 1-1) to simulate the error surface and subtract it from the output. In the past, such a suggestion could be considered a speculative fantasy, but with the low cost of today's multipliers and the availability of high-powered computing techniques to test conceptual models, the cost and time involved may be by no means prohibitive.

KINKS

Power-supply decoupling capacitors are often built into discrete-component modules; but it is not practical to include them on an IC chip. As in the case of IC operational amplifiers, it is good practice to use them routinely, especially for designs involving high overall gain in the circuit. Bypass capacitors, at each device, should be located as close to the device as practicable, but never through a switch or circuit-board edge connector, which can introduce undesirable series inductance. A typical configuration is a 0.01μF ceramic, in parallel with a 1-10μF tantalytic, from each side of the supply to common.

Some care may be necessary in locating offset-adjustment potentiometers. Since long lead lengths may introduce undesirable effects, the adjusting potentiometer should be as close to the device as feasible.

Integrated circuits can be particularly sensitive to capacitive loading. While good design dictates that general-purpose devices should be (and ADI's are) capable of driving 500 to 1000pF worth of capacitive loading at any output level without modification,* it may be wise to include some resistance in series with the load (Figure 6a) or to add a suitable buffer (Figure 6b) whenever

---

*Not all manufacturers see this in the same light

larger capacitive loads are anticipated. For the AD530, 531, 532, and 533, the series resistance should be of the order of 100Ω. (Caution: when the nonlinear function cell and its amplifier are connected in a feedback configuration to obtain the inverse function, for example, the square-root (AD530, 532, 533), the capacitive loading capability may be adversely affected because the loop gain is doubled at high levels.)

A word about warmup shifts! The large thermal mass and greater power-dissipation capability of modular function circuits generally results in cooler operation and smaller warmup shifts. IC function circuits, on the other hand, may have a large power-to-volume ratio, depending on the circuit complexity, and can run quite warm. Thus, the user should allow a 5-10 minute warmup time (in free air) before performing final trim adjustments or measurements of device accuracy to specifications. Warmup effects can be reduced by perhaps an order-of-magnitude if the device is mounted in a heat sink.

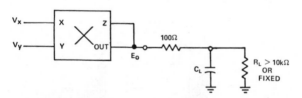

a. Isolating I.C. multiplier from large capacitive load by fixed resistance

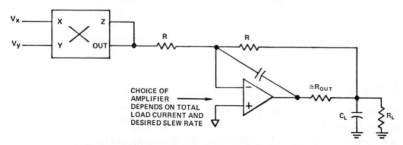

b. Isolating multiplier from load by buffer op amp

Figure 6. Isolating large capacitive load to prevent multiplier from oscillating

PITFALLS

Unlike discrete modular devices, the inherently high reliability of integrated circuits can be easily compromised by applications abuses. As shown in Figure 7, the "circuit board" for IC's consists of a P-doped substrate with P-doped isolation barriers that are biased at the negative-supply voltage level to provide the necessary isolation between circuit elements. When the power supply is *off*, the isolation is reduced and the chip is left vulnerable to the voltage stresses present at its input(s) and output. It is for this reason that most IC manufacturers state that the absolute maximum input and output ratings are $\pm V_s$, which strongly implies 0V with $\pm V_s = 0$.

Some IC's, such as the AD531, incorporate input-circuit protection. Many don't. Worse than failing, unprotected input transistors may simply change their parameters, a change which can be manifested in terms of increased bias current (decreased beta) or voltage offset.

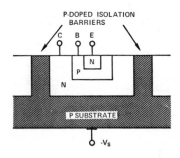

Figure 7. Bipolar I.C. construction, using junction isolation

Although failures are not generally a problem when the inputs are less than ± 1V, and are never a problem when V_{IN} is less than $V_s \pm 0.5$V, higher input levels can be accommodated by using input/output protection circuits, such as those shown in Figure 8. While the 1kΩ series resistors alone will often suffice, the diode clamps to the supplies provide certain protection, except when the supplies have been switched off.

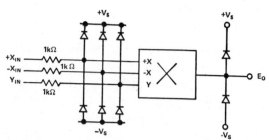

Figure 8. Protection of I.C. device against overvoltage. Diodes protect against voltage spikes, and resistors protect against power-off stresses.

ON THE HORIZON — AND BEYOND

The field of IC nonlinear devices is a largely virginal field within the rapidly-expanding analog ("linear") IC industry. As recently as one year ago, only one self-contained 4-quadrant multiplier[1,2] (AD530) and several multiplier cells had been introduced. It was noted earlier that while monolithic matched-transistor techniques are ideally suited to transconductance and log-antilog circuit techniques, the missing complement — on-chip precision resistors — have only been available for the last three years. The utilization of precision thin-film monolithic resistors, in conjunction with the superbly-matched devices that are characteristic of silicon semiconductor technology will in the long run make available scores of new circuits, both linear and nonlinear, in addition to the three basic families of multipliers (AD530, AD531, AD532) now available, and the AD7520 multiplying D/A converter. Laser trimming (exemplified by the AD532, which permits operation to rated accuracy without *any* additional components) adds a key capability to the nonlinear IC device-manufacturing process, *adjustability*. With effective utilization of this capability, the only deficiency that appears to pose fundamental limitations is speed.

Even the speed frontier can be crossed by going to dielectrically-isolated, dynamically-trimmed IC's. With such combined process-

[1]"Self-Contained I.C. Multiplier/Divider," by R.S. Burwen, *NEREM 1970 RECORD*, p. 56, Boston Section IEEE.

[2]"A Complete Multiplier/Divider on a Single Chip," by R.S. Burwen, *Analog Dialogue*, Vol. 5, No. 1, January, 1971.

ing capabilities, 10MHz, dynamically-trimmed, <0.5% multipliers (complete-on-a-chip) become possible.

However, even before such combined techniques are ultimately exploited, circuit- and device-design ingenuity will make available a wide range of analog circuits that rely for precision on monolithically-matched active devices and dynamically-trimmed deposited thin-film resistors. For instance, it appears likely that careful exploitation of currently-available technology will soon make available multipliers with errors in the 0.1% range, using variable-transconductance techniques.

Other more-specialized multipliers will also appear for limited-polarity, wide-dynamic-range, high-accuracy applications. New IC device structures, not yet put into practice commercially (and which depart radically from conventional circuit concepts), hold out the prospect of marked improvement of one of the fundamental limitations of transconductance techniques − noise.

Along with refinements in the multiplier/divider art, IC's will soon provide high-accuracy log-antilog operations, with the potential of growth comparable to that witnessed by the industry in multiplier/dividers (and still in the accelerating stage). The first of such products will be available at the time this book sees the light of day. Such circuits are eminently integratable. Also, more-complex functions that depend on multiplication, division, or log operations, e.g., true-rms, vector-sum, and precision AGC loops, will soon appear as single chips.

Thus, the future of "nonlinear" "linear integrated-circuits" is exceptionally bright. A rough analogy can be established between the development of these basic nonlinear circuits and the history of op amps (the basic linear circuit) as follows: the "multiplier cell" techniques brought us to the 709 level, while the AD532 has brought us to the 741 stage of development, the point at which op amps became practical for widespread application. It is not unsafe to predict that the wave of op amp development that followed the 741 and led to hundreds of different IC amplifiers for virtually every application will be paralleled by a similar rapid growth in the usage and proliferation of nonlinear IC's.

Discontinuous Approximations

Chapter 5

In this chapter, circuits used for obtaining piecewise-linear input-output relationships will be discussed. Examples of such relationships include piecewise-linear function fitting, absolute value, dual-mode linear responses, and sign-magnitude-to-bipolar conversion. There will also be a brief discussion of digital aids to analog function fitting.

Often, a nonlinear response can be conveniently simulated or linearized by a circuit with a response that is neither curvilinear nor linear, but is built up into an approximation of the desired response by changing the gain of the circuit as input (or output) thresholds are crossed.

Piecewise-linear approximations to curvilinear functions were introduced in Chapter 2-1 (Figures 15 and 16), and an application to a linearization problem was plumbed in Chapter 2-3 (Figure 7). To ensure high accuracy and sharp breakpoints (the latter is not necessarily an advantage), operational amplifiers and diodes in an "ideal diode" configuration were suggested; but they are not the only way. This chapter considers the problem in a somewhat wider circuit perspective.

BREAKPOINTS

If a diode can ideally be considered as a polarity-sensitive switch having zero voltage drop when positive voltage is applied, and zero leakage current when negative voltage is applied, it can be used with bias voltages and resistance networks to construct an op-amp circuit with a controlled nonlinear response.

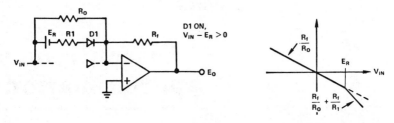

a. Series breakpoints — input

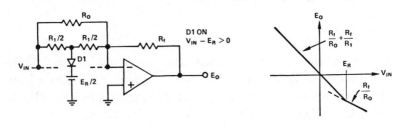

b. Shunt breakpoints — input

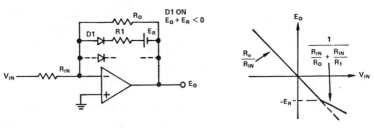

c. Series breakpoints — feedback

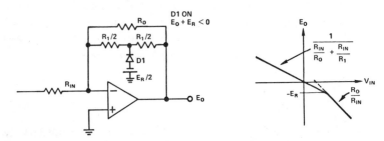

d. Shunt breakpoints — feedback

Figure 1. Series and shunt breakpoints generated in input and feedback circuits of single op amp

Figure 1 shows the input output-relationships for circuits having ideal series and shunt breakpoints, in the forward and the feedback paths. In all cases, the diode and reference are biased and circuited to cause the diode to conduct when the input has increased beyond the threshold.

The threshold, referred to the input, is equal to E_R for the diode in the forward path, and E_R, divided by the gain, for the feedback path (in the plots, it is simply referred to the output).

The degrees of freedom include: series vs. shunt circuits, choice of the input vs. the feedback path, and choice of polarity of both reference voltage and diode connections, as well as the choice of resistance ratios.

For additional breakpoints, similar circuits, with graduated values of E_{Ri}, are connected between the summing point and either the input or the output. The series connection is often preferred, because it is easier to get extremely high resistance in the *off* condition than negligibly low resistance in the *on* condition. However, the shunt circuit has the advantage that the reference source can be grounded (hence driven by an op amp output if a variable threshold is desired).

If the desired function does not go through the origin, an additive bias can be summed at the amplifier's summing point, via a resistor, to relocate it. Its effect can be viewed as a translation of the function along the input or the output axis, depending on the location of the breakpoints.

Additional monotonic shapes are available by changing the polarities of the reference and/or the diodes. Figure 2 shows the ideal responses available with different combinations of polarity in the series input case. Figure 3 shows the effect of reference polarity with multiple slopes. Note that zero output at the origin is obtained when the reference and the diode are of opposite polarity (input zero); if the reference and the diode are of the same polarity, the *extrapolated* R_f/R_o response passes through the origin, but the break occurs before it reaches zero. If zero output is desired, a bias can be added at the summing point, as mentioned above. A complete array of shapes can be catalogued by considering all combinations of polarity, series vs. shunt, and input vs. feedback.

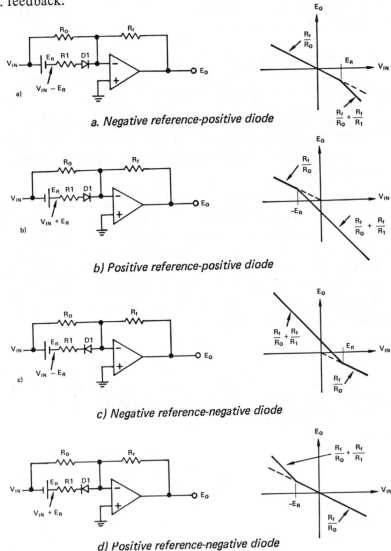

a. Negative reference-positive diode

b) Positive reference-positive diode

c) Negative reference-negative diode

d) Positive reference-negative diode

Figure 2. Effect of reference and diode polarity on input-output relationship; series breakpoints, input circuit

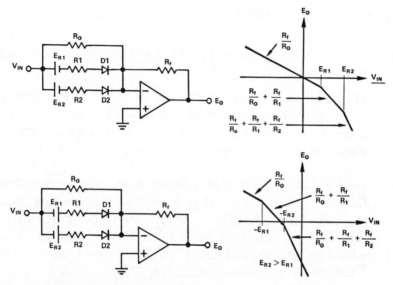

Figure 3. Effect of reference polarity on multiple slopes

The above-mentioned shapes all have monotonic derivatives, i.e., they are all concave-up or concave-down. However, by combining degrees of freedom, it is possible to obtain reversals (and ideally, many reversals, depending on the references and gains).* Figure 4 shows a single reversal, employing series input and feedback. Note that an input bias is employed to obtain zero output for zero input.

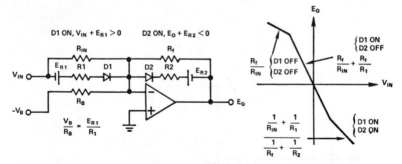

Figure 4. Use of input and feedback breakpoints to obtain reverse slopes with single amplifier

---

*But the incremental slope is always negative unless the + input is driven.

PRACTICAL CIRCUITRY

There are several problems in putting the ideal circuits mentioned earlier into practice:

1. The references are shown floating. It would be desirable to have them derived via resistance networks, starting with a single (preferably grounded) reference. This would allow a good, stable power supply to also serve as the reference.

2. Diodes have appreciable voltage drop when conducting (about 0.5^+ V). Instead of their having a clear-cut threshold, the behavior is logarithmic. Over a 100:1 range of current variation, the threshold shifts by at least 120mV at 25°C, with a thermal shift of about 2mV/°C. From 1mA down to 10μA, series resistance increases from 25Ω to 2500Ω.

Figure 5a shows the basic elements of a circuit that can function in either the forward or the feedback path of an operational amplitier. It answers objection 1. The ideal threshold of conduction is at $V_{IN} = E_R \, R_1/(R_1 + R_A)$ for the first break, $V_{IN} = E_R \, R_2/(R_2 + R_B)$ for the second break, etc. The incremental gains are R_f/R_o, R_f/R_1, R_f/R_2, etc. Thus, the break points depend on the reference. Diodes D_A and D_B limit the reverse swing of D1, D2, etc.

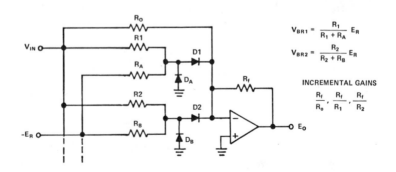

a. Uncompensated-input circuit with grounded reference

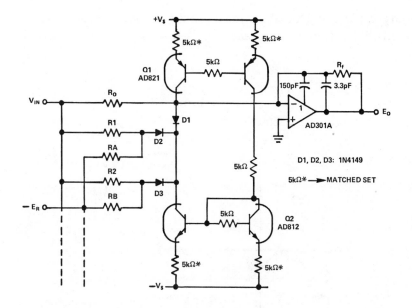

b. Compensated high-speed piecewise-linear function fitter

Figure 5. Shunt-biased, series diode piecwise-linear function fitter a) in principle, b) with first-order temperature compensation

Figure 5b shows the same basic circuit, but with the added benefit of first-order compensation for the diode drops and their variation with temperature.

If the breakpoint must be sharp and located precisely, with near-zero drift, an "ideal diode" feedback circuit, employing diodes in the feedback path of an operational amplifier for polarity sensing only, may be used as a dual-mode circuit (Figure 6). In the circuit of Figure 6, if the net input current $V_{IN}/R_1 + V_R/R_4$ is positive, diode D1 is turned on, D2 is turned off, and the output (1) is

$$V_1 = -\frac{R_3}{R_1} V_{IN} - \frac{R_3}{R_4} V_R \qquad (1)$$

If output (2) is unloaded, or loaded only to common (or a virtual ground), its output is zero, since D2 is turned off and no current flows in R2.

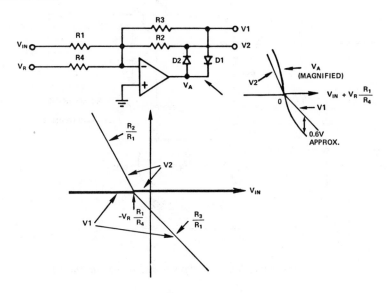

Figure 6. Ideal diode circuit

If the net input current is negative, D1 is turned off (output (1) = 0), D2 is turned on, and output (2) is

$$V_2 = -\frac{R_2}{R_1} V_{IN} - \frac{R_2}{R_4} V_R \qquad (2)$$

Thus, the output of the circuit at (1) is zero for all V_{IN} less than $-V_R (R_1/R_4)$ and proportional to positive values of the difference; output (2) is zero for all V_{IN} greater than $-V_R (R_1/R_4)$ and proportional to negative values.

Either output can be used as an incremental "ideal-diode" breakpoint in a system which sums a number of breakpoints;[1] the

[1] See Figures 15 and 16, Chapter 2-1 and Figure 7, Chapter 2-3.

two outputs ($R_2 \neq R_3$) summed together provide a two-slope response, with the break at $-V_R(R_1/R_4)$. A bias can be summed into the summing amplifier to offset the break point (Figure 7).

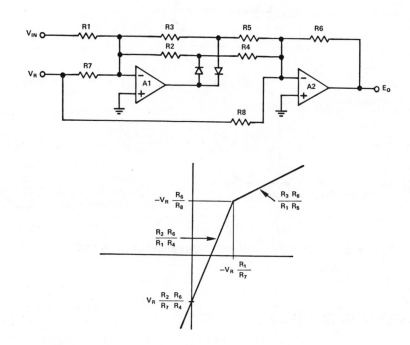

Figure 7. Ideal diode in high-accuracy dual-mode circuit

CURRENT SWITCHING

The circuit of Figure 8 is an "ideal diode" with current output. When V_{IN} is greater than V_R, the amplifier output drives the gate of the enhancement-mode FET with whatever voltage is necessary to maintain the voltage across R equal to $(V_{IN} - V_R) = I_R$. When V_{IN} is less than V_R, the FET shuts off, and the diode conducts away the reverse current from V_R through R to maintain the voltage at R equal to V_{IN}.

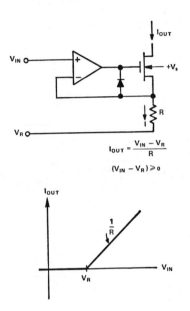

$$I_{OUT} = \frac{V_{IN} - V_R}{R}$$

$$(V_{IN} - V_R) \geqslant 0$$

Figure 8. Unidirectional linear voltage-to-current switch

SPEED IMPROVEMENT

All of the above "ideal-diode" circuits are relatively slow: at the break point, the output must slew through a dead band of two diode drops; and charge stored in a capacitive load requires time to discharge through the feedback resistor in the *off* condition. Though the circuit has a low dynamic impedance while conducting, it must contend with greatly-reduced loop gain while switching at high speeds—first, because of the amplifier's reduced open-loop gain, second, because of the increased loop attenuation due to the high diode impedance near zero.

In the circuit of Figure 9, the switching diodes have been replaced by transistors Q2 and Q1, which saturate when the net input current becomes positive or negative, respectively, and maintain the output near ground. The circuit is considerably faster than the diode circuit, because the output impedance is always quite

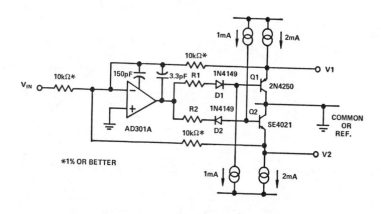

Figure 9. Low-output impedance "ideal-diode" circuit

low—in the saturated condition, it is about 5Ω, and in the active region, it is even lower, about 0.01Ω. D1, D2, R1, and R2 can be adjusted to minimize the dead zone. The output offsets of Q1 and Q2 in the saturated condition will be about 10mV.

This circuit can be connected as a full-wave mean-value circuit by connecting a differential low-pass (R-C-R) filter across the two outputs, and subtracting to obtain the average of the rectified input. It will provide reasonably accurate measurements at frequencies as high as 100kHz.

ABSOLUTE-VALUE CIRCUITS

Many uses of absolute-value circuits have been discussed in earlier chapters. They include measurements of magnitude, conditioning of signals for input to one-quadrant multipliers and other nonlinear devices, measurements of mean absolute deviation, full-wave rectification, vector computation, etc. Absolute-value voltages can be made available in either positive or negative polarity (Figure 10).

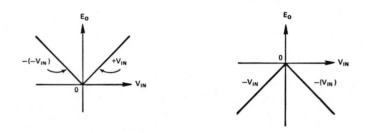

a) Positive absolute value + | V_{IN} | b) Negative absolute value − | V_{IN} |

Figure 10. Positive and negative absolute value

The circuit of Figure 11 is a typical absolute-value circuit. It comprises an ideal-diode and a differencing circuit. For positive input voltages, $V_B = 0$, $V_A = -V_{IN}(R_3/R_1)$, and the output is

$$E_o = \frac{R_3 R_5}{R_1 R_4} V_{IN} \qquad (3)$$

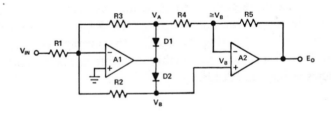

Figure 11. Practical absolute-value circuit

if all resistors are equal. For negative inputs, D1 is open, D2 conducts

$$-\frac{V_{IN}}{R_1} = \frac{V_B}{R_2} + \frac{V_B}{R_3 + R_4} \qquad (4)$$

and

$$E_o = V_B \left(1 + \frac{R_5}{R_3 + R_4} \right) \qquad (5)$$

Again, it can be seen that if all resistances are equal, the right-hand sides of both equations will be $1\frac{1}{2}V_B$ (multiplying (4) through by R_1); hence $E_o = -V_{IN}$. Therefore, the output will be equal to V_{IN} if positive and $-V_{IN}$ if negative, or $+|V_{IN}|$.

POLARITY DETECTION

Quite often, it is useful to have a polarity signal, as well as the absolute-value output. For example, to use a one-quadrant multiplier as a four-quadrant multiplier, it is possible to take the absolute-value of both inputs, and compare the polarities in an exclusive-or gate ("0", first and third quadrants (+), "1", second and fourth quadrants (−)). A sign-magnitude-to-bipolar circuit may be used to restore the polarity, if desired. Another application is in sign-magnitude A/D conversion, using a unipolar converter.

The circuit of Figure 11 may be followed by the polarity detection circuit shown in Figure 12, which makes use of the 2-diode-drop transition region at the output of A1 to derive a logic signal as a function of polarity. The output of the circuit will be positive-true "1" when the input signal is positive, and "0" when the input signal is negative.

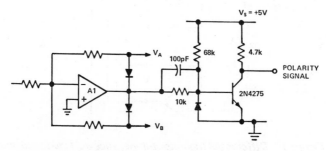

Figure 12. Polarity output signal from ideal-diode circuit

SPECIAL-PURPOSE ABSOLUTE-VALUE CIRCUITS

Figure 13 shows a high-input-impedance absolute-value circuit, making use of an ideal diode circuit and an adder-subtractor. The negative inputs of both amplifiers must follow the positive inputs.

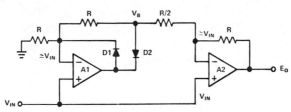

Figure 13. High-input-impedance absolute-value circuit

When V_{IN} is positive, the current V_{IN}/R to ground is supplied by D1; D2 is off; therefore no current flows through any of the other three resistors, and E_o must follow V_{IN}. When V_{IN} is negative, D2 conducts, D1 is off, and $V_B = 2V_{IN}$. The output is $(+3V_{IN} - 4V_{IN})= -V_{IN}$, a positive voltage; therefore $E_o = |V_{IN}|$.

Figure 14 is a circuit that takes the absolute value of a differential input voltage and converts it to a current. It operates in similar fashion to the current-switching circuit of Figure 8, but responds to full-wave inputs. It is useful as an input to devices that call for current inputs, such as integrating meters and logarithmic multiplier/dividers.

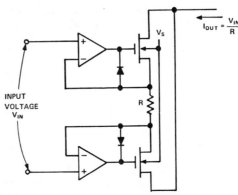

Figure 14. Differential high-input-impedance voltage-to-current absolute-value circuit

SIGN-MAGNITUDE TO BIPOLAR

Figure 15 depicts a circuit that will accept a positive *magnitude* signal and a negative-true *polarity* signal; at the output, it provides a bipolar signal with polarity determined by the inputs.

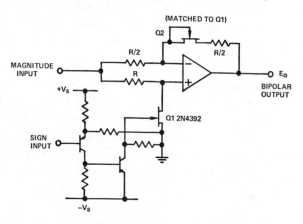

Figure 15. Sign-magnitude-to-bipolar circuit

When the FET switch, Q1, is *off*, the amplifier A1 operates as a unity-gain follower, with performance limited by the common-mode performance of the amplifier, and the impedance level of the input resistors (bias current and noise pickup). When the FET switch is *on,* the positive input signal is shunted to ground, and the amplifier functions as an inverter. Since Q1 does not have zero *on* resistance, there is some leakage of signal to the + input terminal. However, it is almost exactly compensated for (over a wide temperature range) by the resistance of Q2, the FET in series with R2. Improved accuracy can be obtained using 3 equal R's, and adding a second matched FET in series with Q2.

A NOTE ON DIGITAL TECHNIQUES

The read-only memory can be thought of as a "digital function fitter," since it can be programmed to fit (in pointwise fashion— $2^n - 1$ points) an arbitrary function of an input digital number ("address").

It is natural to consider the possibility of converting to digital, applying the digital number to a ROM, and (if necessary) converting the modified digital number occurring at the output of the ROM back to analog. It may (or may not) be more cumbersome than purely analog function fitting, and the discreteness of the fit may (or may not) be undesirable. Costs for both digital and analog techniques are coming down, and the issues are not clearly definable, since experience and inclination can weigh equally with marginal cost differences.

However, there are applications that involve conversion anyway, where consideration of a ROM for linearizing may be especially appropriate. For example, in a conventional dual-slope integrating system (used, typically, in digital panel meters), the signal input is integrated for a fixed number of counts. Then a reference voltage is applied to the integrator in the opposite polarity, and the number of counts required for the return to zero from the initial condition established by the integrated analog voltage, is a measure of the average value of the analog voltage.

For compensation and linearization of transducers, it is possible to use a D/A converter to adjust the reference-voltage level (Figure 16). The D/A converter is fed by a ROM, which has as its input

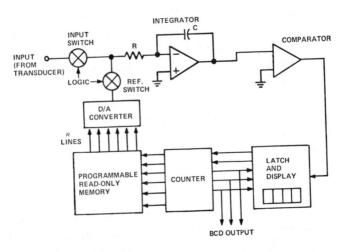

a. Block diagram of linearized dual-slope A/D converter

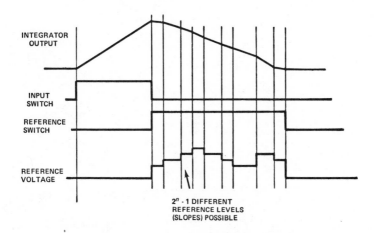

INTEGRATOR
OUTPUT

INPUT
SWITCH

REFERENCE
SWITCH

REFERENCE
VOLTAGE

$2^n - 1$ DIFFERENT
REFERENCE LEVELS
(SLOPES) POSSIBLE

b. Waveforms of linearized dual-slope converter

*Figure 16. Analog-digital transducer linearizing circuit and
A/D converter*

the counter output. As the count increases during integrator discharge, the "address" (i.e., input number) to the ROM changes. The ROM's programmed output changes, depending on the program, and feeds the D/A converter a number, which is converted to an analog reference level, which modulates the rate at which the integrator is discharged. Using this approach, there can be as many as $2^n - 1$ integrator rates, where n is the number of bits used by the ROM and the D/A converter.

With this technique, individual transducers can be linearized and gain-compensated. First, the transducer output can be calibrated, by feeding its output and the true measurements into a computer, which can generate a paper tape of the program needed by the ROM. The tape can then be used to program a field-programmable ROM. The ROM can then be shipped with the transducer (it can even bear the same serial number), much as calibration curves have been in the past. At the same time the transducer is installed, the ROM can be installed in a compatible A/D converter; in this way, the output of the A/D converter will be completely matched (linearized and gain-compensated) to the individual transducer to which it is committed (or multiplexed).

Multifunctional Devices:
Powers & Roots

A perhaps unfamiliar, but nevertheless very useful, nonlinear device is the *multi-function circuit*. It combines multiplication and division with the ability to raise a voltage (or voltage ratio) to an arbitrary positive power or negative (reciprocal) power. The magnitude of the power may be greater than unity ("power") or less than unity ("root"). A good example of such a multifunction circuit (and the first to appear on the market) is the Analog Devices Model 433. Its transfer function is

$$E_o = \frac{10}{E_{REF}} V_y \cdot \left(\frac{V_z}{V_x}\right)^m \tag{1}$$

where

E_o = Output voltage, $\geqslant 0$

V_y, V_z, V_x = Inputs, $\geqslant 0$

E_{REF} = Constant $\cong +9V$

m = Any number in the range 0.2 to 5, set by the ratio of two resistors, e.g., m = 2.72, or m = 0.318

A reference voltage, equal in magnitude to the constant, E_{REF}, is available as an input for operations involving only one or two variables, or for other purposes involving a reference voltage.

This combination of the three functions, multiplication, division, and exponentiation, within one small (38 × 38 × 16mm) module, yields computing power comparable to that of a small algebraic analog computer, or a log-log slide rule.

The multifunction circuit can be easily connected to perform many different functions, as determined by the choice of inputs, jumpers, and (for exponents) resistors. Among the available functions are:

1. Multiplication $\qquad E_o = K\,V_y\,V_z$

2. Division $\qquad E_o = K(V_z/V_x)$

3. Squaring $\qquad E_o = K\,V_y\,V_z\;(V_y = V_z)$

   or $\qquad E_o = K\,V_z^2$

4. Square-Rooting $\qquad E_o = K\,V_y/E_o$

   or $\qquad E_o = K\,V_z/E_o$

   or $\qquad E_o = K\,V_z^{0.5}$

5. Root of ratio $\qquad E_o = K\,V_y\left(\dfrac{V_z}{V_x}\right)^m \qquad m < 1$

6. Power of ratio $\qquad E_o = K\,V_y\left(\dfrac{V_z}{V_x}\right)^m \qquad m > 1$

7. Reciprocal power $\qquad E_o = K\,V_y\left(\dfrac{V_z}{V_x}\right)^m = K\,V_y\left(\dfrac{V_x}{V_z}\right)^{-m}$

In addition, there are other functions that can be performed by a single multifunction circuit and a modicum of external components

8. True rms (Chapters 2-3, 3-7) $\qquad E_o = \overline{K\,V_y\,V_z}/E_o = \sqrt{\overline{K V_{in}^2}},$

$$(V_y = V_z \geqslant 0)$$

9. Vector sum (Chapter 2-3)

$$E_o = \sqrt{V_1{}^2 + V_2{}^2}$$

$$V_1, V_2 \geqslant 0$$

10. Trignometric Functions
(Chapters 2-1, 2-3, and 2-5)

$$E_o = K \tan^{-1}(V_2/V_1)$$

$$E_o = K \sin \theta$$

Two factors account for the versatility of the multifunction circuit:

1. *Log-antilog operating principle*: Powers and roots can be easily generated by adjusting the gain of either the log or antilog portions of the circuit (or both). In the divider connection ($m = 1$, V_y constant), the log-ratio input section provides good accuracy over a much wider dynamic range than is possible with linear ratio circuits. A comparison of the multifunction circuit, as a divider, and a conventional "inverted-multiplier" divider circuit (Chapter 3-3) is shown in Figure 1.

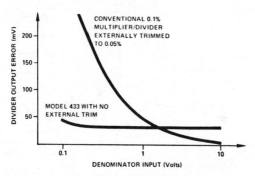

Figure 1. Divider error as a function of denominator level: multifunction circuit (m = 1) compared with an "inverted-multiplier" circuit.

2. *Three variable inputs are useful*: The ZY/X transfer function makes possible implicit solutions, through the use of feedback, of equations such as those for root-mean-square and vector addition. The additional variable also makes possible two-variable multiplication or division with voltage-adjustable scale factors, and direct implementation of such equations as the ideal gas relationship, involving pV/T.

CIRCUIT DESCRIPTION

The multifunction circuit is shown as a simplified schematic diagram in Figure 2 and an operational block diagram in Figure 3. It is in many ways similar to the log-antilog multiplier-divider circuits described in Chapter 3-2 (Figure 20) and Chapter 3-3 (Figure 20).

Amplifiers A1 and A2 and dual-transistor Q1 are connected so that the log of the ratio of input voltages V_z and V_x (the difference of their logarithms) appears at the base of Q1B, multiplied by the usual kT/q. The antilog circuit is basically the log circuit in reverse: the input is applied at the base of Q2B, and the output, at A3, is proportional to the exponential of the input-times-the reference-current established by D1, R4, and A4. Since the argument of the exponential is multiplied by q/kT, the temperature effects are cancelled out.

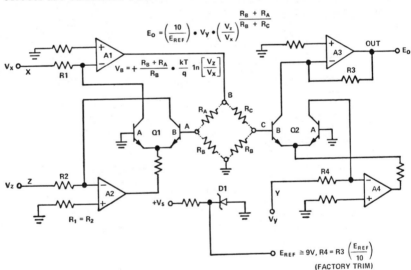

Figure 2. Simplified schematic of the multifunction circuit

The symmetrical arrangement of the circuit provides very low scale-factor and offset drift, independent of the denominator (V_x) input level. The scale-factor stability is limited primarily by the zener reference, D1, and the relative drifts of resistors R1 through R4.

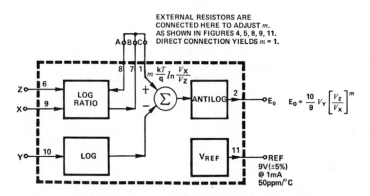

Figure 3. Functional block diagram of Model 433

POWERS AND ROOTS

The distinguishing feature of the multifunction circuit is its ability to take roots or powers of voltage ratios (with one or both voltages variable).

Any exponent between 1/5 and 5 (e.g. 1/4.73, 1/3, 1/2, 1/1.7, 1.05, 2.0, 2.1, etc.) can be obtained by attenuating or amplifying the log of the ratio of the Z and X inputs that appears at point "A" in the circuit of Figure 2. Negative exponents are obtained by interchanging the signals applied to the X and Z inputs.

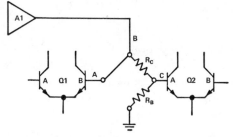

Figure 4. Connections for $1 > m = \dfrac{R_B}{R_C + R_B}$

Connections for the "root mode" (exponents less than unity) are shown in the partial circuit of Figure 4. The resistive divider R_C, R_B, attenuates the log of the ratio of Z and X before the antilog is taken in the output section. The exponent m is equal to the attenuation, $R_B/(R_B + R_C)$.

$$E_o = K\,V_y\,\epsilon^{\,m\,\ln(V_z/V_x)} \tag{2}$$

$$E_o = K\,V_y\left(\frac{V_z}{V_x}\right)^m = K\,V_y\left(\frac{V_z}{V_x}\right)^{\frac{R_B}{R_B+R_C}} \tag{3}$$

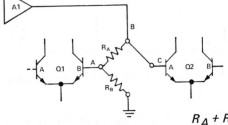

Figure 5. Connections for $1 < m = \dfrac{R_A + R_B}{R_B}$

Connections for the "power mode" (exponents greater than unity) are shown in the partial schematic of Figure 5. In this case, since the feedback voltage is attenuated by $R_B/(R_B + R_A)$, the gain, m, is $(1 + R_A/R_B)$. Therefore

$$E_o = K\,V_y\left(\frac{V_z}{V_x}\right)^m = K\,V_y\left(\frac{V_z}{V_x}\right)^{1+\frac{R_A}{R_B}} \tag{4}$$

Figure 6 shows normalized plots of the function for several representative values of m, $V_x = 10$, $V_y = E_{REF}$. As might be expected, for m = 1, the response is linear with a slope of unity. For m > 1, the slope increases from zero at $V_z = 0$ to m at $V_z = 10V$. For m < 1, the slope decreases from "infinity" at zero to m at $V_z = 10V$. Thus, while "powers" are quite stable, with a maximum gain of 5 at full scale, "roots" become less stable as the input approaches zero (resulting in greatly magnified noise and drift), but are quite stable for larger values of the input ratio. Fortunately, the multifunction circuit has very low noise and drift, referred to the input, so that over a 100:1 range of input ($V_z = V_x$), the errors at the extreme exponents (1/5 and 5) tend to be small. This can be seen in Figure 7, a plot of output noise vs. denominator at constant ratio.

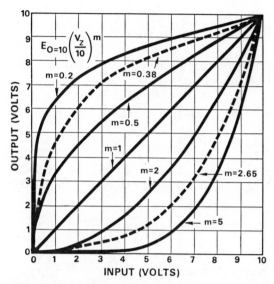

Figure 6. Output vs. input for several values of the exponent, m

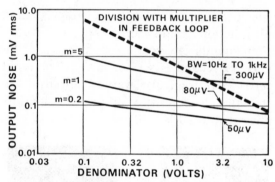

Figure 7. Output noise vs. denominator voltage (unity ratio) for the 433, compared with noise output of inverted-multiplier divider (log scales)

Since incremental input noise, referred to the output, is approximately proportional to the slope of the function, it is useful to know the input level at which a given slope (of the curve in Figure 6 corresponding to the desired m) occurs. This is determined by differentiating the exponential relationship, setting the derivative, dE_o/dz (where $z = V_z/V_x$), equal to a specific value of G (the slope or gain), and solving for the corresponding value of z, i.e., z(G), at which it occurs.

$$z(G) = \left(\frac{G}{m}\right)^{1/(m-1)} \tag{5}$$

The values of z corresponding to various orders-of-magnitude of G are listed below:

m	z(G=1)	z(G=10)	z(G=100)
1/5	0.134	0.008	0.0004
1/4	0.157	0.007	0.0003
1/3	0.192	0.006	0.0002
1/2	0.250	0.0025	0.00002
1	all values		
2	0.500	5.0	50.0
3	0.577	1.826	5.77
4	0.630	1.357	2.92
5	0.669	1.189	2.115

A convincing demonstration of the efficacy (and repeatability) of the multifunction circuit can be seen in Figure 8. Two 433 multifunction circuits are connected in cascade; the first is set for m = 1/5, the second is set for m = 5. The input to the first is a 0 to +10V 5Hz triangular wave. The output of the first, as expected, has the characteristic 5th-root relationship sketched in Figure 6; it is equal to $10(V_{in}/10)^{1/5}$.

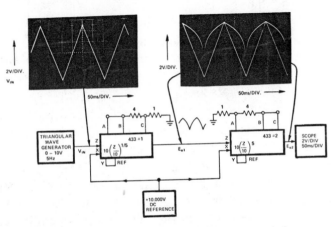

Figure 8. Fifth root and fifth power cascaded

The output of the second 433 is a triangular wave with little apparent distortion, except in the vicinity of zero. Its response of $10(E_{o_1}/10)^5$ inverts the fifth-root response of E_{o_1} to achieve a linear overall response.

Tailoring the Range of Exponent-Adjustment

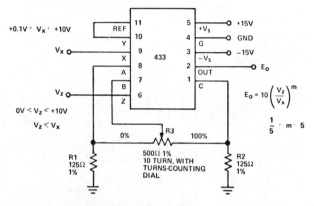

ADJUST POTENTIOMETER TO SET M TO
DESIRED VALUE USING GRAPH IN FIGURE
10 OR EQUATION (8)

Figure 9a. 433 multifunction module computes continuously-adjustable power or root of ratio

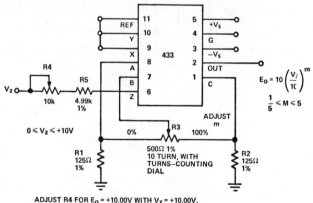

ADJUST R4 FOR E_O = +10.00V WITH V_Z = +10.00V.
ADJUST M POTENTIOMETER FOR DESIRED VALUE
OF M USING EQUATION 8 OR GRAPH OF FIGURE 10.

Figure 9b. 433 multifunction module takes continuously-adjustable power or root of single variable

Figure 9. Wide-range exponent adjustment

For some purposes, it is useful to allow the exponent to be adjusted through unity without rewiring or switching. This can be easily accomplished with a single potentiometer, connected as shown in Figure 9a (V_z and V_x both variable) or Figure 9b (V_z only variable), to allow a range of continuous variation of m from 1/5 to 5. Sucn an arrangement is especially useful for linearizing and curve-fitting, since the pot can be adjusted until the output waveform has the desired shape. The exponent may then be read to within ±5% (or better) by the use of the calibration curve (Figure 10).

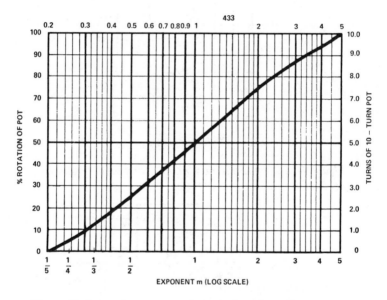

Figure 10. Potentiometer rotation as a function of exponent, m, for the circuits of Figure 9, using 433

For example, suppose that the setting of the pot that produces best linearity (in a linearizing application) is experimentally found to be 3.20 turns. Referring to Figure 10, one finds that for 32% rotation, the calibration curve shows an m of 0.6. Thus, the empirically-arrived-at optimum exponent turns out to be 0.6:

$$E_o = 10(V_z/V_x)^{0.6} \qquad (6)$$

The calibration curve can also be used to preset a desired exponent. Suppose, for example, that a transfer function of

$$E_o = 10(V_z/10)^{1.5} \qquad (7)$$

is desired. First, set up the circuit of Figure 9b, then determine the setting of the exponent potentiometer from Figure 10. In this case, m = 1.5 corresponds to about 64%. Set the pot to 6.40 turns.

If an accurate calibration of m is desired, a high-quality 10-turn pot with less than 0.1% nonlinearity and an accurately-calibrated turns-counting dial should be used. Resistors R1 and R2 should be matched to 0.1%, and should exactly equal 1/4 of the total resistance of the pot. Under these conditions, if α is the fractional rotation of the pot, the relationship between α and m is

$$m = \frac{1 + 4\alpha}{5 - 4\alpha}, \qquad \alpha = \frac{5m - 1}{4(m + 1)} \qquad (8)$$

as plotted in Figure 10.

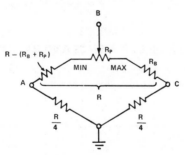

Figure 11. Configuration for arbitrary range of adjustment of the exponent, m

For smaller ranges of exponent adjustment, the potentiometer circuit may include resistances on either or both ends of the pot to tailor the range of adjustment to within the desired limits. Figure 11 shows how this may be accomplished. The design equations for R, R_B, and $(R - R_B - R_P)$, in terms of the minimum and maximum values of m and a chosen value of pot resistance are:

$$\frac{R_B}{R} = \frac{5 - m_{max}}{4(1 + m_{max})} \tag{9}$$

$$R = R_P \frac{1 + m_{min}}{\dfrac{5}{4} - \dfrac{R_B}{R} - m_{min}\left(\dfrac{1}{4} + \dfrac{R_B}{R}\right)} \tag{10}$$

The ratios R_B/R and R/R_P are calculated from the maximum and minimum values of m. Then, a suitable (low) value of R_P is chosen, R is computed, R_B is computed, and $(R - R_P - R_B)$ is established.

Table of Exponents and Ratios

Table 1 shows a set of calculated values of $(V_z/V_x)^m$, for the integral (and their reciprocal) powers from 1 to 5, to facilitate checks on the accuracy of the power/root setting.

TABLE 1

TABLE OF EXPONENTS AND RATIOS

$(V_z/V_x)^m$ as a function of V_z/V_x and m [in $E_o = KV_y(V_z/V_x)^m$]

$KV_{y_{max}}$ = 10V, except as noted for values of $V_z/V_x > 1$

	m → V_z/V_x ↓	1/5	1/4	1/3	1/2	1	2	3	4	5
$\frac{V_z}{V_x} < 1$	0.01	0.398	0.316	0.2155	0.100	0.010	0.0001	–	–	–
	0.025	0.478	0.3975	0.2925	0.158	0.025	0.0006	–	–	–
	0.05	0.5495	0.473	0.3685	0.2235	0.050	0.0025	0.0001	–	–
	0.1	0.631	0.5625	0.464	0.316	0.100	0.0100	0.0010	0.0001	–
	0.25	0.758	0.707	0.630	0.500	0.250	0.0625	0.0156	0.0039	0.0010
	0.5	0.8705	0.841	0.7935	0.707	0.500	0.250	0.125	0.0625	0.0312
	1.0	1.	1.	1.	1.	1.	1.	1.	1.	1.
$\frac{V_z}{V_x} > 1$	2.0 ($KV_{y_{max}}$)	1.1485 (8.706)	1.189 (8.409)	1.260 (7.937)	1.414 (7.071)	2.000 (5.000)	4.000 (2.500)	8.000 (1.250)	16.000 (0.625)	32.000 (0.312)
	5.0 ($KV_{y_{max}}$)	1.3795 (7.248)	1.4955 (6.687)	1.710 (5.848)	2.236 (4.472)	5.000 (2.000)	25.000 (0.400)	125.00 (0.080)	625.00 (0.016)	3125 (0.003)
	10.0 ($KV_{y_{max}}$)	1.585 (6.310)	1.7785 (5.623)	2.1545 (4.642)	3.1625 (3.162)	10.000 (1.000)	100.00 (0.100)	1000.0 (0.010)	10,000 (0.001)	–

Square-Root Circuit

While the square root may be computed by setting m = 1/2, it is also possible to compute the square root with m = 1, saving the cost of a precision resistor-pair. This is done by an implicit solution, with the output voltage fed back to the denominator input, as shown in Figure 12.

$$E_o = 10(V_z/V_x)^1 = 10(V_z/E_o) \qquad (11)$$

$$E_o^2 = 10 V_z \qquad (12)$$

$$E_o = (10V_z)^{1/2} \qquad (13)$$

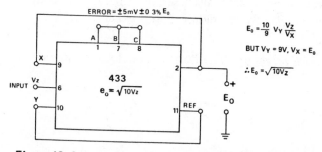

Figure 12. Square root circuit using divider with feedback

Figure 13 is a plot of the error of the 433 connected in this configuration, compared with the "inverted-multiplier" square-root circuit, using a conventional 0.1% multiplier/divider. For low values of input ($\leqslant 0.1$V), the multifunction circuit is considerably more accurate.

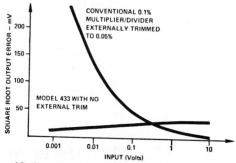

Figure 13. Square root error of 433 compared with square-root error of inverted-multiplier divider connected for square-rooting

In general, the multifunction circuit can obtain the $m/(m + 1)$ power of V_z by this feedback technique, but its usefulness is limited, since it "uses up" the denominator input, and, in any event, requires a pair of resistors that could be used to implement the power $m' = m/(m + 1)$ directly.

SPECIFICATIONS

Specifications of a representative modular multifunction circuit, the Analog Devices Model 433J/B are listed in Table 2. The general format is similar to that shown for the multiplier and divider specifications discussed in Chapters 3-2 and 3-3; it will not be reviewed in detail here, except for specific pertinent comments.

General Expression: The transfer function is listed as

$$E_o = \frac{10}{9} V_y \left(\frac{V_z}{V_x}\right)^m \tag{14}$$

The coefficient "9" is initially set at 9.0V ±5% to match the reference output voltage V_{REF}. Thus, if the reference voltage (or a calibrated equivalent voltage) is applied to V_y,

$$E_o = 10\left(\frac{V_z}{V_x}\right)^m \tag{15}$$

The transfer gain may be trimmed to a value less than (10/9 ±5%) by adding resistance in series with the Y input; a 25kΩ variable resistor is the usual vehicle for settings near unity.

Input Range

The input signal range to the Y, Z, and X terminals is 0 to +10V, positive only; i.e., the 433 is a one-quadrant (I) device that will not respond to negative input voltages or produce negative output voltages. Nevertheless, the *Maximum Safe* input voltage rating is

TABLE 2. SPECIFICATIONS OF MULTIFUNCTION CIRCUITS
(Typical @ +25°C Unless Otherwise Noted).

Model	433J	433B
General Expression	$E_O \triangleq + \dfrac{10}{V_{REF}} V_y \left(\dfrac{V_z}{V_x}\right)^m$	•
Rated Output[1]	+10.5V @ 5mA	•
Input		
Signal Range	$0 \leq V_x, V_y, V_z \leq +10V,$	•
Max Safe Input	$V_x, V_y, V_z \leq \pm18V$	•
Resistance		
X Terminal	100kΩ ±1%	•
Y Terminal	90kΩ ±10%	•
Z Terminal	100kΩ ±1%	•
External Adjustment of the		
Exponent, m		
Range for m < 1 (Root)	$1/5 \leq m < 1, m = \dfrac{R_B}{R_C + R_B}$	•
Range for m > 1 (Power)	$1 < m \leq 5, m = 1 + \dfrac{R_A}{R_B}$	•
	$(R_1 + R_2) \leq 200\Omega$	•
Accuracy (Divide Mode)[2,3]		
Total Output Error @ +25°C		
(for specified input range)		
Typical (RTO)	±5mV ±0.3% of output	±1mV ±0.15% of output
Max Error (RTO)	±50mV	±25mV
Input Range ($V_z \leq V_x$)	0.01V to 10V, V_z	•
	0.1V to 10V, V_x	•
Over Specified Temp. Range	±1%	±1% max
Output Offset Voltage		
(Not Adjustable)		
Initial @ +25°C max	±5mV	±2mV max
Offset vs Temp.	±1mV/°C	±1mV/°C max
Noise, 10Hz to 1kHz		
$V_x = +10V$	100μV rms	•
$V_x = +0.1V$	300μV rms	•
Bandwidth, V_y, V_z		
Small Signal (-3dB), 10%		
of DC Level V_y or V_z		
$V_y = V_z = V_x = 10V$	100kHz	•
$V_y = V_z = V_x = 1V$	50kHz	•
$V_y = V_z = V_x = 0.1V$	5kHz	•
$V_y = V_z = V_x = 0.01V$	400Hz	•
Full Output (V_y or $V_z = 5VDC \pm5VAC$)	$(V_x) \bullet (5kHz)$	•
Reference Terminal Voltage[1]		
V_{ref} (Internal Source)	+9.0V ±5% @ 1mA	•
vs Temp (0 to +70°C)	±0.005%/°C	•
Power Supply Range		
Specified	±(14.7 to 15.3)VDC @ 10mA	•
Operating	±(12 to 18)VDC	•
Temperature Range		
Specified	0 to +70°C	-25°C to +85°C
Storage	-25°C to +85°C	-55°C to +100°C
Package Outline	FA-7	•
Case Dimensions	1½" x 1½" x 0.62"	•
	38 x 38 x 16mm	

•Same specifications as 433J.
[1] Terminals short circuit protected to ground only.
[2] Accuracy is specified in divide mode which is a worst case condition.
  Input range is 10mV to 10V for specified accuracy when connected
  as a multiplier.
[3] Error is defined as the difference between the measured output and
  the theoretical output for any given pair of specified input voltages.
Specifications subject to change without notice.

±18V; though voltages outside its normal range are computationally irrelevant, they will not harm the unit.

Accuracy, Divide Mode

Accuracy as a divider is specified in two ways,

1. as a worst-case, maximum error of ±50mV (0.5% of full scale output)

2. as a small fixed error, plus a fixed percentage of output; ±5 mV ±0.3% of actual output, typical

(2) is more closely descriptive of the actual error of the device. The worst-case ±50mV, or 0.5% of full-scale, the customary overall accuracy specification, is useful near full scale, but it is overly conservative at the lower levels.

TESTING THE MULTIFUNCTION CIRCUIT

In the multiplier and divider modes (m = 1), the multifunction circuit may be tested with the circuits and techniques described in Chapters 3-2 and 3-3, as they are applied to 1-quadrant devices. Figures 14 through 17 illustrate tests that are of particular interest for the multifunction circuit. They test, in this order: divider errors, multiplier errors, dynamic errors, and exponent errors.

The *divide mode* accuracy test, Figure 14 shows a single dc reference driving the denominator directly, and the numerator through

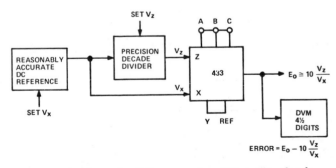

Figure 14. Measuring DC errors of multifunction circuit in divide mode

a precise voltage divider. This approach provides a true-ratio input, independent of the absolute accuracy of the dc reference. It is especially important that this approach be used for high-accuracy dividers, such as the 433, since the seemingly–negligible absolute errors that can exist between two independent dc references can cause large absolute errors at the divider output.

For example, suppose that we want to test the 433 at V_z = 10mV, V_x = 100mV, and that two separate references are used, both having reasonable accuracy — say, 0.01% of full scale. The theoretical output voltage should be

$$E_o = 10\frac{0.010}{0.100} = 0.1 \times 10V = 1V \tag{16}$$

If the Z reference is high by 0.01% × 10V = 1mV, and the X reference is low by 0.01%, or 1mV, the actual test that is performed is

$$E_o = 10\frac{0.011}{0.099} = 10/9V = 1.11V \tag{17}$$

This means that a perfect divider will indicate an output error of 11% (of reading) from the value that would be ideally given by two perfect references! The numerator is 10% (of setting) high and the denominator is 1% low, despite the 0.01% specification on the references.

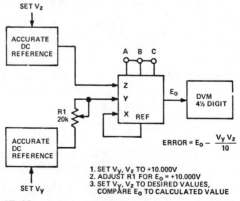

Figure 15. Measuring multiplication errors of multifunction circuit

The error due to imperfect references can be found and accounted for by measuring the X and Z inputs directly with a DVM with 100μV (or better) resolution and good linearity, but it is usually easier and more satisfactory to use a precise voltage-divider or very accurate (100μV-or-less absolute error) reference for the divider inputs.

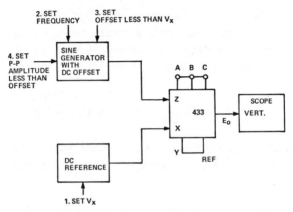

Figure 16. Measuring frequency response to numerator signals, as a function of V_X. Offset voltage keeps the AC signal in the first quadrant.

Accuracy as a Function of Exponent Setting (m). This is the most interesting (and difficult) test to perform, unless you have access to means of mechanizing the measurement, and an adequate table of ideal output values for comparison. The best non-mechanized approach is to measure the output as a function of the input (or input ratio V_z/V_x), point by point, using the test setup of Figure 17. As noted above, the X and/or Z input voltages (or their ratio) must be known to an accuracy greater than that of the device under test. Assuming that the inputs are known, the actual output can be compared to the theoretical output at various values of the inputs, or their ratio.

Semi-automated testing, using analog waveforms, may be performed if a unit having greater accuracy than that specified for the units under test is available for comparison. A slow triangular wave is fed into the inputs of both devices and the outputs are compared via a difference amplifier. Another means of automating the point-by-point error measurement is to apply the inputs via a D/A converter, programmed by a computer. The output of the unit at each point is compared with the computed theoretical value, after conversion to digital form.

The apparent dynamic range of the 433 depends greatly on the exponent, m, as discussed earlier and indicated in Table 1. For example, let $V_x = 10V$, $V_z = 1V$

If $m = 1$, $E_0 = 10 (1/10) = 1V$

If $m = 2$, $E_0 = 10 (1/10)^2 = 0.1V$

If $m = 5$, $E_0 = 10 (1/10)^5 = 100\mu V$!

At the other extreme, $m = 1/5$, the gain of the 433 $\rightarrow \infty$ as the ratio $V_z/V_x \rightarrow 0$. For the above values,

If $m = 1/2$, $E_0 = 10(1/10)^{1/2} = 3.16V$

If $m = 1/5$, $E_0 = 10(1/10)^{1/5} = 6.31V$

From these examples, it is evident that forethought and care should be employed when measuring (and interpreting) the errors of the 433 in the power and root modes, especially for very low values of V_z/V_x. Careful testing will show that multifunction devices based on logarithmic circuitry can give surprisingly good results, even at such outlandish powers as Z^5 and $Z^{1/5}$.

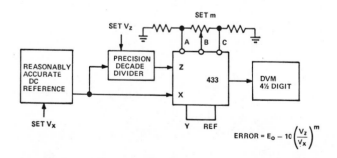

$$ERROR = E_0 - 10\left(\frac{V_z}{V_x}\right)^m$$

Figure 17. Measuring DC error of multifunction circuit in power or root mode.

Root Mean–Square

Chapter 7

In the introductory chapter and in Chapter 2-3, there were brief discussions of some properties and applications of the root mean-square, and typical circuits for implementing the function. This chapter, after a brief review of the nature of rms measurement, provides an in-depth discussion of the circuitry used by manufacturers of modular rms devices, errors and specifications of rms circuits, and configurations for testing rms devices.

The root mean-square of any voltage over an interval T is

$$\text{rms} = \sqrt{\frac{1}{T} \int_0^T [f(t)]^2 \, dt} \qquad (1)$$

The computing process is to square the voltage, $f(t)$, instant by instant, integrate it over the period T, divide by T to obtain the mean, and compute the square root. The integration indicated here is a true time integration, starting with initial conditions at $t = 0$, and ending with a measurement at $t = T$ (or after T, if the final value is retained in the *hold* condition). The desired initial conditions must be reset (e.g., to zero) before another measurement is made.

While this measurement gives a reading of true rms, it is somewhat cumbersome for all but one-shot phenomena. More typically, the waveform being measured is random or periodic and has stationary properties (including, among other things, a constant rms). If the rms is reasonably constant, the mean (of the squared signal) may be measured by a circuit that responds to the running average. In its simplest form, it is a low-pass filter consisting of a simple RC unit-lag circuit, with the RC time constant chosen to be considerably longer than the longest period present in the signal, but short enough to follow variations in the signal's rms value without introducing excessive delay errors.

Three fundamental properties of rms quantities are important to the instrument designer:

1. The rms is a measure of the heating value of a voltage or current applied to a resistor; over the interval, T, all waveforms having the same rms voltage or current will dissipate exactly the same amount of energy in the resistor, irrespective of their variation with time. This is true whether the waveform is constant, sinusoidal, biased-ac, random, or a train of pulses. The rms is a fundamental physical measurement.

2. The rms value of any stationary zero-mean random process is equal to the standard deviation of that process.[*] Whether the distribution measured by the electrical waveform involves electrical random noise or the size of apples on a conveyor belt, the rms measurement is a valid measurement of the standard deviation for large sample size. The rms is a fundamental statistical parameter.

3. If orthogonal or uncorrelated quantities are summed, the rms of their sum is equal to the square-root of the sum of the squares of their individual rms values.

The standard deviation is the square-root of the *variance* from the *mean value* of a set of samples. For an infinitely large sampling,

$$\sigma = \sqrt{\int_{-\infty}^{\infty} (X - m)^2 p(X) dX},$$

where the mean,

$$m = \int_{-\infty}^{\infty} X p(X) dX$$

and $p(X)$ is the probability that X has a given value.

Until recently, accurate rms instrumentation was not feasible for compact, wideband, low-cost applications. A widely-used substitute was the mean absolute value (or statistically, the mean absolute deviation). The mav, or mad, is obtained by simply full-wave-rectifying a signal and averaging the resulting waveform. The table of Figure 14 in Chapter 2-3 lists a number of waveforms, their rms and mad values. Table 1 shows similar data for several other waveforms.

TABLE 1. CHARACTERISTICS OF SOME COMMON WAVEFORMS.
See Chapter 2-3, Figure 14 for Sine, Square, Triangle, Sawtooth, Gaussian Noise, Zero-Based Rectangular Pulse Train.

WAVEFORM	RMS	MAD	$\dfrac{\text{RMS}}{\text{MAD}}$	CREST FACTOR
AMPLITUDE–SYMMETRICAL RECTANGULAR	1	1	1	1
SINE–SQUARED (RAISED COSINE)	$\sqrt{\dfrac{3}{8}}$ $= 0.6124$	$\dfrac{1}{2}$	1.225	1.633
SAWTOOTH PULSE	$\sqrt{\dfrac{\eta}{3}}$	$\dfrac{\eta}{2}$	$\sqrt{\dfrac{4}{3\eta}}$	$\sqrt{\dfrac{3}{\eta}}$
OFFSET PULSE $\left(\begin{array}{c}\text{IF AVE} = 0\\ \eta = A\,(1-\eta)\end{array}\right)$	$\sqrt{\eta\,(1-A^2)+A^2}$ $\sqrt{A}$	$\eta\,(1-A)+A$ $\dfrac{2A}{1+A}$	$\dfrac{\text{RMS}}{\text{MAD}}$ $\dfrac{1+A}{2\sqrt{A}}$	$\dfrac{1}{\text{RMS}}$ $\dfrac{1}{\sqrt{A}}$
EXPONENTIAL PULSE	$\sqrt{\dfrac{\tau}{2T}\left(1-\epsilon^{-2T/\tau}\right)}$ $\cong \sqrt{\dfrac{\tau}{2T}}$	$\dfrac{\tau}{T}\left(1-\epsilon^{-T/\tau}\right)$ $\cong \dfrac{\tau}{T}$	$\sqrt{\dfrac{T}{2\tau}}$	$\sqrt{\dfrac{2T}{\tau}}$

Because measurements on sine waves were widely used, meters were calibrated to read the rms value of a sine wave, while the mad was the actual voltage that was measured. Thus, such meters would read $\pi/2\sqrt{2}$ (= 1.111) times the mean absolute value. For waveforms other than undistorted sine waves, this ratio could be greatly in error: for dc or symmetrical square waves, the error is 11% high; for triangular or sawtooth waves, the error is 4% low; for Gaussian noise, it is 11.3% low. But these waveforms could be calibrated, if their nature was known. Much worse is the inability of such devices to measure waveforms of unknown form or of variable "duty cycle." For example, for a train of zero-based rectangular pulses, the ratio of rms/mad is 2 for a 25% duty cycle, and *10* for a 1% duty cycle. Figure 1 shows the error of measurement of rms over one-half cycle of a sinusoidal waveform as a function of firing angle in a SCR circuit, if 1.111 times the mean is used.

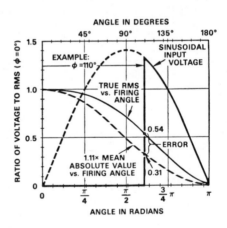

Figure 1. RMS and mean value of ideal full-wave SCR output as a function of firing angle ϕ

Since the mean square (hence the rms) measures, with accuracy and consistency, the power in a signal of known or unknown shape, averaged over a given time interval, it provides valid, universal, and repeatable measurements of the intensity of random phenomena, such as acoustic noise, mechanical vibration, and electrical noise, as well as phenomena characterized by waveforms with more determinate shapes.

Although, surprisingly, even some of today's "better" voltmeters still use rms-calibrated mad, a pronounced and rapid swing toward true-rms instrumentation is now in process. This swing is a result of two trends: the increased use of time-domain measurements, and the greatly decreased cost and increased availability of compact true-rms instruments. One can buy rms-to-dc converter modules with accuracy (errors) in the 0.1% to 0.5% class for about $50. Previously, the cost of classical thermal rms-to-dc converters (of the order of $1000) had restricted their use to the most-expensive digital voltmeters or specialized true-rms analog meters.

This restriction to expensive laboratory meters will vanish during the next few years. The growing market for low-cost portable digital multimeters and ac transfer standards will undoubtedly stimulate the demand for accurate and reliable ac measurements. Users of digital instruments naturally expect good accuracy and resolution, and they will want the ac-measurement capability of an instrument to be as good as its dc capability. Portable meters are generally used to measure complex waveforms in locations where auxiliary instruments (oscilloscopes and laboratory DVM's) are not available; therefore the ac accuracy must be independent of the waveform.

As noted earlier, rms converters are useful in industrial measurement and control: SCR waveforms, noise and vibration analysis, and power dissipation in fixed resistors are but a few applications. However, for variable or reactive loads, average power measurement is performed by instant-by-instant multiplication, followed by averaging, to obtain the average power (Chapter 2-3, Figure 17). (In such applications multiplication of rms values can lead to wrong answers.)

RMS-to-DC CONVERTER CIRCUITS

There are, today, three electrical techniques in common use for performing rms measurements:

    1. Thermal, based on the conversion of an unknown voltage or current to heat in a known value of resistance.

2. *Direct computing*: the use of analog-computing techniques to straightforwardly compute the rms value of an input waveform by squaring, averaging, and rooting in an open-loop configuration,

3. *Implicit computing*: a variation of (2) in which the square-root operation is performed implicitly (i.e., by feedback).

Of these three techniques, the thermal converter is satisfyingly basic by its nature, but it is difficult to realize. Computing types, especially those employing implicit square-rooting, can provide accuracies and bandwidths equalling the best of the thermal converters, usually at lower cost.

THERMAL RMS-DC CONVERTER CIRCUITS

The simplest thermal RMS-dc converter circuit that is useful for low-frequency ($<$10MHz) measurements is the fixed-gain, variable-temperature converter, shown in Figure 2. The input, applied to

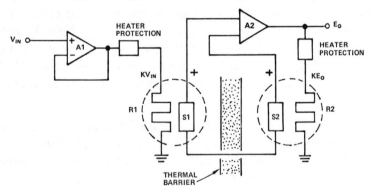

Figure 2. Thermal true rms-to-dc converter (fixed gain)

resistor R1, causes it to heat up. The circuit applies a dc voltage to R2, sufficient to cause it to heat up to the same temperature as R1, and continuously adjusts it to maintain the temperature difference at zero, as measured by the sensors S1 and S2. If both R1 and R2 have identical thermal paths to the environment, the power dissipated in both is identical, and, since $(KV_{IN})^2/R_1 = (KE_o)^2/R_2$, the output

$$E_o = \sqrt{\frac{R_2}{R_1}} \cdot (V_{RMS}) \qquad (2)$$

The input buffer amplifier, A1, should provide a high impedance for the input signal, sufficient output current to drive the low-resistance (10 to 100Ω) heater, R1, and adequate bandwidth (the amplifier is usually the limiting factor on input bandwidth in thermal converters, but thermal lag causes settling of the rms to new values to be slow, especially for decreasing input.

The structure of the heater-sensor assemblies is critical to the accuracy and bandwidth of the thermal rms converter. Until recently, the best converters used vacuum-sealed resistor-sensor assemblies, containing thin-wire-wound resistors and thermocouples. It is essential that the wire maintain constant resistance as a function of temperature, so that the voltage across R2 will vary *linearly* with the rms value of V_{IN}. It is also important that R1-S1 and R2-S2 be thermally isolated from one another. If a significant amount of heat from R1 reaches S2 (or S1 from R2), then the sensitivity of the converter will be diminished, and non-linearities may be introduced.

If thermocouples are used for S1 and S2, then A2 must be quite stable — perhaps chopper-stabilized — because of the low (e.g., $40\mu V/°C$) sensitivity of thermocouples. Some detectors use several thermocouples in series, but the signal levels are still submillivolt.

Recently, balanced thermal detectors have been developed that use thin-film resistors for the heaters, and transistors for the sensors. Since the base-to-emitter voltage of a transistor has approximately a $-2mV/°C$ temperature coefficient at 25°C, it is nearly two orders of magnitude better as a sensor than a thermocouple. While this reduces the performance requirements for A2, the thermal balance, input amplifier, and settling time problems still remain.

Performance of the "Fixed-Gain" Thermal Converter

Errors can be quite low — typically less than 0.1% of reading for input signals over a narrow range of amplitudes (usually less than 3:1). Since the power dissipated in the heater, R1, is proportional to the square of the input rms, a 3.2:1 change of input amplitude will change the power dissipation by 10:1, which can result in a comparable temperature rise. If the input amplitude is increased too much, R1 may be burned out. On the other hand, if the signal level is too low, the temperature rise of R1 will be too small for satisfactory operation.

The bandwidth of the thermal converter (in terms of input-signal response) is limited by the bandwidth of the input amplifier A1 at the upper end, and by the thermal time constant of R1-S1 and R2-S2 at low frequencies (i.e., 1Hz to 10Hz). The converter works perfectly at dc, since no averaging is required. The thermal converter can therefore be calibrated against an accurate dc reference, and then used to measure an ac signal (within its bandwidth) of about the same rms level (within $\pm 50\%$) as the reference.

The limited dynamic range implies a limitation on crest factor (the ratio of peak input to rms). Since the heaters must be operated in the upper portion of their dynamic ranges under steady-state low-crest-factor conditions (such as sinewave inputs, c.f. $= \sqrt{2}$), there is little "headroom" for peaks. For instance, if a heater-detector works best with 10mA to 30mA into the heater, and a crest-factor capability of 3 is required, the input buffer amplifier must be able to supply 30-90mA linearly. At a crest factor of 5, the requirement is 50mA to 100mA. Even if the amplifier can supply the current, the heater might burn out if its instantaneous power rating is exceeded (power is proportional to the *square* of voltage or current).

Variable-gain thermal rms-dc converter: The limited dynamic range and crest factor, and long settling time of the fixed-gain thermal converter can be substantially bettered by operating the heaters, R1 and R2, at constant power (temperature). As shown in Figure 3, the gain of the input buffer-amplifier is manipulated by the null-sensing amplifier A2 to bring the power in R1 into equilibrium with the power in R2, which is driven from a constant voltage V_{REF}. If the input amplifier has a gain that is inversely proportional to the control voltage, E_o, then the control voltage will be proportional to the rms of the input voltage. At null,

$$\overline{\left(K \frac{V_{IN}}{E_o} \right)^2} \frac{1}{R_1} = \frac{V_{REF}^2}{R_2} \tag{3}$$

Therefore,

$$E_o = \sqrt{K\frac{R_2}{R_1}\frac{\overline{V_{IN}{}^2}}{V_{REF}{}^2}} = K'\sqrt{\overline{V_{IN}{}^2}} \qquad (4)$$

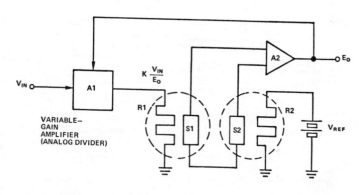

Figure 3. Variable-gain thermal rms-to-dc converter

The variable-gain thermal rms-dc converter has better dynamic range and accuracy than the fixed-gain converter, but it shares some of the same weaknesses. Since it requires fairly large (10mA to 100mA) currents in the heater resistors, good ground-return practices should be followed to avoid causing significant voltage drops in the ground paths.

Since the heater-sensor pairs operate above ambient temperature, a significant warmup time (5 minutes or more) is usually required for a thermal converter to reach usable accuracy. The thermal time constant determines the averaging time, and thus the lowest frequency (excluding dc) for which rms can be measured accurately. The averaging time cannot be increased by the use of low-pass filters (as is the case for computing-type converters). For this reason, thermal converters do not usually work well at frequencies below 10Hz.

DIRECT COMPUTATION

Computing the root mean-square of a waveform requires three mathematical operations: squaring, averaging, and square-rooting. They can be implemented in a straightforward fashion, using multipliers and operational amplifiers, as shown in Figure 4.

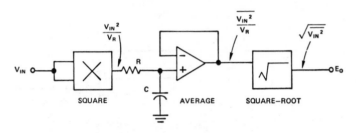

Figure 4. Explicit RMS circuit

This scheme, which embodies explicit computation is unsatisfactory, for several reasons.

1. Cost and complexity: It requires two multipliers, or a square-rooter, plus an op amp.

2. Limited dynamic range: The output of the squarer will vary over a 10,000:1 dynamic range (1mV to 10V) for a 100:1 instantaneous-input range (0.1V to 10V). Since the input multiplier will have errors greater than 1mV, the error will strongly depend on signal level, resulting in an overall dynamic range less than 100:1, probably only 10:1. In this respect, the direct-computing rms circuit shares similar dynamic-range limitations with the fixed-gain thermal converter. Despite these limitations, the direct computation of the rms value can be made quite accurate over a 10:1 range of input. If a 0.1% multiplier is used as the squarer, and a high-accuracy square-rooter (e.g., 434B) is used, an error level of ±0.1% of full scale can be achieved.

IMPLICIT COMPUTATION

Perhaps the best approach to computing the rms value of a signal is to use a circuit that implements an implicit solution to the rms equation:

$$V_{RMS}{}^2 = \overline{V_{IN}{}^2} \qquad (5)$$

by the use of the identity

$$V_{RMS} = \frac{\overline{V_{IN}{}^2}}{V_{RMS}} \qquad (6)$$

Figure 5 is a block diagram of a circuit that performs the indicated operations of squaring, averaging, then dividing by the output. Since the output is essentially constant over the period of the signal being averaged, it may perform the division *before* the average is taken.

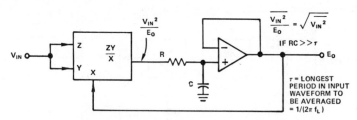

Figure 5. Implicit RMS-computing circuit

This scheme overcomes all the limitations of the direct computing approach, and it has much greater dynamic range than the variable-gain thermal converter. Also, it can be designed to handle very slow waveforms, because the choice of RC is essentially arbitrary, within the constraints that it be much longer than the longest period to be measured, and short enough to provide adequate settling time.

There are two ways in which the scheme can be implemented: direct multiplication-division, and via a specially-designed true-rms module employing log-antilog operations.

Direct multiplication-division with implicit feedback can be implemented using such 3-variable devices as the AD531 IC multiplier-divider (see Figure 15, Chapter 2-3), the 433 multifunction module, or the 434 multiplier-divider. The AD531, used in the rms-circuit, requires careful trimming, plus at least 1 external amplifier. The circuit is accurate to within only 1% or 2%, because of device limitations. The 433 or 434 can be used to make a very high accuracy (within 0.1%), wide-dynamic-range (1000:1) rms circuit. However, it is a 1-quadrant device and must be driven by a high-precision rectifier if the rms of bipolar signals is required. Neither the 433 nor the AD531, with external circuitry, can compete with specialized rms modules on the basis of price or performance.

Special-purpose log-antilog rms-dc conversion combines logarithmic and implicit computing techniques to achieve overall errors of less than 15mV + 0.2% of expected value over a 1000:1 dynamic range. It consists of a log-antilog squarer-divider, with an absolute-

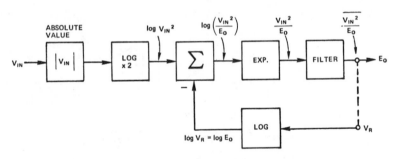

Figure 6. Log-antilog RMS-to-DC converter

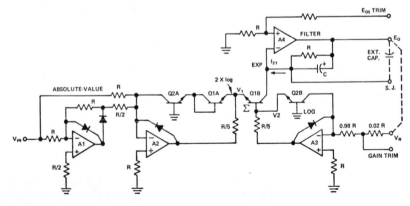

Figure 7. Schematic diagram of log-antilog RMS-to-DC converter

value (full-wave rectifier) front end, and an internally-connected filter. A block diagram of the circuit is shown in Figure 6, and a schematic appears in Figure 7.

The bipolar input signal, V_{IN}, is converted to a current representing the magnitude by an absolute-value circuit (amplifier A1 and the associated components). The unipolar current is transformed into a voltage proportional to twice the logarithm of the input (i.e., $2 \log x = \log\{x^2\}$) by two junctions in series (A2, Q1A, and Q2A). The log of the *output* is obtained by A3 and Q2B and subtracted from the log-of-the-square-of-the-input; the result is antilogged by Q1B and A4, averaged by the filter RC, in the feedback path of R4, and transformed to the output voltage by R.

The derivation shows how the temperature-sensitive terms are cancelled, if the transistors are matched and isothermal.

$$V_1 = -2\,\frac{kT}{q}\,\ln\!\left(\frac{V_{IN}}{RI_{ES}}\right) \tag{7}$$

$$V_2 = -\frac{kT}{q}\ln\!\left(\frac{E_o}{RI_{ES}}\right) \tag{8}$$

$$I_{21} = I_{ES}\,\epsilon^{\,q(V_2 - V_1)/kT} \tag{9}$$

$$V_2 - V_1 = \frac{kT}{q}\left(2\ln\frac{V_{IN}}{RI_{ES}} - \ln\frac{E_o}{RI_{ES}}\right) \tag{10}$$

$$= \frac{kT}{q}\ln\!\left(\frac{V_{IN}^{\,2}}{E_o\,RI_{ES}}\right)$$

$$I_{21} = I_{ES}\,\frac{V_{IN}^{\,2}}{E_o\,RI_{ES}} = \frac{V_{IN}^{\,2}}{RE_o} \tag{11}$$

$$E_o = \overline{I_{21} R}\bigg|_{RC} = \overline{\left(\frac{V_{IN}^2}{E_o}\right)} \qquad (12)$$

For signal frequencies that are high compared to $1/2\pi RC$ $\overline{E_o} \cong E_o$; hence

$$E_o = \sqrt{\overline{V_{IN}^2}} \qquad \text{(cf. (5) and (6))} \qquad (13)$$

Thus, the output voltage is equal to the rms of the input voltage assuming that the corner frequency of the low-pass filter is much lower than the lowest-frequency component of the input signal. The circuit responds accurately to dc inputs, which require no averaging. Because it can respond to dc, this rms-dc converter can be calibrated with ease, since a dc reference can be used for comparison. The circuit can also be used for *mean-square* output if the denominator input, which controls the scale factor, is supplied by a constant voltage instead of feedback from the output.

For dc and low-frequency inputs, errors can be trimmed to very low values ($\sim$ 0.02%). The primary sources of static errors are the voltage and current offsets of the operational amplifiers. At high crest factors (3 to 5), log-conformity error of the transistors introduces nonlinearity; even so, the error of the circuit will increase by only 5%-of-reading for c.f. = 10.

The dynamic response of the rms circuit depends on the signal level. In the 440, at 20Vp-p input level, the slewing rate of A1 limits the overall −3dB sine-wave bandwidth to 500kHz and the 1%-of-reading error-bandwidth to 50kHz. At 2Vp-p, the 1%-of-reading error bandwidth is typically about 150kHz. However, the bandwidth decreases as the signal level is reduced further, because of the reduction of current through the log transistors Q1A and Q2B. The bandwidth for signals in the range of 1 to 2Vrms can be increased to the order of 5MHz by using fast amplifiers for A1 and A2.

The offset and scale-factor drift with temperature or power-supply variations are negligible sources of error. The symmetrical arrangement of the log and antilog transistors results in complete cancellation of the temperature-dependent terms, kT/q and I_{ES} The result is that the scale-factor drift is determined primarily by the temperature coefficient of the resistors, which can be 10ppm/°C or less. The major source of output drift is the offset voltage- and current-drift of the output amplifier, A4, and the feedback amplifier, A3. The input offset is 1mV or less, and the drift is about $20\mu V/°C$.

RMS-TO-DC CONVERTER SPECIFICATIONS

The most salient feature of a true-rms-to-dc converter is that it ideally has no error due to an indirect approximation to the rms. Static errors are due only to scale factor, linearity, and offset errors: dynamic errors are due to insufficient bandwidth at the high end of the frequency range, insufficient averaging time at the low end, and linearity errors that affect crest factor in mid-band.

Salient specifications of rms modules are summarized in an example: the set of specifications for the Model 440 general-purpose (low-cost) rms module (Table 2). Employing a circuit similar to that of Figure 7, it has provisions for output-offset adjustment, scale-factor trim, and for the addition of external capacitance to increase the averaging time. The Appendix to this chapter illustrates means of reducing the ripple by the use of an external filter, and of obtaining a better approximation to a true time-average over any time period through the use of gated integration and sampling, and incremental summation.

The specifications can be interpreted as follows:

Maximum Error

A catchall specification for quick reference, this is the maximum deviation of the dc component of the output voltage from the theoretical output value at full-scale output.

Accuracy

Maximum error, no adjustment is the amount by which the output will differ from the theoretical value. It is the sum of a fixed error and a component proportional to the theoretical output.

TABLE 2. SPECIFICATIONS OF A TYPICAL GENERAL-PURPOSE
RMS-TO-DC CONVERTER
(typical at 25°C unless noted otherwise)

Parameter Model →	440J	440K
MAXIMUM ERROR	0.35%	0.15%
ACCURACY		
Maximum Error, No Adjustment	±15mV±0.2%	±5mV±0.1%
Maximum Error, Externally Adjusted	±10mV±0.1%	±2mV±0.05%
Typical Error, Externally Adjusted	5mV±0.05%	1mV±0.05%
TEMPERATURE COEFFICIENTS		
Output Offset, maximum	0.2mV/°C	0.2mV/°C
Scale Factor, maximum	0.02%/°C	0.02%/°C
DYNAMICS		
Frequency for Specified Error, Minimum	10kHz	10kHz
Frequency for 1%-of-Reading Error		
Sine Wave, 20Vp-p, minimum	50kHz	50kHz
Sine Wave, 2Vp-p, minimum	100kHz	100kHz
Sine Wave, 0.2Vp-p	8kHz	8kHz
-3dB Bandwidth		
Sine Wave, 20Vp-p	500kHz	500kHz
Sine Wave, 2Vp-p	500kHz	500kHz
Sine Wave, 0.2Vp-p	100kHz	100kHz
CREST FACTOR @ 1Vrms Output		
For Specified Error	2	2
For 1% Additional Error	3	3
FILTER		
Time Constant (internal)	10ms	10ms
Time Constant Increase vs. External	50ms/μF	50ms/μF
INPUT		
Voltage Range, Specified Operation	±10V	±10V
Voltage Range, Maximum	±V_s	±V_s
Resistance	10kΩ	10kΩ
OUTPUT		
Voltage Range, Specified Operation	0 to +10V	0 to +10V
Current, minimum available	10mA	10mA
POWER SUPPLY		
Error Sensitivity	0.2mV/V	0.2mV/V
Range for Specified Performance	±14V–±16V	±14V–±16V
Operating Voltage Range	±6V–±18V	±6V–±18V
Quiescent Current	±10mA	±10mA
TEMPERATURE RANGE	0°-70°C	0°-70°C

7557   1/21   1

1 60916550110   5.25≑NTXT
                          5.25 STL
                           .00 TAX
                          5.25 CASH
                           .00 CNG

4/12/84   1 CASH   5.25 TTL

Maximum error, externally adjusted is the amount by which the output will differ from the theoretical value when the output offset and scale factor have been trimmed.

Temperature Coefficients

Output offset: the maximum sensitivity of the output to temperature with zero input, or the maximum displacement of the average error vs. output plot over a temperature range, divided by the temperature range.

Scale factor: the maximum sensitivity of the slope of the output-vs.-input (dc) to temperature

Dynamics

Frequency for specified error is the minimum value of frequency (at the high-frequency end) at which the error is guaranteed to be equal to or less than the specified midband value (sine-wave input). Error at the low-frequency end is governed by the choice of filter, both internal and external. Figure 8 shows a typical error-vs.-frequency plot. Low-end error behavior is shown with no added capacitance, and with 1 and $10\mu F$ added (60 and 510ms filter time constants). At very low frequencies, the output of the circuit follows the instantaneous ($\rightarrow$ dc) value of the input.

Frequency for 1%-of-reading error is the minimum value of frequency (at the high end) at which the error (except for offset) is guaranteed to be equal to or less than 1% of reading. It is a function of peak-to-peak amplitude.[*]

–3dB Bandwidth is the minimum value of frequency (at the high end) at which the error is guaranteed to be equal to or less than 30%.

Crest Factor: (a property of the signal) the ratio of peak signal voltage to the ideal value of rms; the value of crest factor for which the error is maintained within specified limits at a given level of

---

[*]For non-sinusoidal waveforms, the attenuation of harmonics increases dynamic errors at lower fundamental frequencies than are specified here. Since square waves and pulses are familiar waveforms and, at the same time, rich in harmonics, Figure 9 shows the error (for such inputs) caused by frequency-response rolloff in the input stages, as a function of pulse width (seconds) and the specified –3dB bandwidth (Hz).

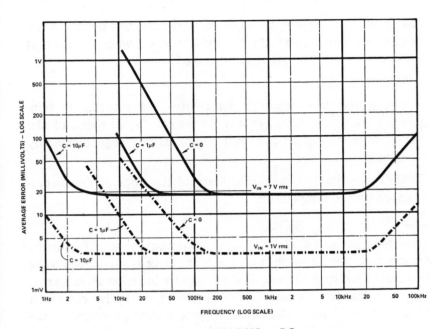

a. Average error of the 440J RMS-to-DC converter as a function of input amplitude, frequency (sine waves), and externally-connected capacitance

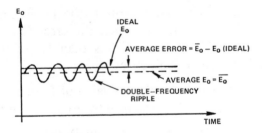

b. Average and ripple components of error of rms-to-dc converter at low frequencies (sinusoidal input). The finite averaging time of the filter produces both a double-frequency ripple component and an offset of the average output value (see Appendix A to this chapter).

Figure 8. Output error of rms-to-dc converter as a function of frequency. The error values plotted are the average errors measured after filtering out the ripple.

rms for a midband-frequency signal. Crest factor, rms, and average are plotted in Figure 10 as a function of duty cycle, for rectangular pulses.

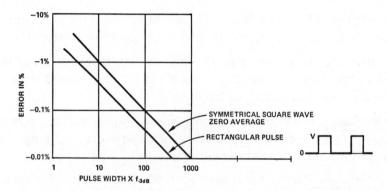

Figure 9. Error of rms-to-dc converter due to finite bandwidth, for pulse and square-wave inputs.[1]

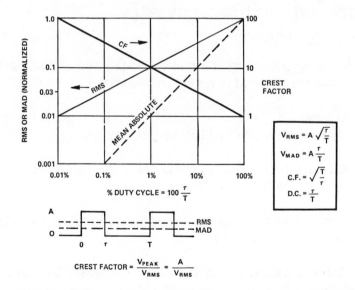

Figure 10. RMS, average and crest factor for rectangular pulse train

[1] Source: "RMS Voltage Measurements — Which Method Works Best?" by Roy Chapel, *Electronic Products Magazine,* January 15, 1973, p. 36.

Filter time constant and Ext. Cap. The time constant of the internal averaging filter, and the increase of time constant per μF of added external capacitance.

Input: the voltage range over which specified operation is obtained, the absolute-maximum voltage, and the effective input resistance.

Output: the maximum output range for rated performance, the minimum current guaranteed available at full-scale output voltage, and the source resistance of the output.

Power Supply: Output-error sensitivity to supply voltage, power-supply range for specified performance, power-supply range for operation, and quiescent current drain.

Temperature Range: the range of temperature variation for operation within specifications. Temperature coefficients are determined by 3-point measurements $(T_H - 25°C)$, $(25°C - T_L)$, when measured.

TESTING THE RMS-TO-DC CONVERTER

USEFUL EQUIPMENT

The following equipment is useful for testing and calibrating rms-to-dc converters. Starred items (*) are essential.

1. * Accurate dc reference: ±1mV (or less) absolute error, adjustable from 0V to ±10V in ±10mV steps

2. * Accurate ac reference: 0.05% (or less) absolute error, rms output 1 to 2V from 100Hz to 10kHz

3.   High-prevision voltage divider with buffered output, adjustable from 1:1 to 100:1

4. * Digital voltmeter, dc-reading, 4½ or 5½ digits, 0.01% error

5. * Digital voltmeter, ac-reading, 4½-digit resolution; can respond to either mean-absolute value or true rms; used for monitoring signal-generator amplitude and making comparative accuracy tests for (clean) sinewaves

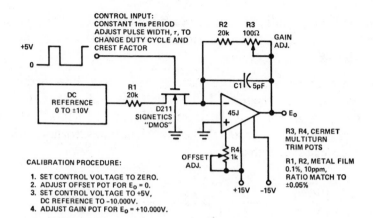

Figure 11. Pulse generator with accurate, adjustable amplitude

6.   Accurate pulse generator (see Figure 11)

7.   True-rms digital voltmeter, capable of responding to ac and ac + dc signals, 4½ digit

8.   Oscilloscope, 5MHz bandwidth

9. *   Sine-wave generator with low distortion ($<0.1\%$), 10mVrms to 7Vrms, 10Hz to 5MHz

TEST PHILOSOPHY AND PROCEDURES

The primary objective in testing an rms converter is to determine how accurately it can convert an ac signal (sine-wave or complex, *including* a dc component) to a dc voltage that is equal to the rms value of the input waveform.

The basic test method is to apply a signal of known rms value to the input, as illustrated in Figures 12 through 17, and measure the resulting dc output with an accurate meter.

This sounds easy, but it isn't all that easy in practice. First, the error of the rms converter depends on the properties of the input waveform: amplitude, frequency, wave shape (crest factor); second, it is difficult to obtain an ac signal, for example a sine wave, of accurately-known rms value.

For these two reasons, it is usually easier to calibrate the rms converter with an accurate dc reference (assuming that the converter responds to dc), as illustrated in Figure 12. (It will also be found useful to calibrate the input signal waveforms with an accurate rms-to-dc converter.)

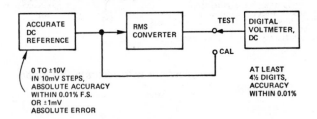

Figure 12. Measurement of absolute accuracy at DC

TABLE 3. TEST CONDITIONS FOR CHECKING DC ACCURACY OF RMS-TO-DC CONVERTER

	V_{IN}	E_o (IDEAL)	Description of Test
1.	0V	0V	Total zero offset, referred to output
2.	+10mV	+10mV	Input offset: + offset will cause output to read $>$ 10mV; − offset vice versa
3.	−10mV	+10mV	Input offset: + offset will cause output to read $<$ 10mV; − offset vice versa
4.	+100mV	+100mV	Low-end accuracy
5.	−100mV	+100mV	Low-end accuracy
6.	+1.00V	+1.00V	Mid-scale accuracy, check for agreement with specification
7.	−1.00V	+1.00V	Mid-scale accuracy, check for agreement with specification
8.	+10.00V*	+10.00V	Full-scale accuracy and symmetry (compare + and − readings)
9.	−10.00V*	+10.00V	Full-scale accuracy and symmetry (compare readings)

*Or ± specified full-scale input for the device under test

The input voltages listed in Table 3 include the most critical points on the dc response function. Other intermediate voltages may be used, as desired, to obtain more-detailed information for a plot of the rms converter's error.

The dc measurements provide information for adjusting the scale factor, the output offset, and (externally, if not available internally) the input offset, which affects symmetry.

AC Measurements

After the converter has been tested for dc errors, and any necessary adjustments have been performed, the ac error may be checked, using a sine wave as the input source. If an ac standard, with 0 offset, is available, the configuration of Figure 13 may be employed. If an ac standard is not available, but a good ac DVM is instead, Figure 14 may be used. The response to ac signals should be tested at a number of different input levels and frequencies to check for nonlinearity, bandwidth limits, and limitations at the low end due to the averaging time of the device (including the intended value of externally-connected capacitance).

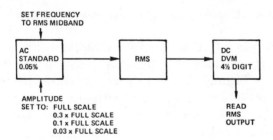

Figure 13. Measurement of absolute accuracy relative to an AC standard

Figure 15 shows a configuration for testing the linearity of the converter over a range of input amplitudes. The amplitude and frequency of the source are set once. Then a low-distortion precision attenuator is used for amplitude adjustment (a) to ensure that the shape does not change, and (b) to obtain precisely-calibrated ratios of input to full-scale.

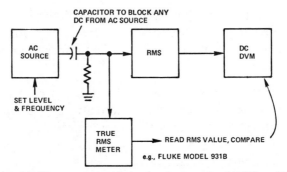

Figure 14. Measurement of absolute accuracy of RMS-to-DC converter relative to accuracy of true RMS meter

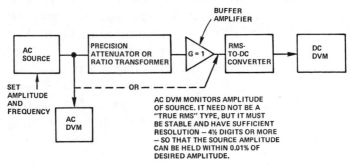

Figure 15. Measurement of nonlinearity for AC input

Figure 16 shows a configuration for measuring error as a function of crest factor, employing a precision pulse source, such as that illustrated in Figure 11. It is easiest to check the crest-factor handling capability with a rectangular pulse of known amplitude and duty cycle. The relationship between crest factor and duty cycle is shown graphically, and in equation form, in Figure 10. Crest factor can be set in terms of the output rms or the input amplitude (which should be less than rated peak input).

The minimum pulse repetition rate, and therefore the frequency at which the crest factor —for rectangular pulses— (for a given accuracy) decreases in terms of the midband value, will be determined by the averaging time-constant. As a rule, the error of the rms-to-dc converter will increase in direct proportion to the crest factor, if the pulse width and repetition rate are within the high and low bandwidth limits of the converter.

Figure 17 shows how the frequency response of the rms-to-dc converter may be measured.

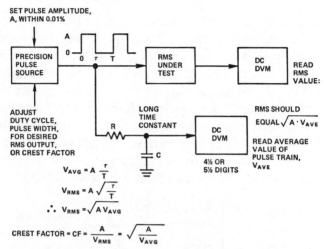

Figure 16. Measuring accuracy of the rms-to-dc converter as a function of crest factor and pulse width (duty cycle)

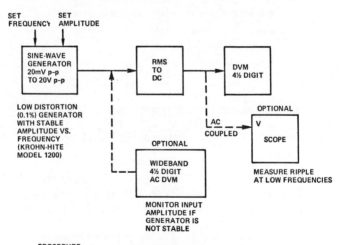

PROCEDURE:
SET GENERATOR AMPLITUDE FOR FULL-SCALE INPUT TO RMS TO DC CONVERTER. SET FREQ. TO MIDDLE OF RMS-DC FREQUENCY RANGE, THEN SWEEPUP TO FREQUENCY WHERE RMS OUTPUT CHANGES ±1% FROM MIDBAND VALUE, THEN FOR ±30% (±3dB) FROM MIDBAND. REPEAT FOR INPUT AMPLITUDES OF 10% AND 1% OF FULL SCALE.

Figure 17. Measuring frequency response of rms-to-dc converter function of crest factor and pulse width (duty cycle)

APPENDIX TO CHAPTER 3-7

A. *Use of a Low-Pass Filter to Reduce Ripple at Low Frequencies*

With externally-connected capacitance, the 440 can be used to compute accurately the rms of signals having components at quite low frequencies. While the ripple content is not important if the output is read on an analog meter (which provides mechanical averaging) or "eyeballed" on an oscilloscope, it can lead to errors if the output is converted to digital form, and to annoyance if a digital panel meter is used for observations. A low-pass filter may be used following the rms circuit, to add an additional stage of averaging and attenuate ripple components.* The illustration shows a low-cost circuit configuration for accomplishing this.

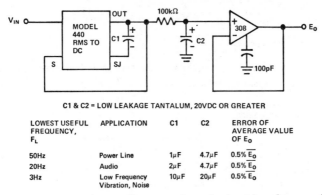

C1 & C2 = LOW LEAKAGE TANTALUM, 20VDC OR GREATER

LOWEST USEFUL FREQUENCY, F_L	APPLICATION	C1	C2	ERROR OF AVERAGE VALUE OF E_O
50Hz	Power Line	1μF	4.7μF	0.5% $\overline{E_O}$
20Hz	Audio	2μF	4.7μF	0.5% $\overline{E_O}$
3Hz	Low Frequency Vibration, Noise	10μF	20μF	0.5% $\overline{E_O}$

A) Adding a low-pass filter to reduce ripple at low frequencies

B. *Use of Controlled Integration and Digital Averaging to Shorten the Averaging Time*

For many applications, the running average provided by an RC low-pass filter (or thermal time constants) does not conveniently provide a close-enough approach to the mathematical average over

---

*Since this averaging is outside the loop, it is only useful for removing ripple components. It is not a substitute for an increased *inside-the-loop* time constant to reduce low-frequency errors. If the inside-the-loop time constant is insufficient, the *average level* of the output will be seen to be in error when the ripple has been satisfactorily filtered out. As a rough guide, if the p-p ripple is less than 10% of the output level, the average error component will be negligible, and external filtering may be used to reduce the ripple to the desired level. If the ripple is greater, it is likely to be accompanied by significant average error; an increased value of C1 may be necessary, in addition to the external filter.

a fixed time period, especially if that period is *very* long, because of the long settling time (many cycles) and the difficulty of obtaining stable capacitances and resistances of adequate magnitude.

The running average is also inadequate if the mean square changes at a rate comparable to the averaging time constant required to smooth out the ripples in the signal (for example, in the case of an amplitude-modulated waveform).

For such cases, controlled integration, sample-hold, and even the use of digital averaging can greatly improve the accuracy of the rms- or mean-square-to-dc conversion and provide an effective averaging time that is arbitrary, ranging from seconds to hours, or even days.

A direct computation scheme, employing all of the above, is shown in illustration (B). The input signal is squared and integrated repeatedly over an interval T, with a characteristic time (RC) equal to T. Thus, the output of the integrator after interval n is the mean square of the nth interval.

At the end of each interval, the integrator output is acquired and held by a sample-hold. Its value (X_n) is compared with the average of the mean-squares up to that time, Y_{n-1}. The difference, divided by n, is added to Y_{n-1} to produce the new average, Y_n, which is stored in a sample-hold for use in computing the next value of Y, Y_{n+1}. The formula for the average over n intervals is

$$Y_n = Y_{n-1} + \frac{X_n - Y_{n-1}}{n} \tag{14}$$

The average of the mean-squares is then square-rooted to compute the true rms over the entire period. For example, if each integration period is 15 minutes, 1000 counts will provide the rms over a 10-day period.

The division by n can be performed by a multiplying DAC, such as the AD7520, connected as a divider, with the digital input, n, supplied by a clock-driven counter, at the appropriate rate. (The count should start with a preset count of 1, to avoid the possibility of dividing by zero. The divider is scaled for unity gain at n = 1, decreasing for higher values.)

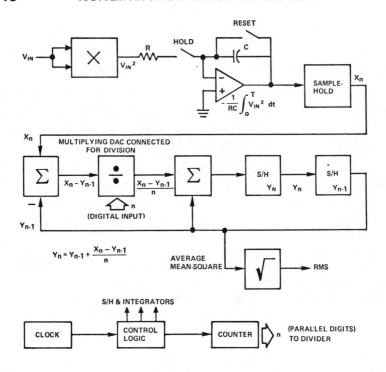

B) Long-term rms computation

The integrator and sample-holds are sequenced by control logic signals derived from the basic clock pulses. During the computing interval, the Y_{n-1} sample-hold is in *hold* and both the X_n and Y_n sample-holds are in *sample*, tracking the integral and continuously computing the next value of Y (count = n) right up to the end of the interval. At the end of the interval, the following events ensue: X_n switches to *hold;* the integrator is quickly reset and immediately starts integrating over the next interval; Y_n switches to *hold*; Y_{n-1} samples the final value of Y_n and returns to *hold*; then, X_n and Y_n revert to *sample*, the counter is incremented to n + 1, and the next value of Y is continuously computed right up to the end of the next interval.

The Y_{n-1} sample-hold, which is quite critical, since its errors, including drift, are cumulative, is perhaps a "sample-infinite-hold" circuit (an A/D converter with output taken from its internal DAC).

Log Circuit Applications

Chapter 1

In Chapter 3-1, the design and testing of logarithmic devices have been discussed and explained in some detail. This chapter will consider the various tradeoffs a designer is confronted with when applying a log device with a predetermined reference. The specifications of log modules, which were listed in Chapter 3-1, will be discussed in greater depth here, in the context of the manner in which they affect performance in an application.

SELECTING THE APPROPRIATE LOG DEVICE

There are many options available to the designer. These range from performing one's own design — from the ground up — using the most appropriate approach (linear approximations, diodes, monolithic-dual transistors, or other components characterized by logarithmic transfer characteristics) to purchasing one of the several types of log module available on the market today. Available standard devices, with guaranteed performance, range from modules containing a simple transistor-pair packaged with a temperature-compensating resistor network to complete log-amplifiers containing op amps, a reference-current source, and all the necessary temperature- and frequency-compensating components.

THE BASIC LOGARITHMIC ELEMENT

The simplest module available for temperature-compensated log circuits is the basic log element. A schematic of the Analog Devices Model 751 is shown in Figure 1.

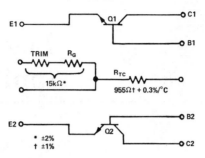

Figure 1. Schematic diagram of 751N basic log element

At first glance, one may think that this is too trivial a module to consider for purchase; it contains only two transistors and three resistors. Superficially, this is true, but it is also true (though less obvious) that an essential characteristic of such a device is its isothermal environment. Also, despite the many advantages of monolithic dual transistors, a matched discrete-transistor pair is better-suited to the purpose of the device. Since the log element is the heart of a circuit that provides a logarithmic output, a thorough understanding of these two points is essential to successful design.

The transistors used in Model 751 are discrete transistors especially selected for log characteristics and carefully matched. There is a subtle, but important, advantage for discrete transistors: the combination of base spreading resistance and contact resistance, which can be lumped together and considered as a small resistor in series with the emitter, is the major cause of log conformity error at current levels from $100\mu A$ to $1mA$. The discrete transistors used in the 751 have a much lower value of series resistance, hence three to ten times better performance at these levels (without compensation circuitry) than do available high-β monolithic pairs.

In order to capitalize on this advantage (as well as the temperature-tracking capability of matched transistors), it is imperative that the two transistors be located in an isothermal environment. At constant current, the V_{be} of a transistor has a temperature-coefficient of approximately -2mV/°C. If the junctions are allowed to be as little as $0.5°C$ apart, the corresponding voltage difference is $1mV$, which corresponds to an input-current change of 4% (If $1mV = kT/q \ln(1 +\lambda) = 26mV \ln(1 +\lambda)$, then $\lambda = 0.04$). If errors from this source are to be held at less than 0.5%, the junctions must be held within $130\mu V$, or $0.06°C$, under all conditions of ambient or internally-generated temperature change.

To obtain an isothermal environment, conductive epoxy and "heat clips" are used, and particular attention is paid to the layout. This environment is also essentially isothermal as regards the temperature-compensating resistor. To appreciate this advantage, one need only consider the effect of a small amount of air circulation over similar components sitting on an open circuit board. Although the compensation resistor may be precisely trimmed to eliminate the temperature-dependence of the scale factor, its effect can be nullified (and even detrimental) for differences greater than $1°C$. These points are illustrated in Figure 2.

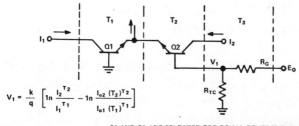

$$V_1 = \frac{k}{q} \left[\ln \frac{I_2^{T_2}}{I_1^{T_1}} - \ln \frac{I_{o2} (T_2)^{T_2}}{I_{o1} (T_1)^{T_1}} \right]$$

IF $T_2 = T_1 = T$
AND $I_{o1} (T) = I_{o2} (T)$

$$V_1 = \frac{kT}{q} \left[\ln \frac{I_2}{I_1} \right]$$

$$E_0 = \left[1 + \frac{R_G}{R_{TC}} \right] \frac{kT}{q} \ln \frac{I_2}{I_1}$$

Q1 AND Q2 ARE SELECTED FOR EQUAL REVERSE-SATURATION CURRENTS AT A GIVEN TEMPERATURE. THIS MATCH HOLDS WELL FOR LARGE TEMPERATURE EXCURSIONS, BUT BOTH TRANSISTORS MUST BE AT THE SAME TEMPERATURE. IF $T_1 - T_2 \neq 0$, THE ERROR BETWEEN V_{be1} AND V_{be2} WILL BE APPROXIMATELY -2mV/°C. R_{TC} IS A TEMPERATURE-SENSITIVE RESISTOR WHICH IS TRIMMED TO CANCEL THE EFFECT OF THE kT/q TERM. ITS TEMPERATURE (T_3) MUST BE THE SAME AS THAT OF T_1, T_2, OR THE DESIRED CANCELLATION OF TEMPERATURE EFFECTS CANNOT BE ACHIEVED.

Figure 2.Importance of an isothermal environment for the basic log element.

When to Apply the Basic Log Element

The simplest of all log modules, it contains only the log transistors, calibrated temperature-compensating resistors, and the isothermal environment necessary for reliable and predictable log operation. By itself, the basic log element can perform no useful function, but when coupled with two or three op amps, a reference-current source, and frequency compensation, a complete log amplifier can be built.

The basic log element should be considered primarily for special-purpose designs, calling for considerable design flexibility, e.g., physical form, current ranges, choice of reference or gains, etc. Though allowing considerable freedom for the circuit designer, it requires the greatest amount of care in the external wiring, circuit layout, and choice of components, to obtain best results.

The designer must take especial pains to obtain the best compromise of speed and dynamic stability. Familiarity with the stability considerations in Chapter 3-1, especially Figures 5 through 7 in that chapter, is essential. The *Log Circuits* Application Note[1] will also be of value.

THE LOG TRANSCONDUCTOR

This term describes a device that combines a log element, reference-current source, op amp for the reference transistor, and the related frequency-compensation components. Figure 3 shows the basic circuitry of a log transconductor (Model 752) and its relationship to the basic log element (751).

Besides offering a completely-tested circuit, the log transconductor eliminates a number of application problems relating to stabilization: it is dynamically stable at all rated levels of input current.

To complete the log amplifier circuit, the designer needs only to provide an external operational amplifier and the two trim

---

[1] "Design of Temperature-Compensated Log Circuits Employing Transistors and Operational Amplifiers," by W. Borlase and E. David, Analog Devices, Inc., March, 1969.

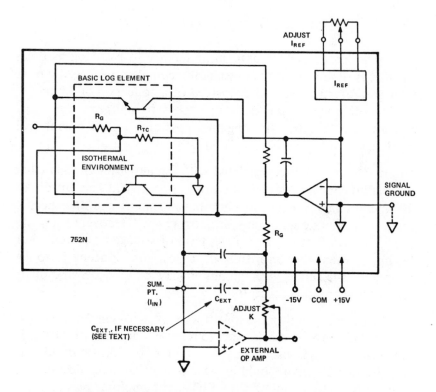

Figure 3. Functional diagram of a log transconductor and its relationship to the basic log element.

adjustments for I_{ref} and the scale factor, K. When currents beyond the specified range of the log transconductor are to be "logged," it may be necessary to add capacitance (10-20pF) in the feedback circuit of the external amplifier (Figure 3) for stability, at the cost of reduced bandwidth.

It should be noted that Figure 3 is a *functional* diagram of the 752 transconductor. For a wholly-packaged circuit, a separate basic log module is not used; in many cases a monolithic dual transistor is used to obtain improved temperature-tracking. To overcome the series-resistance problem mentioned above, a feedback compensation circuit is used internally to reduce the series resistance.

When to Apply the Log Transconductor

Log transconductors are used in preference to basic log modules when it is desired to use an essentially-complete packaged circuit with guaranteed performance instead of a collection of parts; yet, on the other hand, the designer wishes to optimize the price/ performance mix by choosing the most appropriate operational amplifier to fit his needs.

The major error terms associated with log amplifier circuits (that are readily controllable) are the offset voltage and bias current of the operational amplifier. By choosing an amplifier to fit a specific application, the designer may, in most cases, effectively eliminate amplifier-caused errors at a given level of voltage or current. For example, when the logarithm of a low-level current is to be computed, the Model 42K amplifier may be used because of its 100fA bias-current specification.

If 100fA is compared to the lowest-level of signal current specified for the 752, 1nA, it can be seen to represent an error of 0.01%. If one starts by choosing this low value of bias current, its contribution to error could be completely ignored, despite the characteristic doubling/10°C of FET-input-amplifier bias current. At 65° C, the bias current would have increased by a factor of $2^4 = 16$, which is still insignificant compared to the other errors.

If the input signal is a voltage, the external op amp would be chosen for minimal offset voltage and drift. For ultimate low-level performance, "spikeless" chopper-stabilized amplifiers are perhaps the best choice; but one should also consider such low-drift chopperless types as the 184L and the AD504M, and especially, the 52K (I_b = 2pA, $\Delta E_{os}/\Delta T = 1\mu V/°C$).

When economy is of prime concern, the log transconductor and a low-cost IC op amp can be the best answer. To achieve good performance when using a low-cost (modest-performance) amplifier, the input signal should be scaled so that its geometric mean falls in the center (i.e., geometric mean) of the range determined by the log transconductor's upper limit and 100X the bias voltage or offset current of the amplifier. With this approach, excellent results can be obtained over a limited range. Figure 4 shows how the log transconductor is connected with an external op amp. Table 1 provides a brief selection guide.

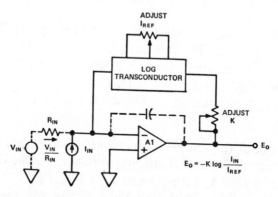

Figure 4. Application of log transconductor. If the input signal is a current, connect it directly to the summing junction, as shown. If the input signal is a voltage, connect it to the summing junction through a resistance as shown by the dashed lines. Note that R_{IN} includes the source resistance of V_{IN}.

TABLE 1. SELECTION CRITERIA FOR A1 (FIGURE 4)

INPUT SIGNAL	SELECT A1 FOR	RECOMMENDED OP-AMP TYPE
VOLTAGE	LOW OFFSET VOLTAGE	CHOPPER (234) CHOPPERLESS LOW-DRIFT (184L, 504M, 52K)
CURRENT	LOW BIAS CURRENT	ELECTROMETER (42K)
VOLTAGE OR CURRENT; LIMITED DYNAMIC RANGE	PRICE-REASONABLE PERFORMANCE	LOW-COST BIPOLAR OR FET, I.C. OR MODULE: AD301A, AD540, MODEL 40, AD308

THE COMPLETE LOG AMPLIFIER

The complete log amplifier contains a log transconductor, all the necessary trim circuitry, and a high-quality FET-input op amp. This is the most convenient type of log module, since the designer has only to connect power (usually ±15V @ ±10mA), input and output leads. All the necessary frequency-stabilization and trimming have been performed by the manufacturer.

When to Apply the Complete Log Amplifier (and When Not to)

The complete log amplifier is the starting point for all new log-circuit designs. By freeing himself from the problems of designing, building, and trimming a log amplifier, the designer has more time available to deal with the problems involved in the instrument, apparatus, or system that led to the use of a log amplifier.

Performance of the complete log amplifier is adequate to deal with a wide range of input currents and voltages; and the fixed choices of reference ($10\mu A$, $100mV$) and scale factor ($2/3$, 1, 2) provide convenient scaling, which can be modified for the system by external gains and biases.

If wider ranges of current or voltage, greatly-different local scaling, or paired operations (such as the use of 752P and 752N in hyperbolic-sine operations), are necessary, then the need for the flexibility inherent in the log transconductor (or basic log devices) will become apparent (see Page 102).

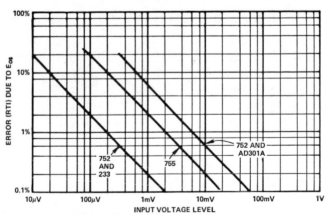

Figure 5. Effect of offset on dynamic range in terms of error vs. input level. Errors are for $+2°C$ temperature change. Greater temperature changes move curves up proportionally (e.g., 755 at 10mV has 0.2% change for $2°$, 2% for $20°$).

As an aid to determining the performance tradeoffs between the complete log amplifier (755) and the log transconductor-plus-external amplifier, the graphs of Figures 5 and 6 will be found helpful. They show the error as a function of input-voltage and current levels for the 755 and for the 752 with-external-op-amps of several types. The voltage error used in the plots is that caused by a 2°C variation from an initially-zeroed condition. The current error is essentially that caused by the amplifier's bias current.

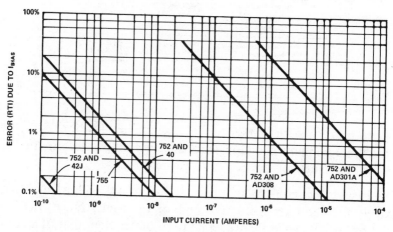

Figure 6. Effect of op-amp bias current on dynamic range in terms of error vs. input level, for various combinations of op amps and log transconductors

Selecting the Best Log Device for the Application

Table 2 provides general guidelines as to the choice of approach. The choice will be determined by the designer's principal objective: best performance, lowest cost, easiest-to-apply.

If cost is the criterion, then the basic log element may be the answer, but the saving is somewhat marginal; the designer must honestly evaluate the cost of designing, assembling, and testing the finished device. If the design is to be used in quantity, costs of drawings, parts inventorying, and production engineering must also be anticipated.

TABLE 2. SUMMARY OF SALIENT FEATURES OF CONSTANT-REFERENCE LOG MODULES (SUMMER, 1973)

Log Module	Description Contents and Applications	Advantages	Disadvantages
Basic Log Element Model 751	{ 2 Matched Log Transistors { Scaling and Temperature-Compensating Resistors For Special-purpose log. designs	Lowest Cost Greatest Flexibility	Most Complex to Apply Requires at least 2 external op amps plus dynamic stabilization in conventional log application.
Log Transconductor Model 752	{ Basic Log Element { Reference-current source To optimize operation at low levels	Best performance obtainable through op amp choice	Requires external op amp, gain trim I_{REF} trim
Log Amplifier Model 755	{ Log Transconductor { FET-Input Op Amp The initial choice for all fixed-reference log applications.	Easiest to apply. Meets specs with no trimming or external components. Best performance over a wide range.	Op amp is optimized for most (but not all) applications

To optimize performance for a specific application, the log transconductor and a high-performance op amp selected for the application can offer the best performance.

The easiest log module to apply is the complete log amplifier. Except for the extremely low end of the signal range, the log amplifier offers performance equal to or better than that of any of the other choices.

While the generalities given above are helpful in selecting the best log device for the application, the proper choice can be made only when specific information regarding signal level, source impedance, and acceptable error has been developed.

Once this is known, the limitations on dynamic range caused by the input parameters of the op amp associated with the log modules can be determined.

STEP RESPONSE

The dynamic parameters of log modules are highly dependent on signal level, as Chapter 3-1 has demonstrated. Perhaps the most useful parameter to discuss in detail is the step response, since the response to a step of a given magnitude is usually a matter of prime concern. The time required for the output to change (or "slew") from one level to another is dependent on the magnitudes of the input currents and on the direction, i.e., whether the current is increasing or decreasing.

Slewing rate of a log transistor's base-emitter voltage can be explained in terms of the effects of current level on the transistor's base-charging capacitance (C_b), transconductance (g_m), and incremental space-charge-layer capacitance (C_{je}). Base-charging capacitance, C_b is defined as

$$C_b = \tau_F g_m = \tau_F \frac{q}{kT} |I_c| \tag{1}$$

where τ_F is the average charge-replacement time in the base and I_c is the collector current.

Transconductance, g_m, and — in turn — C_b are proportional to I_c, provided that the base current is much greater than the reverse saturation current. This condition is met for all log modules operated within the specified range.

The capacitance of the transistor hybrid Π model, C_Π, directly controls the common-emitter current gain at high frequencies. It is equivalent to

$$C_\Pi \equiv C_{je} + C_b \tag{2}$$

which, from (1), may be written as

$$C_\Pi \equiv C_{je} + \tau_F \frac{q}{kT} |I_c| \tag{3}$$

The dependence of ω_t, the frequency at which current-gain is unity, on C_Π is

$$\frac{1}{\omega_T} \equiv \frac{C_\Pi + C_\mu}{g_m} \tag{4}$$

where C_μ is the feedback capacitance of the hybrid model.

Substituting (2) and $g_m = \frac{q}{kT} |I_c|$ into (4),

$$\frac{1}{\omega_T} = (C_{je} + \tau_F \frac{q}{kT} |I_c| + C_\mu) \left[\frac{kT}{q|I_c|} \right] \qquad (5)$$

$$= \tau_F + (C_{je} + C_\mu) \left[\frac{kT}{q|I_c|} \right] \qquad (6)$$

Although base-charging capacitance, C_b, is directly proportional to collector current, its effect on ω_t is nullified by the proportionality of g_m to collector current.

The net effect of these expressions is to show that the admittances of C_{je} and C_μ are controlled by collector current; an increase in collector current results in greater bandwidth and faster slewing rate.

The above discussion relates only to the effect of the signal level on speed through its effect on the parameters of the log transistor. In a practical situation, there are many other factors that influence the speed of response of the circuit. The most important of these factors in log modules are the added feedback capacitance for stabilization, the stray capacitance at the amplifier input, and the bandwidth of the amplifier.

With proper stabilization circuitry, the amplifier and its stabilizing feedback capacitor should restrict speed only at the high end of the allowable input range. Stray capacitances modify the values of C_{je}, C_μ, and τ_F in the expression defining ω_t. The stray capacitance from summing junction to common is a constant that modifies τ_F. Stray capacitance from collector to base is feedback capacitance and affects C_μ.

Table 3 lists response times for steps of differing magnitude and current level for the 755 log amplifier. By comparing them with the equation for the radian period $1/\omega_t$, it is clear that the predicted linear relationship does exist at the lower levels, with a limit determined by the feedback capacitance and strays at the higher levels. In this region (1μA to 1mA), other considerations, such as the bandwidth of the amplifier and the size of the stabilizing capacitor, dominate the response, and further improve-

TABLE 3. RESPONSE TIME (755) FROM 10 TO 90%

I_{IN} INCREASING		I_{IN} DECREASING	
I_{IN}	TIME	I_{IN}	TIME
1nA → 10nA	1ms	10nA → 1nA	4.5ms
10nA → 100nA	100µs	100nA → 10nA	400µs
100nA → 1µA	7µs	1µA → 100nA	30µs
1µA → 1mA	4µs	1mA → 1µA	7µs

ments in response time become marginal. Even if an external amplifier having near-infinite gain and requiring no stabilization capacitor were used, the improvements in slew rate would be slight. C_{je} and C_{μ} are not quite constant, as assumed, but increase with signal level for currents of 0.1mA and more).

If steps of current are applied to the log module in the direction of increasing magnitude, a faster slew rate will be achieved than for steps in the direction of decreasing magnitude. Since the time to charge or discharge a capacitor is dependent on the available current, it is to be expected that steps which increase quiescent current will have a faster slewing rate and a shorter final exponential "tail." Because of the decreased time constant at high current levels, responses that end at higher levels will be completed faster than those ending at lower levels, even though the latter start faster than the former in traversing a given current range.

For changes over combinations of the ranges listed in Table 3, the response time will be determined by the final value of current. For example, in traversing the entire range of 1nA to 1mA, the total time will be dictated by the final value of 1mA, resulting in a total response time of about 6µs. Conversely, when slewing from 1mA to 1nA, the final value is 1nA, and the total response time will correspond approximately to that for 1nA, or 4.5ms.

The dependence of slew rate on signal level will cause varying degrees of distortion for input square waves at various signal levels. Shown in Figure 7 are several input signals and the corresponding outputs from the log amplifier. By the choice of V_{IN}, R1, and R2, and the position of electronic switch S1, the various input signals are readily obtainable. Alternatively, the square-wave-plus-bias of Figure 20, Chapter 3-1, may be used.

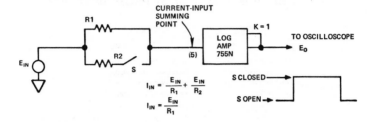

a) Test setup for measuring response time (10% – 90%)

E_{IN}	R_1	R_2	S Operation	Magnitude Change	Output Waveform Fig. 7c ( )
100mV	100MΩ	11MΩ	CLOSED	1nA → 10nA	7c (1)
100mV	100MΩ	11MΩ	OPEN	10nA → 1nA	7c (2)
100mV	100MΩ	100Ω	CLOSED	1nA → 1mA	7c (3)
100mV	100MΩ	100Ω	OPEN	1mA → 1nA	7c (4)
100mV	1MΩ	1kΩ	CLOSED	0.1µA → 100µA	7c (5)
100mV	1MΩ	1kΩ	OPEN	100µA → 0.1µA	7c (6)

b) Table of input values

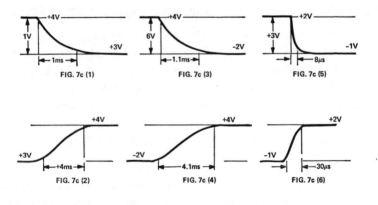

c) Responses (10% – 90%)

Figure 7. Response time (10% – 90%) for steps of various magnitudes and polarities

ANOTHER LOOK AT SPECIFICATIONS OF LOG DEVICES

The transfer equation for a log module is

$$E_o = -K \log \frac{I_{IN}}{I_{REF}} \text{ or } -K \log \frac{V_{IN}}{E_{REF}} \qquad (7)$$

I_{REF} is a dimensional constant necessitated by the fact that logarithms exist only for pure numbers (the expression $\log I_{IN}$ by itself would imply a reference current of 1A). When $I_{IN} = I_{REF}$, the logarithm of the ratio is zero. I_{REF} for practical devices can be chosen as one or the other extreme of the range of I_{IN}, or it may be approximately in the middle of the range (geometric mean). I_{REF} for Model 755 and similar devices is $10\mu A$, which corresponds to 0.1V for E_{REF}, the geometric mean between 1mV and 10V, hence mid-scale. Since the logarithm is real only for positive arguments, the input and reference must be of the same polarity, i.e., if V_{IN} is negative, V_{REF} must also be negative.

The gain, or scale factor, K is also a dimensional constant (volts). The typical log amplifier (755) can be connected for K = 1, K = 2, or K = 2/3. -K = -1 gives an output decrease of 1 volt for each 10X increase in the input ratio. $K \log_{10}$ (ratio) can also be interpreted in terms of other logarithmic bases (B), according to the relationship, $E_O = -K' \log_B$ (ratio), where

$$K' = K \log_{10} B \qquad (8)$$

For K = 1, BASE (B)	K'
100	2.0
10	1.0
8	0.903
5	0.699
3.16	0.50
ϵ	0.434
2	0.301

K' is the output change corresponding to an input ratio change, B.

The 755 and 752 modules are really families having devices of both polarities. 755N and 752N utilize NPN transistors (751N); 755P and 752P utilize PNP transistors (751P). When used for log operations, the "N" versions accept positive voltage or current, and K is positive (that is, the output becomes less positive or more negative as the input becomes more positive — the characteristic response of any circuit involving an inverting operational amplifier); the P versions accept negative voltage or current for logarithmic operation, and K is negative (that is, the output becomes less negative, or more positive, as the input becomes more negative — again, an inverting response). The output voltage of both devices is plotted against current input (log scale) in Figure 8. All three inherent values of the scale factor, K, are shown.

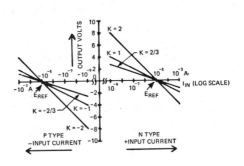

Figure 8. Plot of output voltage vs. input current for model 755

INTERPRETING LOG-DEVICE TERMS

Log Conformity Error: Log conformity error is the difference between the actual output voltage and the output voltage predicted by the log-transfer equation. A plot of output vs. input, when plotted on semilog paper (linear output scale, log input scale), should be a straight line. Any deviation from this straight line is log conformity error; in this sense, it is analogous to linearity error for linear devices. For most log amplifiers, the best linearity is obtained in the middle 4 decades (10nA to 100μA). In this range, for 755, log-conformity error is ±0.5% referred to the input (RTI) or 2.17mV referred to the output (RTO) for |K| = 1. To obtain best results, the input data, if possible, should be centered within this range.

Log conformity error is the one irreducible error of the log amplifier. It appears as an error in the log operation and cannot be compensated-for with internal circuitry.

Offset Voltage (E_{os}): The offset voltage of the log module is the offset voltage of the internal amplifier. This voltage acts as though it were a small dc offset voltage in series with the input terminals. For voltage-logging operations, best performance is obtained with the offset voltage trimmed.

Since E_{os} appears in series with V_{IN}, the effect of E_{os} depends on the level of V_{IN}. Referred to the input, the error contribution of E_{os} is

$$\% \text{ RTI} = (E_{os}/V_{in}) \, 100\% \tag{9}$$

Determining the output error corresponding to a given RTI error is straightforward. For example, for a RTI error of ½%, (1± 0.005), the respective values of the logarithm are 0.002166 (1.005) and -0.002177 (0.995), or about ±0.0022. Multiplying by K = 1V yields an output error of ±2.2mV for an input error of ±½%.

An alternative method of determining RTO errors from RTI (or vice versa) is to use an abbreviated table, such as Table 4,* with linear interpolation, if necessary.

---

*Note: Table 4 differs from the table in the Specifications section of Chapter 3-1; in the latter table, the errors are based on low-side error (e.g., 1−0.1% → 0.999), hence give greater errors referred to the output. In table 4 on this page, the errors are based on high-side error (1 + 0.1% → 1.001), hence give greater errors referred to the input. While not significantly different, worst-case error computations should take into account the more-important error from the standpoint of the application.

% Error Referred to Input (δ_i)	Millivolt Error Referred to Output (δ_o)		
	K = 1	K = 2	K = 2/3
0.1%	0.43mV	0.87mV	0.29mV
0.5	2.17	4.33	1.44
1.0	4.32	8.64	2.88
3.0	12.84	25.67	8.56
4.0	17.03	34.07	11.36
5.0	21.19	42.38	14.13
10.0	41.39	82.79	27.60

This table gives representative examples of input errors and the corresponding error at the output for common values of K. For a given output error, the input error can be computed by linear interpolation. The curves may also be helpful in obtaining approximate error conversions instantly. Example shown: 2% error ≅ 8.5mV.

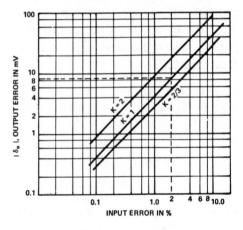

Table 4. Error conversion: $|\delta_o| = K \log_{10}(1 + \delta_i/100)$

For example, to determine the RTI error equivalent to 14mV out for K = 1, the nearest value to 14mV is 12.84mV, at 3.0%. To this should be added $(4.0 - 3.0)(14 - 12.84)/(17.03 - 12.84) = 0.28$, for a total error of 3.3% RTI.

Offset Current (I_{os}): The offset current, I_{os}, of the log amplifier is the bias current of the internal amplifier. This parameter can be a significant source of error when processing signals in the nanoampere region. For this reason, it is important to select a log amplifier having bias current much less than the smallest signal to be processed. Error contribution of I_{os} is:

$$\% \text{ RTI} = (I_{os}/I_{IN})\,100\% \qquad (10)$$

Reference Current (I_{REF}): I_{REF} is the internally-generated current source to which all input currents are compared. Tolerance errors in I_{REF} appear as a dc offset at the output. Other offsets appearing at the output cannot be differentiated from the effects of I_{REF} error. This is easily demonstrated by considering the transfer equation with an added offset

$$E_o = -K \log \frac{I_{IN}}{I_{REF}(1 \pm P/100)} \pm E_1 \qquad (11)$$

where P represents the percent tolerance in I_{REF} and E_1 is the output offset.

This equation can be rewritten

$$E_o = -K \log \frac{I_{IN}}{I_{REF}} + K \log (1 \pm P/100) \pm E_1 \qquad (12)$$

Since both the second and the third terms on the right-hand side of (12) are constants, they can be combined to form a constant E_2:

$$E_o = -K \log \frac{I_{IN}}{I_{REF}} \pm E_2 \qquad (13)$$

Table 4 can be used, as before, to determine the RTI equivalent of the combined offset, E_2, and this new tolerance (P′) can be part of the error in I_{REF}:

$$E_o = -K \log \frac{I_{IN}}{I_{REF}(1 \pm P'/100)} \qquad (14)$$

The specified tolerance for I_{REF} of log amplifiers includes dc output offset errors, since they are inseparable from the effects of I_{REF}.

The effect of I_{REF} tolerance errors can be compensated for by adding a constant at the output or in a stage following the output, or by trimming the input scale factor ahead of the log amplifier.

Reference Voltage (E_{REF}): E_{REF} is the effective internally-generated voltage to which all input voltages are compared. It is related to I_{REF} through the equation,

$$E_{REF} = I_{REF} \, R_{IN} \qquad (15)$$

where R_{IN} is the total input-circuit resistance, including the input resistor, the signal source resistance, and any other appreciable series resistance. Virtually all the tolerance in E_{REF} is due to I_{REF}, provided R_{IN} is made up of stable precision resistors. Consequently, variations in I_{REF} cause a shift in E_{REF}.

The effects of E_{REF} tolerance errors are compensated for in the same manner as the similar effects of I_{REF}.

Scale Factor (K): Scale factor is the voltage change at the output for a decade (i.e., 10:1) change at the input. Scale-factor error is equivalent to a change in gain, or slope, and is specified in percent of the nominal value. An external adjustment may be performed to fine-trim the scale factor (usually preset) or it may be locally adjusted to a value that is several times the initial value by adding series resistance at the feedback terminal. It can also be manipulated by adjusting the gain of a stage following the log amplifier. (Its effect is the same as that of manipulation of the exponent of the log-amplifier input.)

ADJUSTING THE PARAMETERS OF LOG DEVICES

Adjusting E_{os}: The amplifier's offset voltage may be adjusted to near-zero by a very simple but unconventional procedure. Most users of op amps are accustomed to "zeroing" an op amp's offset voltage by adjusting for a zero-volt output. For log modules, this is not applicable, since a zero output corresponds to the log of 1 ($I_{IN} = I_{REF}$), and log (0) is not defined.

There is, however, a quite convenient method of adjusting E_{os} without disturbing the log circuit arrangement (except for grounding the input). The method is shown in Figure 9, and, while specifically applied to the 755, it is applicable to all log amplifiers. Under the conditions shown in Figure 9, the output is

$$E_o = -K \log \frac{V_{IN} - E_{os}}{E_{REF}} \qquad (16)$$

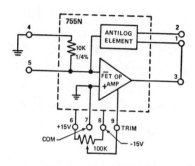

Figure 9. Trimming E_{OS}

Since the input terminal is grounded through the 10kΩ resistor, $V_{IN} = 0$. The equation for output voltage then becomes

$$E_o = -K \log \frac{-E_{os}}{E_{REF}} \qquad (17)$$

Remembering that the log is undefined for zero, assume a practical limitation on E_{os}, and then calculate E_o: Since the change in E_{os} with temperature for a good FET amplifier is in the range of $20\mu V/°C$, it is reasonable to assume that adjustment of E_{os} to $10\mu V$ would be reasonable.

Using this value of E_{os} and the value of K and E_{REF} specified for the log amplifier, the output can be calculated. For Model 755N, as shown in Figure 9, the output is

$$E_o = -1V \log \left[\frac{-\pm 10\mu V}{100mV} \right] = 4V, \text{ for } E_{os} = -10\mu V \qquad (18)$$

The amplifier's offset voltage can therefore be adjusted to within $10\mu V$ by adjusting for a +4V output (for 755P, the output would be -4V). By adjusting for a higher output voltage, one can adjust the offset to a still smaller value. Care should be taken to ensure that the amplifier's output voltage specification is not exceeded, since this may lead to a false offset indication.

A volt-ohmmeter is all that is required to monitor the output, since the signal level of 4V is high, and impedances are low. If an oscilloscope is used to monitor the output, high noise levels will be observed when performing this adjustment. This is to be expected, since the sensitivity is extremely high: for the next decade, $(5V - 4V)$ out/$(10\mu V - 1\mu V)$ in = 111,000, gain for input noise. The open-loop gain of the amplifier may, in some cases, limit the closeness with which zero input may be approached, beyond $4V$ out. For 755P, the output is adjusted to $< -4V$.

Reference-Current Adjustment: A shift of reference current results in a dc offset at the output. By adding a dc voltage to the output, one may change the reference current to the desired value. This can be accomplished in the amplifier stages following the log amplifier or by injecting a current into an unused scale-factor-feedback terminal. In the latter case, the impedance of the scale-factor terminal being used must be known. The current inserted into the unused terminal, multiplied by the resistance in series with the used terminal determines the amount of offset change. This point is illustrated in Figure 10.

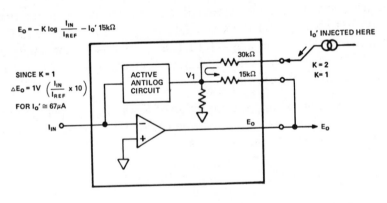

Figure 10. Adjusting I_{REF} by inserting current into unused scale-factor terminal. Since V_1 is determined by input only, I_o' must flow out through $15k\Omega$ resistor, biasing E_o by $(15k\Omega)(I_o')$. If $I_o' \cong 67\mu A$ at terminal 2, I_{REF} is, in effect, divided by 10.

It should be noted that shifting I_{REF} has no effect on the output other than changing the value of input at which zero output is obtained (Figure 11).

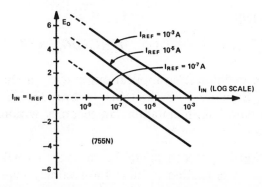

Figure 11. Output vs. input (log scale) as a function of I_{REF}

Reference-Voltage Adjustment: The reference voltage is defined in (15). Provided that R_{IN} is constant over the range of operation, the effects of changes of E_{REF} are the same as for those of I_{REF}, and the adjustment technique is exactly the same.

Scale-Factor Adjustment: Scale-factor may be adjusted by changing the total resistance between the output and the dummy summing junction shown in Figure 10. By adding resistance in series with one of the scale-factor terminals, the scale-factor may be increased from its nominal value. In general, the total resistance required to obtain any value of scale factor is

$$R_T = K R_1 \qquad (19)$$

where R_1 is the resistance of the resistor at the 1V/decade scale-factor terminal and K is the desired scale factor.

Example: To achieve a 5V/decade scale factor for a log amp having a 15kΩ input resistor at the 1V/decade terminal, calculate the external resistance required $(R_T - R_1)$ or $(R_T - R_2)$.

$$R_T = (5)(15k\Omega) = 75k\Omega \qquad (20)$$

The total resistance required is thus 75kΩ. If one uses the 1V/decade terminal, the external resistance required is (75 − 15)kΩ = 60kΩ. Alternatively, one might use the 2V/decade terminal, and connect an external resistance of (75 − 30)kΩ = 45kΩ.

APPLICATIONS

LOG OF VOLTAGE – DATA-COMPRESSION EXAMPLE

One of the more-interesting applications of a log module is in data compression. Suppose it is desirable to introduce a signal to a computer or data link, maintaining an accuracy within 1% of the signal throughout the range.

The conventional approach might be to select an A/D converter having sufficient resolution to meet the accuracy requirement. If the input range were 1V to 10V, the dynamic range would be from 1% of 1V to 10V, or 1000:1. To obtain this resolution, a 10-bit converter might be used,* and 1 LSB (least-significant bit) would represent 1% of the smallest signal. If the range were increased to 10mV to 10V, the converter would require a resolution of 1% of 10mV to 10V, or 1:100,000. To even approach this resolution, a 16-bit converter would be barely sufficient.* An economical alternative, maintaining the required 1%-of-signal accuracy error is to use a log module over a 3-decade range for data compression, and a *12-bit* A/D converter. The proposed scheme is shown in Figure 12.

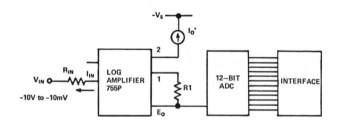

Figure 12. Data compression, using a log amplifier

In order to convert the log output to a unipolar signal, an external current source will be used to shift the reference current. To achieve the proper scale-factor, or gain, an external resistor, R1,

*$2^{10} = 1024$, $2^{16} = 65,536$

has been added in series with the K = 1 output. The input resistance value, R_{IN}, will be chosen for the range of best log conformity, and connected at the current input, pin 5.

The first step in applying the log amplifier is to select the proper polarity. Since the input voltage is negative (-10mV to -10V), the P-type log amplifier is required. A Model 755P complete log amplifier is chosen, in order to minimize the design effort.

Choosing R_{IN}*:* After selecting the log amplifier, the region of best log conformity is noted, and an attempt is made to shift operation to those decades. For Model 755. the best log conformity is 0.5%, specified for the range from 10nA to 100μA. R_{IN} is selected, for use at the current input, to provide the highest input current desired (100μA) at the highest input voltage magnitude (10V).

$$R_{IN} = 10V/100\mu A = 100k\Omega \qquad (21)$$

The lowest value of input current expected is 10mV/100kΩ = 100nA. The input-current range is therefore well within the range of 0.5% log conformity.

Adjusting K: To determine the best value of K for the application, the input requirements of the following stage must be considered. If we assume the 12-bit converter to have an input range of 0 to +5V, K can be calculated. The total output voltage required is 5V, and the input range spans 3 decades. Therefore, K must be 5/3V.

From the data sheet for the log amplifier selected (755P), a value of 15kΩ is given as the input resistance of the 1V/decade terminal. The total resistance required for K = 5/3V is

$$R_T = (5/3)(15k\Omega) = 25k\Omega \qquad (22)$$

A nominal 10kΩ is required in series with the K = -1 terminal. To allow for tolerance of the internal resistor, a 10kΩ 10-turn pot in series with a 5kΩ resistor is adopted as R1.

Shifting I_{REF}: Since I_{REF} determines the point at which a zero output will be obtained, and since a zero output is desired for the smallest input current to be processed, it is this value of current, 100nA, that I_{REF} will be shifted to.

The transfer curves for the 755P (Figure 8) show that a -2V output would be obtained for 100nA input at K = -1V. Since this is the current level for which a *zero-volt* output is desired, the current $I_o{}'$ injected into the unused scale-factor terminal must shift the output in the positive direction. Referring to Figure 10, one can see that a polarity inversion occurs between the unused scale-factor terminal and the output. To cause a shift in the positive direction, the current to be injected must be negative, i.e., derived from a negative voltage.

The amount of current to be injected can be calculated by Ohm's Law, using the total resistance of the scale-factor terminal connected to 1, the output, and the amount of voltage to be shifted.

$$I = \frac{2K}{R_T} = \frac{2 \times 5/3}{25k\Omega} = 133\mu A \tag{23}$$

A resistor to the negative supply (of value $15V/133\mu A = 30k\Omega$) can be used to obtain this current, but shifts in the offset voltage

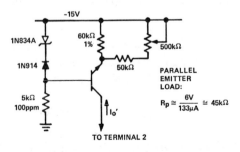

TO TERMINAL 2

Figure 13. Constant-current source to shift reference current ($I_o{}'$ = 133μA) in circuit of Figure 12.

at the dummy summing junction (Figure 10) of 60mV/decade can cause significant errors. For example, if I_{REF} were adjusted at 100nA I_{IN}, a shift of 180mV would occur as the input increased to 100μA (3 decades @ 60mV/decade), resulting in a shift of offset current of (180mV/15V) × 100 = 1.2%.

Therefore, a rudimentary current source, shown in Figure 13, is used.

Trim Procedure

1. E_{os}: The first step in the trim procedure is to adjust E_{os} of the log amp to nearly zero volts. As mentioned earlier, this can be accomplished by leaving pin 5 open (or V_{IN} disconnected), and grounding pin 4. The trim pots for I_{REF} and the scale factor should be set to midrange, in order to reduce interaction. R2 would then be adjusted for

$$E_o = -K \log(10\mu V/E_{REF})$$

$$= (5/3) \log (10\mu V)/(100nA \times 10k\Omega)$$

$$= -10/3 \text{ V} \tag{24}$$

Adjusting for any voltage between −10/3V and −5V will insure that E_{os} has been adjusted to within 10μV.

2. I_{REF}: After E_{os} has been adjusted, I_{REF} can be adjusted by applying a value of input that will cause I_{REF} to flow into the log amplifier. To accomplish this, set V_{IN} to 10mV and adjust the reference-current source for zero volts out of the log amplifier.

3. K: To adjust K, the input signal should be increased to its maximum value of -10V. R1 is then adjusted for 5V output.

4. Because some interaction among the adjustments cannot be avoided, all adjustments should be repeated at least once, and in the same order as initially performed, E_{os}, I_{REF}, K.

Error Analysis

Parameter	Error	Comment
E_{REF}	$\cong 0$	Initial error is trimmed to zero
E_{REF} Drift	0.5% RTI	$\pm 0.1\%/^{\circ}C \times 5^{\circ}C = 0.5\%$
Log Conformity	0.5% RTI	Input was scaled to within range of $\pm 0.5\%$ log conformity
E_{os}	$\cong 0$	Initial error trimmed to zero
E_{os} Drift	0.75% RTI	$(\pm 15\mu V/^{\circ}C \times 5) / 10mV$ worst-case condition; occurs only for smallest input signal $(7.5\mu V/10mV)$ $(100\%) = 0.75\%$
I_{os}, I_{os} Drift	$\cong 0$	Negligible contribution $(10\ pA/100\ nA)$ $(100\%) = 0.01\%$
K	$\cong 0$	Initial error trimmed to zero
K Drift	$\pm 10mV$ RTO	$\pm 5^{\circ}C \times 0.04\%/^{\circ}C \times K \log (V_{IN}/E_{REF})$ $= \pm(3 - 0.33)mv \log (V_{IN}/0.01V) = 10mV$ @ $V_{IN} = 10V$

Total error, referred to input, at constant temperature: $\pm 0.5\%$.

Total error, RTI, over the temperature range, assuming a worst-case condition that all errors are additive:

$$(\pm 0.5\ \pm 0.5\ \pm 0.75)\%\ \text{RTI} + 10mV\ \text{RTO}$$

Converting 10mV RTO to an RTI term by Table 4 yields 1.4%.

Total error RTI over the $10^{\circ}C$ temperature range is 3.15% (1.74% root-sum-squares).

A/D converter error: If the total converter error is kept to 1LSB = 1/4096F.S., the equivalent log-amplifier output error is 5/4096 = 1.22mV. From Table 4, this is equivalent to 0.28%, referred to the input, at any level.

LOG OF CURRENT – PHOTOMULTIPLIER EXAMPLE

In this example, the output current of a photomultiplier tube is to be applied to a log amplifier, as shown in Figure 14.

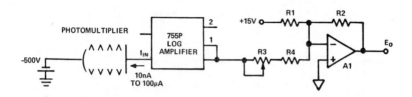

Figure 14. Log of current — photomultiplier input

Since the input is a current source, there is no need to perform the offset-voltage adjustment. This can be verified by determining the source resistance that will produce 0.5% error RTI at the lowest level: $R_S = 500\mu V/(0.005 \times 10nA) = 10M\Omega$, easily achievable with a photomultiplier.

The output of the 755P will be as predicted by the transfer curve (Figure 8). Level shifting, in this case, is accomplished by adding an offset to A1 *via* R1, and gain is trimmed by a small resistor R3 in series with R4.

For 0 to -10V output, the gain of A1 is 10/4, or 2.5, since 4V is the entire output range of the log module when spanning 4 decades at 1V/decade. The ratio of R_2 to $(R_3 + R_4)$ is then 2.5:1.

At the low end of the range (10nA), the corresponding output voltage is -3V. At the output of A1, it is amplified to (-3)(-2.5) = 7.5. In order to offset it to zero at the output of A1, R1 must have the value, determined by the gain equation

$$+15V(-R_2/R_1) = -7.5V$$

$$R_1 = 2R_2 \qquad (25)$$

If R2 is selected to be 25kΩ,

$$
\begin{aligned}
R_1 &= 50k\Omega \\
R_3 &= 10k\Omega \text{ pot} \\
R_4 &= 5k\Omega
\end{aligned}
$$

Error Analysis @ $\pm 10^{\circ} C$ Temperature Range

Parameter	Error	Comment
I_{REF}	$\cong 0$	Trimmed to zero by R1
I_{REF} Drift	$\pm 1\%$	$\pm 0.1\%/^{\circ}C \times 10^{\circ}C = \pm 1\%$
Log Conformity	$\pm 0.5\%$	Not trimmable
E_{os}, E_{os} Drift	$\cong 0$	Can neglect when equivalent (Thevenin) voltage is much greater than E_{os}
I_{os}	$< 0.1\%$	Worst case at lowest input current (10 nA/10 pA) (100%)
I_{os} Drift	$< 0.1\%$	Additional 10pA due to I_{os} doubling per $10^{\circ}C$ increase
K	$\cong 0$	Initial error of 10mV trimmed to zero by R3
K Drift	12mV	$\pm 10^{\circ}C \times 0.04\%/^{\circ}C \times K \log (I_{IN}/10\mu A)$ = 4mV log ($I_{IN}/10\mu A$), worst-case at I_{IN} = 10nA

Total error, referred to input, at constant temperature: $\pm 0.6\%$.

Total error, RTI over the temperature range:

1.7% + 12mV RTO = 4.5% RTI. Referred to the output by Table 4, 4.5% RTI is equivalent to 19.3mV, or less than 1/2% of F.S.

Log Ratio
Applications

Chapter 2

The logarithm is a mathematical function. When it is employed to describe the behavior of a physical entity, its argument must be dimensionless. Accordingly, practical logarithmic devices always compute the log of a ratio of two voltages or currents; the numerator is termed the "signal," the denominator the "reference."

The distinction between "log" devices and "log-ratio" devices is practical, not semantic. It is determined by the requirements of the application on the "reference," and the consequent effects on the circuit design and external connections. If the reference is more-or-less fixed and considered a constant, the subject is a "log" device. If the reference is controlled by an external signal, or is simply considered to be freely variable, a different circuit design is usually employed, and it is called a "log-ratio" circuit. It is to the latter group of applications that this chapter is devoted.

A typical commercially-available log-ratio circuit (Model 756) has its performance defined over a range of 4 decades of signal current and 3 decades of reference current, or a total range of ratio of $10^7 : 1$. While it is convenient to use and by no means expensive, the user can often find it a more practical matter to design a log ratio circuit to meet a set of *specific* needs with the aid of the basic log element (751). Log-ratio circuitry is somewhat simpler to deal with than log circuitry, because a separately-generated reference current is unnecessary.

Figure 1 shows the simplest temperature-compensated log-ratio circuit, driven by two current sources, I_{s1} and I_{s2}. Since the operational amplifier maintains its inputs at essentially the same potential, V_A, the two log-diode-connected transistors are essentially in series-opposing, and the sum of their voltage drops, proportional to the log of their current ratio, appears at the tap of the voltage divider, V_B. If I_{s2} can be kept low enough so that it does not significantly load the voltage divider,

$$E_o \cong \left[1 + \frac{R_G}{R_{TC}}\right] \frac{kT}{q} \ln\left[\frac{I_{s1}}{I_{s2}}\right] \tag{1}$$

The divider incorporates a temperature-sensitive resistor, R_{TC}, which is designed to compensate for the temperature variation of kT/q. The resistance values of R_G and R_{TC} are chosen and trimmed to make

$$\left[1 + \frac{R_G}{R_{TC}}\right] \frac{kT}{q\bullet} \ln(10) \cong 1.00V \tag{2}$$

Therefore,

$$E_o \cong 1\cdot\log\left[\frac{I_{s1}}{I_{s2}}\right] \quad (1V/decade) \tag{3}$$

independently of temperature.

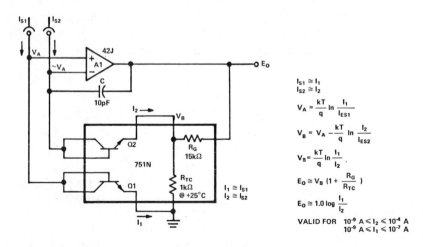

$I_{s1} \cong I_1$
$I_{s2} \cong I_2$

$V_A = \frac{kT}{q} \ln \frac{I_1}{I_{ES1}}$

$V_B = V_A - \frac{kT}{q} \ln \frac{I_2}{I_{ES2}}$

$V_B = \frac{kT}{q} \ln \frac{I_1}{I_2}$

$E_O \cong V_B (1 + \frac{R_G}{R_{TC}})$

$E_O \cong 1.0 \log \frac{I_1}{I_2}$

VALID FOR 10^{-9} A $\leqslant I_2 \leqslant 10^{-4}$ A
$\qquad\qquad 10^{-9}$ A $\leqslant I_1 \leqslant 10^{-7}$ A

Figure 1. Simple temperature-compensated log-ratio circuit

VOLTAGE vs. CURRENT: INPUT-LOADING EFFECTS

Since the basic log elements convert linear current to log voltage, inputs in the form of current from ideal current sources having "infinite" source resistance can be used with any of the circuits to be discussed here. If the input signal is a voltage, it must first be converted to a current. This is, of course, an inherent feature of circuits that employ inverting operational amplifiers operating at zero common-mode potential.* A precise value of series resistance will determine the current scaling. However, in the simple circuit of Figure 1, employing a differential-input operational amplifier, a voltage source and its series resistor "look into" V_A. At the negative input terminal, V_A^- is a variable common-mode voltage; at the positive input terminal, it is a nonlinear load resistance. If the input voltages are V_{s1} and V_{s2}, in series with resistances R_{s1} and R_{s2}, then (Figure 2a)

$$\log\frac{I_1}{I_2} = \log\left[\frac{V_{s1} - V_A}{V_{s2} - V_A} \cdot \frac{R_{s2}}{R_{s1}}\right] \tag{4}$$

If the inputs are imperfect current sources, having non-infinite internal resistances, the effect is similar. The log ratio of the actual input currents is

$$\log\frac{I_1}{I_2} = \log\frac{I_{s1} - V_A/R_{s1}}{I_{s2} - V_A/R_{s2}} \tag{5}$$

V_A is an implicit function of I_{s1}, R_{s1}, V_A, and the reverse-saturation current of D1, I_{ES1}

$$V_A \cong \frac{kT}{q}\ln\frac{I_1}{I_{ES1}} = \frac{kT}{q}\ln\frac{I_{s1} - V_A/R_{s1}}{I_{ES1}} \tag{6}$$

While V_A cannot be determined explicitly, it can be plotted as a function of I_{s1} for various values of R_{s1} (Figure 2b). The plot is based on the assumption that I_{ES1} at 25°C is about 2×10^{-15}A, a realistic value for the best log transistors. For higher values of I_{ES1} or temperature, the plotted value of V_A is high, i.e., conservative, by about 60mV/decade of I_{ES} and 2mV/°C. Quite often, V_A is simply assumed to be about 0.6V; however, for

*See Fig. 8, this Chapter, and Fig. 9 in Chapter 3-1.

low-current operations, this figure could be conservative by a factor of 2 or more.

The output error attributable to V_A and finite source resistance is a term that adds to (3), approximately

$$1 \cdot \log \left[\frac{1 - V_A/R_{S1}I_{S1}}{1 - V_A/R_{S2}I_{S2}} \right] \qquad (7)$$

The difference between the term inside the brackets and 1.00 is the ratio-error (%/100) contribution, referred to the input. The relationship between input and output errors can be seen in Table 4, Chapter 4-1 (K = 1).

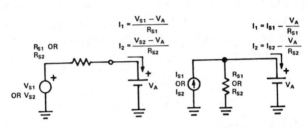

Figure 2a. Input current as a function of source voltage or current, source resistance, and V_A. Amplifier bias current and circuit leakage current are considered negligible here.

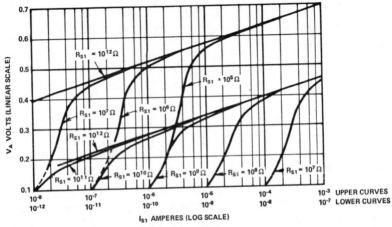

Figure 2b. V_A as a function of input current for various values of source resistance, assuming that $I_{ES} \cong 2 \times 10^{-15}$ A @ +25°C

It has been assumed that I_2 does not load the feedback voltage divider. To investigate the validity of this assumption, note that

$$E_o = V_B + R_G\left(\frac{V_B}{R_{TC}} - I_2\right) = V_B\left(1 + \frac{R_G}{R_{TC}}\right) - I_2 R_G \qquad (8)$$

The error, $I_2 R_G$, if $I_2 = 1\mu A$, is (for $R_G \cong 15k\Omega$) 15 mV, or about 4% referred to the input. For 1% error, RTI, the output error is 4.3mV, corresponding to a maximum I_2 of 290nA, for this circuit.

Because the properties of current sources may be affected by voltage, it is important, in such circuits as Figure 1, to ensure that the "compliance voltage" of the current inputs is greater than V_A, and that V_A does not seriously affect the parameters of the current source.

RESPONSE AND STABILITY

Since the transistors are connected as diodes, D2 is a passive feedback element, with a resistance $= r_E \cong (40I_2)^{-1}$. The 10pF feedback capacitance should be quite adequate to compensate for input strays (Figure 3).

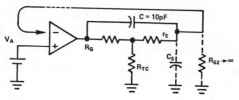

Figure 3. Dynamic model of log-ratio circuit

The effective net resistance of the feedback divider (output to summing point) is

$$R_{eff} = R_G + r_E\left(1 + \frac{R_G}{R_{TC}}\right) \cong R_G + \frac{16}{40I_2} \qquad (9)$$

For this circuit, because of the accuracy requirement implied by (8), I_2 is likely to be less than $1\mu A$, hence $R_{eff} \cong (2.5I_2)^{-1}$, and the small-signal time-constant, $R_{eff}C = C/(2.5I_2)$. For example, if $C = 10pF$ and we are considering small-signal variations of I_2 in the neighborhood of 100nA, $\tau = 40\mu s$, $\omega = 25kr/s \rightarrow 4kHz$.

For small changes in I_1, which may be considered to be changes in V_A/r_E (I_{s2} fixed and having negligible admittance), the output will immediately change by $V_A (1 + C_s/C)$, then continue on to the final value with the same exponential time-constant as discussed above.

For transdiode applications, if feedback to the amplifier's positive input can be considered negligible, the stability considerations are quite similar to those discussed in Chapter 3-1 (Equations 10 to 16 and Figures 5, 6, and 7 are quite relevant). For stability explorations, the open-loop gain of the amplifier may be considered to be reduced by $(1 + R_G/R_{TC})$, thus lowering its Bodé plot by 20 log 16 = 24.1dB, and R_E is, in effect, the parallel combination of R_G and $R_{TC} \cong 940\Omega$.

APPLICATION CONSIDERATIONS

Thus far, there have been a number of tacit assumptions: that the diode-connected log elements are matched and are isothermal with each other and with the compensation resistors, that the diodes operate in the range for which h_{FE} is sufficiently large and the bulk resistance sufficiently small to cause negligible errors, and that the amplifier's offset voltage, bias current, and common-mode errors are negligible. We shall now look more closely at the amplifier requirements, at various elements of log-ratio-circuit performance specifications, and at a number of alternatives to the circuit of Figure 1.

CHOOSING THE OP AMP

In low-current operation, the time constants involved are significantly longer than amplifier time constants and preclude the possibility of operation at high frequency. The bandwidth restrictions on the choice of op amp are therefore minimal, unless log operation is confined to high current levels.

Bias current is the primary specification for all log applications involving current-input signals. The op amp's bias current is added to the input signal (±) and flows through the log element. Errors can be stated directly as percentage of input.

$$\% \text{ error RTI due to } I_{BIAS} = 100\% \ (I_b/I_s) \qquad (10)$$

For log voltage ratio applications, the input signal is converted to a current by the input resistor. The resulting current should be the value of I_s that is compared with I_b in (10).

Offset voltage is of little importance in log current-ratio applications if the current signals are "true current sources." The effect of offset voltage (E_{os}) for either current or voltage applications is

$$\% \text{ error RTI due to } E_{os} = 100\% \ (E_{os}/V_s) \qquad (11)$$

For voltage applications, V_s is the voltage signal to be logged. For current applications, V_s is the equivalent open-circuit voltage source corresponding to $I_s R_s$. If E_{os} is not very much smaller than the smallest value of V_s for which a stated accuracy level is desired, it can be a significant source of error and should be trimmed to zero. (Also, in such applications, the amplifier should be chosen for low thermal drift, as noted below.)

Offset-voltage drift is important for those applications in which the current input is not provided by an ideal current source. To determine the required amplifier-offset specification, the temperature range, the allowable % error over the temperature range, and the lowest voltage for which that % error must be maintained are decided upon. The maximum allowable amplifier offset temperature-coefficient is

$$E_{os} \text{T.C.} = \frac{V_{min}}{\Delta T} \frac{\delta_{max}}{100} \qquad (12)$$

where V_{min} is the lowest input voltage

δ_{max} is the largest allowable % error at V_{min}

ΔT is the change of temperature from the temperature at which E_{os} was trimmed to zero.

For example, an input range of 10mV to 10V is to be applied to one terminal of a log-ratio device (assuming that the common-mode voltage is zero), and it is necessary the E_{os} contribute no more than 1% error over the range 20°C to 40°C. E_{os} might be trimmed to zero at mid-range (30°C). Then the required offset temperature coefficient would be

$$E_{os}T.C. = \frac{0.01 \times 1}{\pm 10° \times 100} = 10\mu V/°C \tag{13}$$

The same procedure would be followed to determine I_b and E_{os} requirements for the second input terminal.

Output-current rating, as with all op amp circuits, would be selected for the capability of supplying the maximum requirements of the load, in addition to the requirements of the feedback circuitry (usually negligible).

OTHER CONSIDERATIONS

Dynamic Range: The dynamic range of the circuitry used in a given application is determined by the logarithmic resolution of the device used as a log element, the dynamic range of the input signal, restrictions caused by characteristics of the specific circuit configuration and the devices used in it, and the desired accuracy level. A typical basic log element, such as Model 751N, can be used over a range of 100pA to 1mA with less than 2% error; over limited ranges, better accuracy can be obtained. For the circuit of Figure 1, dynamic range was limited primarily by the common-mode range for voltage sources and source resistance for current sources, at the low end, and by $I_2 R_G$ at the high end. More-sophisticated circuits are available, as will be shown in a later section, in which the characteristics of the log element are the primary limit to the dynamic range, assuming proper op-amp selection and care in circuiting.

Polarity of the Input Signal and of the Log Device: All log elements are restricted to inputs of a single polarity (but the outputs can be bipolar, with zero occurring at unity current ratio). By definition of the logarithm, the log of zero and the log of a negative number do not have real values. This means that for a given design, the input currents must both be of the same polarity, and of such polarity that the diodes are conducting in the forward direction, i.e., the most favorable direction for log behavior. Of course, input signals may be conditioned in preceding stages, except in the case of extremely low currents, for which immediate conversion may be desirable.

If a transistor is diode-connected, either an N (NPN) or P (PNP) log element may be used, since the transistor is connected as a reversible two-terminal device. (In such cases, direct input currents of either polarity may be applied, and properly dealt-with in post-conversion conditioning.) In general, log-transistor connections (transdiode and follower) are not reversible, and polarity must be specified as "P" type for applications requiring PNP transistors (751P) and "N" type (751N) for applications requiring NPN transistors.

Temperature Range: Nearly all of today's commercially-available logarithmic modules, whether basic log elements, log transconductors, log amplifiers, or log-ratio modules, are temperature-compensated. Specially selected resistors (R_{TC} in the circuits discussed above), having resistance that increases predictably with temperature, are used to reduce the temperature-dependence of the log equation, $(kT/q) \ln (ratio)$, from 0.3%/°C to 0.04%/°C. Although this is an order-of-magnitude reduction of error, there can still be considerable contribution of scale-factor error if large temperature excursions are to be encountered. The offset voltage and bias current of op amps can also contribute significant error over wide temperature ranges. Of concern for measurements at very low currents, high temperatures may increase I_{ES} by several orders of magnitude; though the log transistors may still track, their log-conformity errors will be considerably increased.

For these reasons, the user of log devices who seeks high-accuracy wide-range operation is urged to limit ambient temperature variations to the vicinity of ±10°C. An analysis illustrating the effects of temperature in a typical application, involving the voltage/current log module Model 756, is given at the end of this chapter.

Output Polarity: The output voltage of the log-ratio circuit can be either positive, negative, or bipolar, depending on the ratio I_1/I_2, and whether or not offsets are added at the output. Figure 4 shows the output as a function of the ratio of input currents for two different scale factors. For any given application, only a portion of the entire 14-decade (potential) range would be used. By determining the range of expected input signals and computing their ratios, one can use Figure 4 to predict the expected output-voltage range. I_1 and I_2 may be assigned arbitrarily, to match device performance to current range, but polarity should be observed.

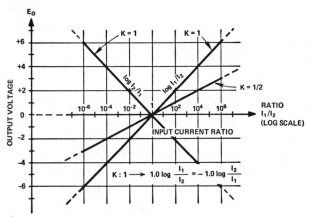

Figure 4. Output of ideal log-ratio circuit vs. input ratio, showing effect of exchanging numerator and denominator

If for any reason* it is necessary to change the effective input scaling, so that the zero-output point occurs at a ratio other than unity, but the input signals themselves are unavailable for scaling, the output can be offset by applying a current to the junction of R_G and R_{TC}, as shown in Figure 5. As noted earlier (Chapter 4-1), the voltage at this point (V_B) is determined by the *inputs*. Therefore, any current added must flow through R_G, offsetting the output by $-I_B R_G$.

Since V_B is a function of the input ratio (60mV per decade-change), the offsetting current source in Figure 5 should have high-enough source impedance to be unaffected by variations of V_B.

*Examples include: locating 0 at the middle or extremities of the range, and measuring deviations of equipment from a fixed gain or attenuation, e.g., determining by how many "dB", an actual gain differs from a nominal gain of 100X.

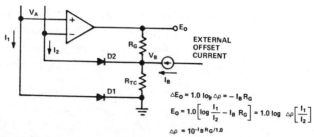

$$\Delta E_O = 1.0 \log \Delta \rho = -I_B R_G$$

$$E_O = 1.0 \left[\log \frac{I_1}{I_2} - I_B R_G \right] = 1.0 \log \Delta \rho \left[\frac{I_1}{I_2} \right]$$

$$\Delta \rho = 10^{-I_B R_G / 1.0}$$

Figure 5. Offsetting the output to shift the zero output (unity ratio) coordinate. The output shift, ΔE_O, corresponds to the input ratio multiple $\Delta \rho$ (K = 1.0).

If the signal is to be processed beyond the log-ratio amplifier, it may be better and easier to simply add a constant voltage or current at an op-amp summing junction in a succeeding stage.

Log Conformity is the specification of logarithmic devices that is akin to *linearity* in linear devices. Log-conformity error is the difference between the theoretical value of the logarithm of a ratio and the actual value that appears at the output of a log-ratio circuit after initial zeroing at unity-ratio, and scale-factor adjustment (either by end-point or best-straight-line* method).

Log conformity is basically limited by degradation of transistor gain at low current levels and by base-spreading resistance at high current levels. The form of connection affects the gain at low current levels; for example, diode-connected log transistors have a useful log range that is limited to about 1nA at the low end. If the log devices are used as transistors ("transdiode") rather than as diodes (base and collector tied together), additional decades of performance are available at the low end.

Base-spreading resistance, for log transistors, is a small resistance of a few ohms that appears in series with the emitter. For an emitter current of 1mA and 1Ω of series resistance, an error of 1mV will be obtained. The feedback attenuator will cause it to be magnified by $(1 + R_G/R_{TC})$ at the output. For 1V/decade operation, a 16mV error will be obtained for every ohm of

---

*But "best straight line" implies the ability to measure log conformity *a priori*. Although "best straight line," usually specified, gives a 2X tighter specification, "end point" is easier to measure, since it is more easily located.

base-spreading resistance. Referred to the input ratio, this is an error approaching $4\%/\Omega$. This source of error can be reduced to a satisfactory level with most of the transistor types used for log operations by restricting input signals to about $100\mu A$ maximum.

Log Voltage-Ratio: Much that has been said about log current-ratio is also pertinent to log voltage-ratio as well. The only additional consideration for voltage applications is the means of converting voltage to the current needed by the log element. If the summing junctions of input amplifiers are available, the problem is trivial, for all that is required is to attach a resistor from the voltage source to the summing junction.

The input current I_{IN} is then

$$I_{IN} = \frac{V_{IN} - E_{os}}{R_{IN}} \qquad (14)$$

The initial amplifier E_{os} should be trimmed to zero, and the amplifier chosen should have low E_{os} drift over the temperature range of concern.

As noted earlier, there are a number of circuit configurations that do not permit voltage to be easily and accurately converted to a current. These configurations are referred to as log current-ratio designs. Fortunately, many of the natural phenomena, for which log-ratio measurements are desired, are measured by transducers that provide current outputs. It is important that the designer assure himself that such transducers have sufficient "compliance voltage" to behave as ideal current sources in the presence of voltages of the order of 0.6-0.7V (V_A in Figure 1).

EXAMPLES OF LOG-RATIO CONFIGURATIONS

The circuit of Figure 6 is similar to that of Figure 1, but Q2 is connected as a "transdiode" rather than as a diode, extending its range at the low end. Currents from a few picoamperes to about $0.3\mu A$ may be converted with good accuracy.

Since I_{s1} must furnish the base current for Q2, while maintaining the base voltage at V_A, the range of I_{s1} is restricted at the low end. For example, if $h_{FE2} = 150$ at $I_{s2} = 0.3\mu A$, then I_{s1} must furnish 2nA of base current to Q2. Even if this is the maximum base current that must be supplied, over the expected range of I_{s2}, it is evident that I_{s1} must always be greater than 400nA to ensure that the contribution of this source of error will be less than 0.5%.

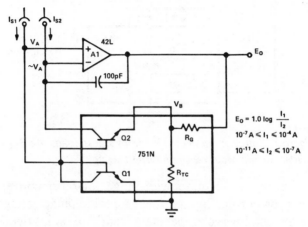

$$E_0 = 1.0 \log \frac{I_1}{I_2}$$
$$10^{-7}A \leqslant I_1 \leqslant 10^{-4}A$$
$$10^{-11}A \leqslant I_2 \leqslant 10^{-7}A$$

Figure 6. Modification of the circuit of Figure 1 to extend the range of I_{s2} at the low end

The amplifier is of the Model 42 family, with bias current considerably less than the smallest signal to be processed.

Even though the dynamic range of I_{s2} has been increased at the low end, the restriction on I_{s2} for currents higher than $0.3\mu A$ still persists, because of the error caused by the flow of I_2 through R_G.

If a follower-connected op amp is used to unload V_A and drive the base of Q2, the current swing of I_{s1} at the low end can be greatly extended.

In the circuit of Figure 7, which happens to have negative input currents but uses an N-type basic log element, the feedback element is a follower-connected log transistor. Since the current that flows through R_G is just the base current of Q1, the dynamic range is greatly improved at the high end. However, the low end for I_{s2} is still degraded, due to the diode connection of Q2. In addition, the log conformity error for the ratio will be increased because the base and collector of Q2 are at slightly-differing potentials.

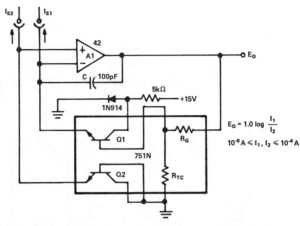

Figure 7. Follower-connected log transistor unloads the $R_G - R_{TC}$ divider, allows wider range of current swing

Figure 8 provides accurate wide-range log-ratio conversion for voltage and/or current inputs. Ideal for wide-range voltage signals, this circuit employs chopper-stabilized 234 amplifiers for low voltage offset and drift, enabling the accurate processing of signals ranging from 1mV to 10V. If higher input voltages are desired, the 100kΩ series input resistors would be scaled proportionally (e.g., 1MΩ for 100V). The dynamic range would be increased, because the low-end resolution would still be 1mV.

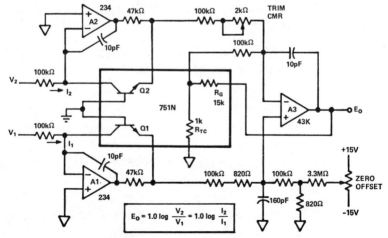

Figure 8. Wide-range log of voltage — or current — ratio

The individual log signals at the outputs of A1 and A2 are subtracted and amplified in A3's circuit, using precision resistors and the 751's temperature-compensating resistance network. As in the other circuits, temperature compensation and gain are provided by R_G and R_{TC}. The high-performance Model 43K was chosen to provide good common-mode rejection for low-level signals, low noise, and low offset drift.

LOG-RATIO MODULE

Figure 9 is a block diagram of a self-contained temperature-compensated log-ratio module, Model 756. Log current-ratio is computed by applying the input currents directly at the input terminals (amplifier summing junctions). Input voltages can be converted to current by applying them in series with external resistances, of appropriate magnitude and stability. The internal log amplifiers convert the input currents to log voltages, which are subtracted to obtain the log-ratio, and the difference is furnished at low impedance by the output amplifier.

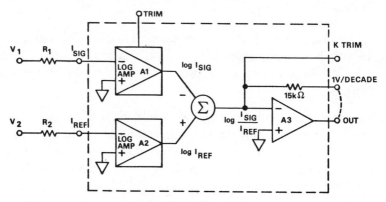

Figure 9. Functional block diagram of Model 756 log-ratio module

The 756 capitalizes on the asymmetry of typical applications that require log ratios, to obtain a near-optimum price/performance ratio: the "signal" input, channel 1, uses a high-performance FET-input amplifier with 10pA maximum bias current, which allows input signals from 1nA to 100μA to be processed accurately. The second input, the "reference," channel 2, can process signals spanning 3 decades of current with good results For best results, if channel 2 is scaled to a geometric mean of $1\mu A$, with variations of ± 1 decade (from 100nA to $10\mu A$), there will be no appreciable error as the input current at channel 1 is varied from 1nA to $100\mu A$ (7 decades of log ratio).

Figure 10 is a typical plot of log-conformity error as a function of I_1 for various intermediate values of I_2. (It must be emphasized that I_2, like I_1, is continuously variable.) The curve is typical for separate or simultaneous variation of I_1 and I_2.

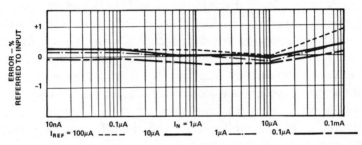

Figure 10. Typical log-conformity plot for Model 756 log-ratio module. (% Error (RTI) vs. I_{SIG})

The performance of the 756 is representative of performance that can be obtained using appropriate log-ratio schemes, involving the basic log element and op amps, provided that good design practice is adhered to.

Best overall operation (except for bandwidth) is almost always obtained when operating in the center of the range (geometric mean of the extremes), avoiding errors due to bias current and reduced transistor current-gain at the low end, and errors due to base-spreading resistance at the high end.

CHOOSING AN APPROACH

The alternatives facing a designer are essentially three:

1. To buy a self-contained log-ratio module

2. To assemble two log amplifiers and a difference amplifier in a configuration similar to that of Figure 9.

3. To design and build a circuit that is an optimum compromise between cost and performance, using basic log elements and operational amplifiers.

For the general run of applications for which its performance is suitable, in laboratory and instrumentation applications, systems, and equipment, option 1 is the best choice. It allows the designer complete freedom to deal with other system problems, once the choice is made. If the potential usage involves large numbers (e.g., more than 100 units), the designer may wish to investigate possible economies via option 3.

If the design problem involves current ranges at one (or both) inputs exceeding the available performance of the log-ratio module, the second option should be considered. It can make available from 12 to 14 decades of log-ratio (240 to 180 "dB"). Naturally, it is somewhat more expensive than the first option. If the potential usage involves large numbers, the designer may again wish to investigate possible economies via option 3.

The designer will consider option 3 where the quantity of devices required (especially if performance requirements are looser than those specified for packaged units) suggest the possibility of economy through a special-purpose design or a more-compact overall package. He should also consider the third option if, in addition, unusual combinations of dynamic range, input signal (voltage or current), polarity, scaling, or combined operations (log products and ratios) are involved.

If possible cost-savings over a standard module are the *sole* motive for considering option 3, the designer should take into account other costs in addition to parts and labor; these include, of course, design and development; they also include costs of parts procurement and inventory, availability of potting facilities (a *must* for reasonable performance of the basic log elements—see Chapter 4-1), test and temperature-performance trimming facilities, the unavoidable scrap, plus appropriate overhead rates. Often these hidden costs increase the attractiveness of complete-module prices in large quantity.

The form in which the input signal presents itself may play a large part in narrowing the selection of designs. If the inputs are true current sources, any of the alternatives presented are suitable. But if V_A is an appreciable fraction of the input voltage, or if the signal is the short-circuit current of a very low voltage device, then amplifier summing junctions must be made available. Such applications suggest options 1 and 2.

A key step in the design is the consideration of polarity. For designs where positive current flow (opposite to electron flow) is from the source toward the input terminal, an N type module (NPN transistor) would be selected. If the current flow is negative (for instance, photomultipliers), positive current must flow toward the source, calling for a P-type module (PNP transistor)*. If the log transistors are both connected as two-terminal diodes, either polarity (N or P) may be used.

DESIGN EXAMPLE

Figure 11 shows a log-ratio module used in a photometer application. Two inputs represent the intensities of light transmitted through space and through a medium that absorbs light. The *absorbance* of the medium is given by the formula

$$-A = \log \frac{I_{signal}}{I_{reference}} \qquad (15)$$

where I_{signal} and $I_{reference}$ are the currents representing the received light intensities.

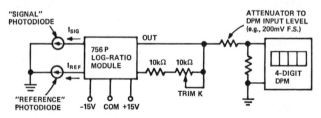

Figure 11. Log ratio applied to absorbance measurement

The transducers used in this application are photodiodes, devices that provide a short-circuit current that is proportional to the intensity of applied light.

*Figure 7 is an exception

The lowest value of absorbance is determined by the value of I_{ref}, since when $I_{sig} = I_{ref}$, $A = 0$. In this case, I_{ref} is assumed to be $10\mu A$. The actual value for the specific application depends on the intensity of light and the characteristics of the photodiode employed.

The output of the log-ratio module is externally trimmed to be precisely 1V/decade and is applied to the input of a 4-digit DPM through the scaling network R1 and R2.

The 756 log-ratio module was chosen for the design, principally because it makes available both amplifier summing junctions. When the photodiodes are connected to the summing junctions, they are operated with zero volts, i.e., in the short-circuit mode.

Short-circuit loading is necessary, because accuracy of the photodiodes can be degraded several percent when operated with as little as 100mV of output voltage.

Error Analysis: Log Current-Ratio analysis for the absorbance measurements of Figure 11

Conditions: I_{sig} = 10nA to $10\mu A$, $I_{ref} = 10\mu A$ ±50%, temperature range = 25°C ± 10°C

Parameter	Error	Comment
Log Conformity	±0.5% RTI	Specified as ±0.5%max for Model 756
Scale Factor, K	$\cong 0$	Initial error of ±1% is trimmed to zero for this application
K Drift	12mV RTO	±10°C x 0.04%/°C x K x log $\frac{I_{sig}}{I_{ref}}$
I_{os}	$\cong 0$	I_{os} at I_{ref} input has negligible error due to the relatively high value of I_{ref}. I_{os} at I_{sig} input is 0.1% of smallest expected I_{sig}.
I_{os} Drift	−0.1%	Additional 10pA due to I_{os} doubling per 10°C
E_{os}, E_{os} Drift	$\cong 0$	This contribution is held to a negligible value due to the low initial offset and offset vs. temperature. The data sheet for the photodiode should be consulted to determine the error in I_{sc} due to E_{os} across the diode.

Total RTI error at $25°C = \pm0.6\%$, or 2.6mV at the output, less than 0.1% of the 3V full-scale output range.

Total RTI error over the temperature range is

$\pm0.7\%$ RTI ±12mV RTO = 3.5% RTI, assuming all errors to be directly additive, a worst-case condition

Referring the 3.5% RTI error to the output results in 15mV of error, or 0.5% of the 3V maximum output.

SPECIFICATIONS OF MODEL 756N/P

Current Log Ratio
Transfer Equation

$$E_o = -K \log \frac{I_1}{I_2} \;,\; I_1 = \text{signal} \quad I_2 = \text{reference}$$

Transfer Equation including
Error Terms

$$E_o = -K \left[\log \frac{I_1 - I_{b1}}{I_2 - I_{b2}} + E_{os3} \right]$$

Voltage Log Ratio
Transfer Equation

$$E_o = -K \log \left[\frac{V_1}{V_2} \times \frac{R_2}{R_1} \right]$$

Transfer Equation including

$$E_o = -K \left[\log \left(\frac{\dfrac{V_1 - E_{os1}}{R_1} - I_{b1}}{\dfrac{V_2 - E_{os2}}{R_2} - I_{b2}} \right) + E_{os3} \right]$$

Parameter	Value
Signal Current, I_1 [1]	10nA to 100μA (4 decades)
Reference Current, I_2 [1]	100nA to 100μA (3 decades)
Log Conformity [2]	$\pm0.5\%$ (2 decades, I_2 constant)
	$\pm1.0\%$ (4 decades, I_2 constant)
Scale Factor, K [1, 3]	1V $\pm1\%$ $\pm0.04\%/°C$
Bias Current, I_{b1}	10pA, doubles/$10°C$
Bias Current, I_{b2}	10nA, max, $\pm1\%/°C$
Offset Voltage, E_{os1} [3]	±1mV, max, 25μV/$°C$
Offset Voltage, E_{os2}	0.5mV, max, 30μV/$°C$ max
Output Offset, E_{os3} [3]	±10mV, max, 85μV/$°C$

Small Signal Response

I_{IN}	f_t
1nA	1kHz
1μA	8kHz
100μA	25kHz

Rated Output

Log Mode	±10V at 5mA
Antilog Mode	±10V at 4mA

Response Time

I_{IN} (increasing)	time	I_{IN} (decreasing)	time
1nA to 10nA	70μs	10nA to 1nA	200μs
10nA to 100nA	25μs	100nA to 10nA	50μs
100nA to 1μA	25μs	1μA to 100nA	25μs
1μA to 100μA	20μs	100μA to 1μA	20μs

Noise in 10kHz B.W.

V_n, INPUT 1	3μV rms
V_n, INPUT 2	3μV rms
I_n, INPUT 1	0.1pA rms
I_n, INPUT 2	20pA rms

Power

Quiescent Current	3mA at ±15V
PSRR	54dB

[1] Positive for positive inputs (N type), negative for negative inputs (P type).

[2] The log conformity specification is referred to input (R.T.I.). Note: 1% error R.T.I. is equivalent to 4.3mV of error at output for K = 1V.

[3] Externally trimmable.

Antilog Applications

The antilog(arithm) is the inverse of the logarithm. It is by definition the *exponential*, in which the logarithmic base is raised to a power. That is,

$$x = \log_B^{-1}(y) = B^y \tag{1}$$

If $y = \log_{10} x$, then $x = 10^y$. If $y = \ln x$, then $x = \epsilon^y$. The same argument can be expressed in terms of exponentials of any base. For example,

$$x = 10^y = (\epsilon^{\ln 10})^y = \epsilon^{y \cdot \ln(10)} = \epsilon^{y/(\log_{10} \epsilon)} \tag{2}$$

Logarithmic devices with the transfer function

$$E_o = -K \log_{10} \frac{V_{IN}}{E_{REF}} \tag{3}$$

are usually available with a connection scheme that allows the input and feedback circuits to be interchanged to compute

$$E_o = E_{REF} \cdot 10^{(V_{IN}/-K)} = I_{REF} \, R_f \cdot 10^{(V_{IN}/-K)} \tag{4}$$

Thus, if $K = 1$, (3) provides an output of 1 volt per-decade of the input ratio, while (4) provides an output that changes by 1 decade for each volt of input. E_{REF} in (3) is interpreted as the starting point (e.g., 0) for the log ratio, the input value at which the ratio is unity and the output is zero. In (4), E_{REF} is interpreted as the normalized value of the exponential, and each volt of input either multiplies or divides E_{REF} by an additional factor of 10.

Figure 1 is a plot of the exponential response of N-type and P-type modules with correct polarity on a linear scale. Because of the wide range of variation of the output, it would be difficult to show the output values accurately without changing scales for each decade. Alternatively, it is possible to compromise and plot output (on a log scale) against the input (Figure 2).

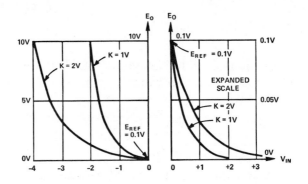

a) E_O vs. V_{IN} for N-type antilog operator. Vertical scale: 10V max, $V_{IN} \leqslant 0$; 0.1V max, $V_{IN} \geqslant 0$

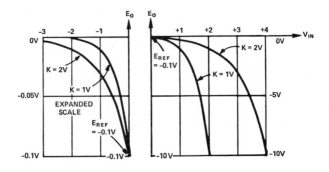

b) E_O vs. V_{IN} for P-type antilog operator. Vertical Scale: -10V min, $V_{IN} \geqslant 0$; -0.1V min, $V_{IN} \leqslant 0$

Figure 1. Antilog-operator response curves (linear scale)

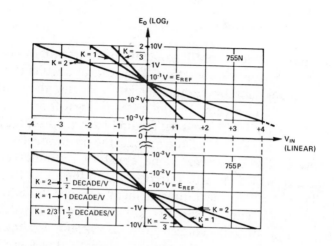

Figure 2. Antilog operator response curves, semilog scale.

$$E_O = E_{REF} \, 10^{V_{IN}/-K}$$

WHERE THEY ARE USED

Exponential devices are used in at least three classes of application

1. When compound multiplications, involving roots and powers, are performed (e.g., $x_1{}^{\alpha} \cdot x_2{}^{\beta} \cdot x_3{}^{\gamma} \cdot x_4{}^{\delta} \cdot ...$), each input is "logged", multiplied by a constant (or variable) exponent of appropriate magnitude and polarity, the terms are summed and/or differenced, then the antilog is taken to convert the result back to the "world of phenomena."

2. If measurements are performed by devices having logarithmic responses they may be linearized, if necessary, by the use of the antilog. An example is given in this chapter.

3. The exponential may be used in function fitting and function generation to obtain relationships or generate curves having voltage-programmable rates of growth or decay. For example, if V_{IN} is a ramp, E_o will be a widely-ranging exponential function of time, with time-constant determined by K, and scale-factor determined by E_{REF}. K can be manipulated by changing the input gain (variably with an IC multiplier, if desired).

AVAILABLE OPTIONS

The options available for antilog operation are essentially the same as those for log-voltage operation. Log-voltage circuits (Chapters 3-1, 4-1) become antilog circuits by interchanging the input and feedback circuit elements. In the case of logarithmic circuitry (Figure 3), the log is achieved by feeding back a current which is the antilog ($I = I_o e^{qV/kT}$) of the output; the operational amplifier forces this feedback current to be equal to the input current, therefore the output must be proportional to the log of the input. For antilog operation, the antilog element is placed in the forward path, and the feedback circuit is closed by a resistor that transduces the output voltage into a current to equal the exponential input current; the output voltage is thus forced to be proportional to the antilog of the input.

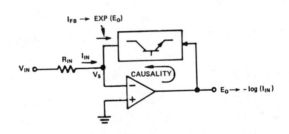

a) Antilog function in feedback path. Amplifier forces $I_{FB} = I_{IN}$ and $V_S = 0$ by making E_O proportional to log (I_{FB})

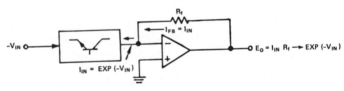

b) Antilog function in forward path converts input voltage to exponential current. Amplifier converts current to exponential voltage via R_f

Figure 3. Using feedback to enforce direct and inverse relationships

There are some log modules that are committed to log operation
alone, but they are in the minority. Most commercially-available
log amplifiers can also serve in antilog applications. Figure 4 shows
how the Model 755 is connected in the log and antilog modes.

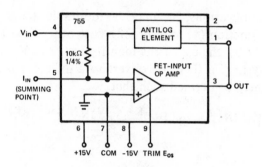

a) Log/antilog amplifier connected in the log mode (K = 1)

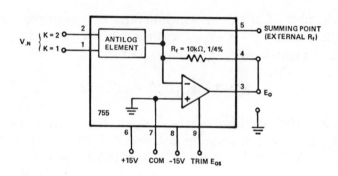

b) Log/antilog amplifier connected in the exponential mode

Figure 4. Log/antilog amplifer connections

The reader may observe that log modules are usually specified to
have considerably greater dynamic range (of accurate operation)
for current inputs than for voltage inputs (6 or 7 decades to 4).
For normal, practical antilog circuitry, however, the output is
from an operational amplifier, the input-offset of which imposes

the low-end output accuracy restriction ($10\mu V$ offset, for example, is 1% of the lowest of 4 decades (1% of $10V/10^4 = 10\mu V$).*

OTHER BASES

For some applications, it may be desirable to interpret the exponential function (4) in terms of a base other than 10. This can be done easily by the use of the identity in (2). If $V_{IN}/-K$ in (4) is the same as y in (2), then to convert to a different base B, a new constant K' may be defined such that

$$10^{V_{IN}/-K} = B^{V_{IN}/-K'} = B^{V_{IN}/(-K \log_{10} B)} \tag{5}$$

from which,

$$K' = K \log_{10} B \tag{6}$$

For example, if B = 1.10, which defines an exponential having a growth rate of 10% as V_{IN} becomes more negative, then K = 1 defines a negative-going input interval of

$$K' = 1 \cdot \log_{10} 1.10 = 0.0414V \tag{7}$$

for each 10% increase of the output. Put another way, it is analogous to a continuously compounded interest rate of 10% per-41.4mV "year." If B = 2, K' = 0.3010; if B = ϵ, K' = 0.4343.

That the input intervals are negative-going is a consequence of the negative slopes that can be observed in Figure 1. This is accounted for by the fact that, like the log circuit, the antilog function is produced by essentially a single inverting operational amplifier,

---

*It should be noted, nevertheless, that the summing-point current from the input element is independent of small amplifier offsets and is imposed at high impedance (the antilog amplifier acts as a unity-gain follower for its offset voltage). In concept, then, if the feedback current could be tapped for a "downstream" element requiring a wide range of exponential current, antilog accuracy could be maintained over a wider dynamic range. In practice, this is difficult to achieve in a simple manner, unless the "load" is, in effect, a two-terminal device that can be connected in series in the feedback path of the amplifier.

which must have a transfer function with a negative slope. If the signal input were held constant and the reference input (if available) varied, the slope would be inverted. An easier way, with standard modules, is simply to use an inverting op amp to precede the antilog circuit; it can also be used to scale K′ to a convenient round-number value.

ERRORS

Although all the specifications for a log module apply to both log and antilog modes of operation, with "RTI" and "RTO" interchanged, the observed effect of each parameter is manifested at the output quite differently. By examining the transfer equation, one can predict these effects.

$$E_o = I_{REF} \, R_f \cdot 10^{V_{IN}/-K} + E_{os} \qquad (8)$$

Offset Voltage (E_{OS}): As noted earlier, it appears at the output as a constant dc voltage. Its percentage error contribution is 100% X (E_{os}/E_o). For small output signals, E_{os} is a major source of error. If output signals in the millivolt range are expected, E_{os} should be adjusted to zero. Since the device is operated in the voltage mode, the contribution of bias current to output offset is negligible (e.g., at 25°C, I_b for the 755 is 10^{-11}A; with a 10kΩ feedback resistance, the contribution is $10^{-11} \times 10^4 = 10^{-7}$V).

Adjusting E_{os} to Zero: To adjust E_{os} to zero, a value of input voltage is applied that will essentially cut off the flow of current to the summing point. Then the offset-adjust potentiometer is adjusted for zero output. For example, if the input terminal of an N-type module (K = 2) is connected to +15V (-15V for P types), the theoretical contribution of input to the output signal is 7.5 decades below 0.1V, well into the noise level. For reasonable sensitivity of output reading, a small resistance can be connected from the summing point to ground (100Ω to ground establishes a gain of 101, resulting in 1 mV out per 10μV of input offset), as shown in Figure 5.

After adjusting to zero, the jumper to 15V and the 100Ω resistor (if used) should be removed.

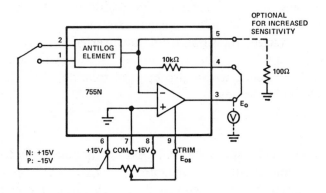

Figure 5. Circuit for trimming E_{OS} to zero in antilog mode

Reference Current (I_{REF}): Errors associated with $E_{REF} = I_{REF} R_f$ appear directly at the output, since I_{REF} acts as a gain factor. For example, if I_{REF} is 1% too high, it will cause the output voltage to be 1% too high. E_{REF} can be calibrated by applying 0 input (grounding the input), then trimming the feedback resistor (by series or parallel resistance) until the output is equal to E_{REF} (0.1V for the 755). It is more convenient to adjust R_f than I_{REF}, if an external means of adjusting I_{REF} has not been provided in the device. I_{REF} can also be adjusted by biasing the input.

Scale Factor (K): Errors associated with K are manifested as errors of the input scale factor; their effect on the output is a modification of the exponent. If the actual value of K is $K/(1 + \delta)$, then the error, defined as the difference between the actual and theoretical value of output, divided by the theoretical value, is

$$\frac{0.1 \times 10^{(1 + \delta)V_{IN}/-K} - 0.1 \times 10^{V_{IN}/-K}}{0.1 \times 10^{V_{IN}/-K}} \qquad (9)$$

$$= 10^{\delta V_{IN}/-K} - 1 \qquad (10)$$

For $\delta = 0.01$ (1%), the fractional output error is $10^{0.01 V_{IN}/-K}$. For $V_{IN}/-K = 2$, the error is $1.047 - 1 = 0.047$, i.e., 4.7%.

The table provides a brief listing of output errors as a function of V_{IN}/K, for several values of δ. (The range of $V_{IN}/$-K for the 755 is -2 to +2.) Linear interpolation may be used for intermediate values.

TABLE 1. OUTPUT ERRORS, AS A FUNCTION OF K ERRORS, IN % OF ACTUAL OUTPUT VALUE

$V_{IN}/$-K	+1%	+0.3%	+0.1%	0	−0.1%	−0.3%	−1%	
				K Error (100% $\times$ δ)				
4	9.6	2.8	0.93	0	−0.92	−2.7	−8.8	
3	7.2	2.1	0.69	0	−0.69	−2.1	−6.7	
2	4.7	1.4	0.46	0	−0.46	−1.4	−4.5	
1	2.3	0.7	0.23	0	−0.23	−0.7	−2.3	Output
0	0	0	0	0	0	0	0	Errors
−1	−2.3	−0.7	−0.23	0	0.23	0.7	2.3	(%)
−2	−4.5	−1.4	−0.46	0	0.46	1.4	4.7	
−3	−6.7	−2.1	−0.69	0	0.69	2.1	7.2	
−4	−8.8	−2.7	−0.92	0	0.93	2.8	9.6	

Log Conformity is specified in %, referred to the input, in the log mode, all other errors adjusted to zero. The relationship between log-conformity errors referred to the input and to the output is discussed in Chapter 4-1, and Table 4 in that chapter, to be found on page 434, provides ready conversion between input and output error. Since input and output are exchanged in antilog applications, the same table may be used here, with the headings exchanged. For example, if a log conformity error is 1%, referred to the input, corresponding to 4.32mV at the output, then in the antilog mode, a 4.32mV error referred to the input is a constant 1% error at any output level. For a given input log-conformity error, γ,

$$E_o = E_{REF} \cdot 10^{(V_{IN}/\text{-}K) + \gamma} = E_{REF} \cdot 10^{(V_{IN}/\text{-}K)} \cdot 10^{\gamma} \quad (11)$$

Since γ is specified independently of V_{IN}, 10^{γ} is a constant multiplier; if γ = 4.3mV, 10^{γ} = 1.01.

APPLICATION EXAMPLE–LINEARIZING LOGARITHMIC OUTPUTS

In pollution monitoring, oxygen detectors that have output voltages that measure $p0_2$, the log concentration of oxygen, are frequently used. If it is desired to determine the actual concentration of oxygen, some form of exponential processing is necessary. This example will consider the design of an analog linearizing circuit employing logarithmic devices.

The objective is to design a circuit that will obtain a linear concentration reading from the logarithmic output of an oxygen detector, the response of which is characterized as follows:

O_2 Concentration	Output Voltage
10%	0mV
1%	−60mV
0.1%	−120mV
0.01%	−180mV

The first step is to reduce the measured data to the form of an equation, if possible. In this example, the output of the transducer varies at -60mV per decade change of concentration. It can be expressed by the equation

$$E_{OD} = 60 \times 10^{-3} \log_{10} \frac{100 C_{0_2}}{10} = 0.06 \log_{10} 10 C_{0_2} \quad (12)$$

$$10^{-4} \leqq C_{0_2} \leqq 10^{-1}$$

The second step is to characterize a circuit that will have a response determined by the inverse of (12). The inverse of (12) is

$$10 C_{0_2} = 10^{E_{OD}/0.06} \quad (13)$$

It is evident that a function of the desired form can be performed by circuits employing either a basic log circuit (751), log transconductor (752), or log amplifier (755). For example, the 755N's transfer relationship is

$$E_o = E_{REF} \, 10^{V_{IN}/-K} \quad (E_{REF} \text{ positive}) \quad (14)$$

The next step is to determine scale factors, and any external circuitry necessary to accommodate the range of the 755 to the range of the oxygen detector.

To scale the exponential device, it is necessary first to determine the range of E_o that will correspond to the range of O_2 concentration and the range of V_{IN} that will correspond to the range of E_{OD}.

Since C_{O_2} ranges from 10% down to 0.01%, a useful maximum value for E_o is 10V, which allows the output to be read directly in percentage concentration (1V/1%). Since at $V_{IN} = 0$, $10^0 = 1$, and $E_o = 10V$, corresponding to $E_{OD} = 0$, at $10C_{O_2} = 100$

$$\frac{E_o}{E_{REF}} = 1 \quad (15)$$

$$E_{REF} = E_o = 10V \quad (16)$$

which happens to be 100 X the nominal E_{REF} for the 755N, equivalent to a -2V input bias (K = 1).

For a factor-of-10 change in concentration, $E_{OD}/0.06 = 1$, and similarly, for a factor of 10 change in E_o,

$$\frac{V_{IN}}{-K} = \frac{GE_{OD}}{-K} = 1, \text{ hence } \frac{G}{K} = \frac{-1}{0.06V} \quad (17)$$

where G is an external coefficient between the output of the oxygen detector and the input of the exponential circuit.

In determining the value of G, it is useful to note that the internal gain developed by the 755 is independent of the choice of K; the K connection simply determines the input attenuation, without affecting any other aspect of performance. Thus, one might make the arbitrary choice of $K = 1$, to simplify the mathematics. If $K = 1$, $G = -16.7$, which could be developed by an inverting preamplifier.

Figure 6 shows a linearizing scheme based on the above discussion; Figure 7 shows a similar scheme employing the 752N log transconductor with an external operational amplifier. If the scheme of Figure 6 is used, the G trim allows the overall coefficient (G/K) of the exponent to be trimmed, taking into account errors in the value of K. E_{REF} is perhaps most easily achieved by biasing the input by -2V, and trimming for 10 volts out with 0 volts in. E_{os} is trimmed by the method mentioned earlier, with V_{IN} at +15V, or by applying a value of V_{IN} that maintains the output of A1 at $\geq$ +10V.

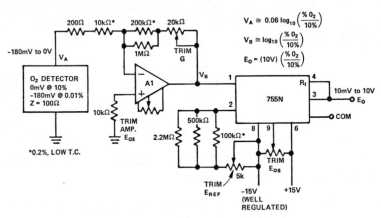

Figure 6. Linearizer for oxygen detector, using log amplifier.
Trim procedure:
*   A. Trim A1 output to zero with zero input*
*   B. Trim E_O to zero with $V_B \geqslant 10V$*
*   C. Trim E_O to +10V (trim E_{REF}) with zero input*
*   D. Trim E_O to +10mV (trim G) with –180mV input*

The transconductor approach (Figure 7) lends somewhat more flexibility to the computation, since I_{REF} is adjustable. For example, in this case, where $V_{REF} = 10V$, I_{REF} can be set at $100\mu A$, and the feedback resistor at $100k\Omega$, without a need for biasing the input or using a greater value of feedback resistance.

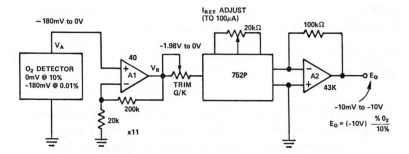

Figure 7. Linearizer for oxygen detector using high-impedance follower-connected op amp and log transconductor. Trim procedure:

    A. Trim A1 output to zero with zero input
    B. Trim E_O to zero (A2 output) with $V_B \geqslant 10V$
    C. Trim E_O to -10V (I_{REF} adjust) with zero input
    D. Trim E_O to -10mV (trim G/K) with -180mV input

As shown here (Figure 6), the input amplifier is connected in the inverting configuration, and consequently loads the input signal source. If high input impedance is necessary, the "P" version of the log transconductor or log amplifier could be used with a non-inverting input-amplifier, but the output polarity (Figure 7) would be negative instead of positive (usually a minor consideration because it can be easily dealt with).

The exponential circuit could also be built "from scratch", using the 751 basic log element, and the principles discussed in Chapters 3-1 and 4-1.

Selecting the Operational Amplifiers

Amplifier A1 is a general-purpose op amp selected for low offset temperature-coefficients. If it is a non-inverting amplifier (with "P" versions), it should also have good common-mode rejection at low levels.

The worst-case error occurs at $(V_{IN}/K) = 3$, which gives the greatest error (% of output); for example, if V_{IN}/K is associated with a 0.1% error, the output error would be 0.7% (from Table 1). If, on the other hand, the maximum output error allowable due to errors in the input circuitry were 0.1%, the input errors would have to be kept at about 1/7 of 0.1%, or 0.014%.

As an aid to considering the effect of millivolt errors at the input on the output, Table 2 may be found useful (K = 1).

TABLE 2. MILLIVOLT RTI VS. % RTO ERROR. Equal Numbers of Millivolts of Error in the Exponent Correspond to Constant Percentage Error at the Output at any Level. (Figures are for K = 1)

RTI Error millivolts	RTO Error percent
0.1	0.02
0.5	0.1
1.0	0.2
2.0	0.5
3.0	0.7
4.0	0.9
5.0	1.2
10.0	2.3
30.0	4.2
100.0	25.9

From the table, it can be seen that noise or drift, resulting in a $167\mu V$ error at the output of A1 ($10\mu V$ error at the input) would create a 0.04% output error.

In choosing A2 for this application, a low-noise, low-drift operational amplifier is required. The output impedance of the current presented to the summing point by the log transconductor varies with current level and may be approximated as

$$Z_O = 1k\Omega + \frac{40}{I_C} \, ,$$

where I_C is the current in mA.

At worst, its minimum value of $1k\Omega$ should be considered as the summing-point load. With $R_f = 100k\Omega$, the "noise gain" is about 100, and consequently the noise and drift of A2 will be magnified by 100. For this reason, the Model 43K was chosen, based on its guaranteed maximum noise of $2\mu V$ rms in a 1kHz bandwidth and its maximum specified drift of $5\mu V/^\circ C$.

Bias current is also an important concern in selecting A2, since for the smallest output (feedback) current level $(0.15\mu A)$, a 1% error would result if the bias current were to exceed 1% of $0.15\mu A$, i.e., 1.5nA. The FET-input Model 43K, with bias current of 20pA maximum at $25^\circ C$, easily meets this requirement over a wide range of temperature.

Multiplying And Squaring

Chapter 4

MULTIPLYING AND SQUARING

In Part 1, the properties and applications of multipliers were described briefly; in Part 2, applications of nonlinear devices were suggested, including a great many that involved multiplication. Chapter 3-2 discussed at length the design, nature, specifications, and foibles of multipliers, and means of measuring their properties. In this chapter, we consider some of the factors involved in choosing and applying a multiplier to perform multiplication and squaring applications.

SELECTION GUIDELINES

It has been determined that multiplication or squaring is required for a given application. Now it is necessary to choose a multiplier that will do the required job at the lowest cost, and to interface it to the rest of the circuit in an optimal manner. Since the choice of the multiplier and the manner of using it are interdependent, they should be considered together. There are a number of questions, the answers to which affect both choice and use. They are grouped here for discussion.

1. *What is the required transfer function? Does it involve simple multiplication, division, or multiplication combined with division? Is the scale factor fixed, adjustable, or variable? What are the polarities of the inputs? What is the polarity of the output? What is the polarity relationship? How many quadrants of operation are involved?*

The usual multiplier transfer function is $E_o = K(V_x \cdot V_y)$, where K represents the constant of proportionality, commonly called the scale factor. For most electronic analog multipliers, $K = 1/(10V)$, in order to obtain 10-volt full-scale output response for 10V full-scale input signals. There are several designs, however (e.g., AD531 and Models 433 and 434), that allow the user considerable freedom in the selection of the scale factor: it may be fixed at some arbitrary value, switched among several arbitrary values, or even varied continuously by an external voltage or current (in effect, combining multiplication and division).*

A multiplication ideally results in a product that has the proper algebraic polarity ($1 \times 1 = 1, -1 \times 1 = -1, -1 \times -1 = 1, 1 \times -1 = -1$). However, not all multipliers accept both input signal polarities or provide both positive and negative output voltage. The number of quadrants of operation (on a plot of one input against the other) is defined in Chapter 3-2.†

In order to make the multiplication operation, as envisioned on a block diagram, correspond to the performance ranges of the hardware to be used, it is often necessary to provide gain scaling, either ahead of one or both inputs, or following the output. All such gains should be taken into account in the overall mathematical transfer equation to ensure that the measured output of the circuit is indeed related to the inputs in the expected manner (Figure 1).

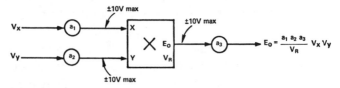

Figure 1. Effects of external scaling (amplification or attenuation) on the overall multiplier scale factor

---

*For division alone, best accuracy and dynamic range are achieved with specialized dividers, such as 434 or 436.

† It is strongly urged that the reader prepare himself for this chapter by ensuring that he is familiar with the general characteristics of multipliers and the definitions of the specifications in Chapter 3-2.

General-purpose 4-quadrant multipliers may be readily used for one- or two-quadrant operations (where one or both input variables is of restricted polarity), often with the advantage that tighter adjustment is possible in a restricted range. However, one-quadrant multipliers are not quite as easily used for multiple-quadrant operations. Chapter 3-2 (Figure 21) shows one way of accomplishing this, by offsetting the input signals. For squaring of 2-quadrant signals, a 1-quadrant multiplier may be preceded by an absolute-value circuit.

Finally, if either input signal, the output, and/or the gain relationship is of an undesired polarity, it can be easily corrected by the use of external unity-gain inverting op amps at the appropriate locations, or by appropriate connections to differential inputs, if available (as in the AD532).

2. What is the range of input magnitudes? What is the range of output magnitudes? What is the input resolution? What is the output resolution?

It is usually desirable to scale inputs and output to full-scale rated input. However, for some applications (e.g., gain control), linearity can be improved by restricting the range of one or both variables. Naturally, offset errors increase in importance, since a given offset represents a larger fraction of a reduced signal.

If one variable has a relatively small range of variation, the output resolution (dynamic range) is the same as that of the other input. However, if both variables have wide swings (e.g., in squaring), the required output resolution is the product of the input dynamic ranges. For example, if both inputs have 100:1 ranges (10V to 0.1V), the output resolution must be 1 in 10,000 (1mV out of 10V). Whether all of it is needed depends on what is to be done with the output signal.

In a frequency-doubler application, all that is needed is a reasonably undistorted sine wave; hence 1mV uncertainty at the low end is irrelevant; in many cases 50mV would be adequate. However if the output of a squarer is to be averaged, then square-rooted (as in a straightforward open-loop rms application), the high sensitivity of the square-rooter at small signals requires that the squarer output actually have the full implied resolution. Fortunately, most commercially-available multipliers have excellent "resolution" near zero — but noise and drifts may mask small ac signal components.

3. Is the application "ac" or "dc?" What properties of the product are important (i.e., instant-by-instant value, or some measure, such as average, peak, peak-to-peak, phase, etc.)? Is the application dc-dc, ac-ac, or ac-dc?

For some "ac" applications, such as modulators, frequency-doublers, and gain-control circuits, output offset may be unimportant. For others, good linearity may be necessary only for one of the inputs (in a modulator, the modulation input; in a gain control, the signal input). In a modulator, dc feedthrough may be important on the modulation signal channel, ac feedthrough on a high-frequency signal channel.

Bandwidth requirements are a function of the job the multiplier performs: for mean-square measurements, average-power measurements, correlation, and phase measurements, where the average value of the output is of interest, output bandwidth can be quite narrow, but the inputs must have negligible phase difference.* For such functions as amplitude modulation, peak-demodulation, frequency-doubling, and AGC, the output amplitude-envelope must be maintained to the desired accuracy (1%, -3dB, etc.). The most-exacting applications are those in which phase must be preserved, that is, the output must follow the input "instantaneously," as predicted by the settling-time specifications or (for sine waves) the "vector" error.

4. What measures of accuracy are important? (overall error? nonlinearity? feedthrough? offsets? noise? gain error?) Over what dynamic range of inputs (output) must accuracy be maintained?

If no adjustments are permissible, overall error may be the key accuracy criterion. On the other hand, offsets, feedthrough, gain

---

*Also, it is desirable to avoid signal amplitudes (rise times) that cause (asymmetrical) limiting or slewing; this will also affect the average value at the output.

error, and − in some cases − nonlinearity may be reduced by external adjustments, allowing a nominally lower-performance device to be used at lower cost, if its drift specifications indicate that the error components will not change excessively over a reasonable range of temperature.

For some applications, the error must be a given fraction of the actual output voltage (say 0.5%) over a range of output values; in other cases, it is sufficient to specify the error as a fraction of full-scale output (say 0.5% of 10V, or 50mV). Nonlinearity and linear feedthrough errors are referred to the inputs; other errors are usually referred to the output (see Chapter 3-2). As noted earlier, external adjustments can be used to further reduce errors over limited ranges of voltage magnitude, and in a restricted number of quadrants.

5. What scope can be allowed for adjustability? Must the device be installed without initial tweaking? Without calibration? Must the installation be factory-adjustable? Field adjustable? To what degree can adjustability be traded for basic device cost?

Without question, performance of the lowest-cost general-purpose devices can be greatly improved by adjustment. However, adjustment has costs of its own, in terms of additional parts, "real estate," procedures, fixtures, and instructions. Higher-cost pre-trimmed devices include these extra costs in their performance guarantees.

Pre-trimmed modular devices are generally close to optimum. Pre-trimmed IC.s, such as the AD532, can often benefit by further trimming, since the automatic laser-trimming process must take into account variations due to warmup, handling, packaging, and aging; this results in a wider specified range of individual variation, but still considerably lower error than untrimmed IC's, and lower cost than most pre-trimmed modular devices. Besides the usual offset, linear-feedthrough, and gain trims, IC multipliers generally can obtain greatly improved performance with cross-feed linearity trimming to reduce the quadratic nonlinearity component (Chapter 3-2, Figures 14, 15, 16, 17, 18).

Multiplier adjustments can often be replaced by overall calibration adjustments to the equipment incorporating the multiplier, since the adjusted variables are input and output bias voltages and overall gain. Gains and additive voltages elsewhere in the system can be manipulated (in the right order, by an appropriate procedure) to compensate for multiplier offset, linear-feedthrough, and gain errors.

Some applications, by their nature, call for fewer adjustments. For example, in squaring, one of the feedthrough adjustments can be eliminated, and quadratic nonlinearity can be eliminated by appropriate gain adjustment. Another example: an amplitude modulator with ac coupling can eliminate *all* local adjustments (dc feedthrough at the modulating input is incorporated in the modulating-signal bias; gain can be controlled in the modulating signal, the carrier, or beyond the output; and dc feedthrough at the carrier input and dc output offset are irrelevant with ac coupling).

For applications where *ac* feedthrough must be reduced (at all costs) at high frequencies, the input signals can in some cases be fed capacitively, with appropriate polarity, into a summing amplifier and summed with the multiplier output out-of-phase, to cancel internal capacitive feedthrough.

6. What are the needs of dynamic response: input frequency range, output frequency range, allowable attenuation and phase shift, slewing rate, settling time, distortion? (See 3.)

7. What are the environmental constraints (temperature variation, humidity, shock, vibration, warmup, power supply, physical size, external circuitry)?

8. What other requirements on the inputs and outputs are not expressed in the transfer function? Differential vs. single-ended? Loading of the signal sources by the multiplier inputs? Output current requirements: boosting, isolation, capacitive load?

9. Finally (and perhaps the most perplexing question) what weight should be assigned to the various costs (what is their relative tolerability?):

A. The cost of the device itself?

B. The cost of adjustments and their procedures, to the degree they are needed?

C. The cost of any other external circuitry peculiar to a given device? Size?

D. Reduced performance or system reliability as a consequence of compromises to A or B?

The investigation of alternatives should start with a translation of the requirements outlined above into a preferred circuit configuration and a set of tentative specifications. The specifications, including size and cost, should be listed in order of priority, taking into account the tradeoffs between cost of an untrimmed device with external trim circuitry and that of a pre-trimmed device.

If, for example, size is of the essence, it will be near the top of the list and will probably result in a restriction of the field to integrated circuits, with no external adjustments. If linearity is the most important criterion, the field will be narrowed to those devices having the desired linearity, or something approaching it.

By listing the specifications in order of priority, one can generally narrow the field of choice very quickly, so that the relative merits of just a few devices may be compared in depth. If the process eliminates *all* available devices, it quickly establishes the need for compromise (or for consultation with a manufacturer).

Most manufacturers (including Analog Devices) group their multipliers into several classes, that make the narrowing easy to accomplish in a general way. For example, integrated-circuit devices are listed in a separate table. Because IC's have minimum size and (often) minimum cost, the rule of the thumb is often used: "choose the cheapest IC that will do the job satisfactorily;" the separate listing allows one to study the whole panoply of IC's at a glance.

If, on the other hand, the priority is given to other considerations, one can find devices under such headings as: High Accuracy, Accurate Low-Drift, Accurate Wideband, Wideband, General-

Purpose, Economy, etc. Table 1 is a thumbnail sketch of the key features and typical applications of each of the Analog Devices multiplier families. Each family has two or more members, graded in terms of one or more key specifications, e.g., nonlinearity, feedthrough, drift (429 vs. 429B; 427J vs. 427K; AD532J, AD532K, AD532L, AD532S, etc.)

TABLE 1. APPLICATION GUIDE BY KEY FEATURE

Key Feature	Multiplier Application	Multiplier Family
Highest precision, lowest noise & drift	Analog computation, dividers, servo multipliers, correlators	424, 427, 436
Low drift, good accuracy, lower cost	Wide temperature range, general purpose multiply/divide	428
Bandwidth, accuracy	Graphic displays, dividers	429, 422
Wide dynamic range, accuracy	Root and power generation, $Y\,Z/X$, $Y(Z/X)^m$	427, 433, 434
External trim for high accuracy	R & D, medical, laboratory, analog computation	424, 425
Economy, size	OEM designs, general purpose multiply/divide	530, 532, 533, 426, 432, 531
MIL spec. available	Military grade design	530, 531, 532, 533, 432, 428

Obviously, if one starts without any prior knowledge of the available devices, the job of choosing a suitable device may involve considerable searching. To make the job somewhat easier, a table (Table 2) has been prepared that selects several salient criteria and divides the specification ranges into convenient increments ("levels"). Then, each and every standard device is listed, in terms of descending desirability for each of the parameters.

For example, considering feedthrough, the best possible choice would be the 424K, with 1mV at the X input, 2mV at the Y input. The worst possible choice would be the AD532J (without external trim) with a possible maximum of 200mV at both inputs. Considering bandwidth, the 429 family would be the best choice, with 10MHz typical -3dB bandwidth, while the 424 or 427 might be the worst choice, with 100kHz bandwidth.

Thus, it is possible to apply the list of priority specifications to Table 2, and, from it to determine whether any of the devices listed there fits the required profile; if none do, the necessary compromises are instantly evident.

Naturally, the utility of the table is limited to "First Resort," for several reasons

1. It represents a tabulation of general-purpose 4-quadrant devices (not including the excellent 433 and 434 YZ/X 1-quadrant-device families).

2. It represents the state of the Analog Devices product line in late 1973. At the time you read this, newer products, lower prices, better specs may be available on the market, from Analog Devices and/or other manufacturers.

3. All specifications are not listed. Not even all salient specifications (for your application). However, the number of data sheets (or catalog tabulations) that must be consulted can be greatly decreased, since the number of devices to be considered is narrowed.

4. "All prices and specifications are subject to change."

5. Some of the figures given are "typical," rather than min/max. The decision to make a specification "typical" is usually an economic decision: certainty of achieving a given performance level is traded for a lower price. In effect, any "typical" specification should be moved down one or more levels, or its price rating should be moved down one level (representing the higher price that might have to be paid to have the spec guaranteed, if feasible).

6. Prices shown are for 1 unit. Since discounts in quantity may be substantial, devices should not be ignored if they are one level too low in their price rating.

NOTES FOR THE USER

1. Improved linearity at low levels. By taking advantage of the nonlinearity specifications, the user can often use a less-costly multiplier to obtain adequate small-signal accuracy. This is done

TABLE 2. 4-QUADRANT MULTIPLIER SELECTION GUIDE

	ERROR (Int. Trim) max. %	ERROR (Ext. Trim) max. %	NONLINEARITY (x, y) max. %	FEEDTHROUGH (Ext. Trim) (x, y) max mV p-p	DRIFT (Total Error) max. %/°C
LEVEL ONE	≤0.3%	≤0.1%	≤0.1%	≤5mV	≤0.01%
	427K .2 427J .25	424K .1 427K .1	424K .04 427K .04 428K .08 424J .08 427J .08	424K 1-2 424J 2-4 427J 4-5 427K 4-5	AD533L* AD530L* AD533S* AD531L*
LEVEL TWO	≤0.5%	≤0.3%	≤0.3%	≤20mV	≤0.02%
	426L 428J 428K 429B	427J* .15 424J .2 428J* .25 428K* .25 429B* .3	429B .2 426L .25 428J .25 AD530L*} AD531L*} .3-.2	428J } 428K } 10-10 429B 10-20 426L 20-20	AD532S* AD530S AD531S 428J/K 424J/K 427J/K
LEVEL THREE	≤1%	≤0.5%	≤0.5%	≤50mV	≤0.03%
	AD532K/S* 426A/K 432K 429A 422A/K	426L* .35 AD533L AD530L* AD531L*	AD530K/S*} AD532K/S*} .5-.2 AD533K/L/S* AD531K/S*} 429A } .5-.3	422A/K 8-35 AD530L 40-30 AD531L 40-30 429A 16-50 AD533L 50-50	AD533K* AD530K* AD532K* AD531K*
LEVEL FOUR	≤2%	≤0.7%	≤0.7%	≤100mV	≤0.04%
	AD532J* 432J	432K* .6 426A/K* .6 429A* .7 422A/K* .7	432K 426A/K 422A/K (.6-.3)	426A/K 20-60 AD530K/S} AD531K/S} 80-60 432K† 50-100 AD533K/S AD532K/S†	AD533J* 432K* 426K/L 429B 422K
LEVEL FIVE		≤1%	≤1%	≤150mV	≤0.05%
		AD533K/S 432J* AD530K/S* AD532K/S* AD531K/S*	AD533J*} AD530J*} AD532J*} .8-.3 AD531J*} 432J* .8-.4	432J† 80-120 AD530J 150-100 AD531J	426A* 429A* 422A*
LEVEL SIX		≤2%		≤200mV	≤0.06%
		AD533J AD530J* AD532J* AD531J*		AD533J 150-200 AD532J† 200-200	AD530J* AD532J* 432J* AD531J*

*Typical
†Int. Trim

OFFSET DRIFT max. mV/°C	TEMP. RANGE Rated Drift Performance	BANDWIDTH small-signal –3dB min. (by family)	COST (1's) $U.S.	SIZE Pkg. or vol. (by family)
≤0.2	–55°C to +125°C	≥10MHz	<$20	IC (TO-100 or TO-116) (or chips)
AD530J/K/L/S* AD531J/K/L/S* 428K 424J*/K 427J*/K	AD533S AD532S AD530S AD531S	429*	AD533J AD533K	AD533 AD530 AD532
≤0.5	–25°C to +85°C	≥5MHz	<$30	IC (TO-116) (or chips)
428J*	426A 429A/B 422A	422	AD530J AD532J AD531J 432J	AD531
≤0.7	0 to +70°C	≥1MHz	<$50	8cm³
AD533J/K/L/S* AD532J/K	AD533J/K/L AD530J/K/L AD532J/K/L AD531J/K/L 432J/K 426K/L 428J/K 422K 424J/K 427J/K	AD533* AD530* AD532* AD531* 432*	AD530K/L AD532K/L AD533L/S AD531K 426A 432K	432
≤1		≥300kHz	<$100	22cm³
432K* 426K/L 429B 422K		426* (400) 428* (300)	AD530S AD531L/S 426K/L 428J	426 428 429 422
≤2		≥100kHz	>$100	47cm³
432J* 426A* AD532S 429A* 422A*		424 427	428K 429A/B 422A/K 424J/K 427J/K	424 427

See Descriptive Notes on next page.

BRIEF DESCRIPTIONS OF 4-QUADRANT DEVICE FAMILIES — NOTES TO TABLE 2

Prefix AD: Integrated Circuits

<u>AD533</u>: lowest cost, general-purpose variable-transconductance, 4-quadrant. Thin-film-on-silicon monolithic construction.

<u>AD530</u>: low-cost, general-purpose variable-transconductance, 4-quadrant. Better specs than AD533. Thin-film-on-silicon monolithic construction.

<u>AD532</u>: internally-trimmed low-cost, general-purpose, variable transconductance, 4-quadrant. Thin-film (laser-trimmed) on-silicon monolithic construction. Both inputs differential.

<u>AD531</u>: similar to AD530, but scale factor is adjustable by an input current (XY/I). Can be used as a three-variable multiplier-divider, to perform vector, square-root, rms, absolute-value operations. Available in TO-116 dual in-line package only. X-input differential.

No Prefix: Modular Packages

<u>427</u>: highest accuracy, lowest noise and drift, pulse-height, pulse-width modulated, internally trimmed.

<u>424</u>: highest accuracy, lowest noise and drift, pulse-height, pulse-width modulated, externally trimmable. Requires external amplifier for division.

<u>428</u>: low-drift, good accuracy, reasonable cost, variable-transconductance type.

<u>429</u>: combines 10MHz bandwidth and good low-frequency performance, variable-transconductance.

<u>422</u>: wideband variable-transconductance type.

<u>426</u>: low-cost general-purpose variable-transconductance type.

<u>432</u>: lowest-cost internally-trimmed general-purpose variable-transconductance in small modular package.

Not Listed in Table 2

<u>434</u>: high-accuracy multiplier-divider (YZ/X), especially useful for dividing and square-rooting. Wide dynamic range, log-antilog type, 1 quadrant.

<u>433</u>: multifunction module $Y(Z/X)^m$, performs 434 operations and also obtains powers and roots from 1 to 5. Wide dynamic range, log-antilog type, 1 quadrant.

<u>425</u>: The 424, mounted on a card, with adjustments.

<u>436</u>: high-accuracy variable-transconductance divider-only — 2-quadrant numerator, positive denominator.

by using the nearly-always-conservative approximation that the nonlinearity error $f(X,Y)$

$$f(X,Y) \cong |\, V_x \,|\, \epsilon_x + |\, V_y \,|\, \epsilon_y$$

where ϵ_x and ϵ_y are the fractional nonlinearities, i.e., % nonlinearity/100, specified for the X and Y inputs respectively.

<u>Example:</u> For Model AD530K, $\epsilon_x = 0.5\%$ and $\epsilon_y = 0.2\%$. What is the maximum error for $V_x = 5V$ and $V_y = 1V$? What happens if the inputs are reversed?

   A. Nominal output is $V_x V_y / 10 = 5 \times 1/10 = 500$mV.

   B. Expected linearity error is $5(0.005) + 1(0.002) = 27$mV, or 5.4% of the output (0.27% of full-scale).

C. Interchanging the inputs, $1(0.005) + 5(0.002) = 15\text{mV}$, or 3% of the output (0.15% of full-scale).

It is evident that for this application, the inputs should indeed be interchanged, if the voltages in the problem statement are the respective maxima. In any event, the errors computed here should be contrasted with the overall 1%-of-full-scale predicted specification: 100mV, or *20%* of the 500mV output, surely belt-and-suspenders conservatism!

This discussion relates only to the nonlinearity specification. All dc errors should have been zeroed, and the range of temperature variation should be small to avoid introducing excessive errors when computing at such low levels.

2. Common-Mode Rejection: Some multipliers have one or more differential inputs, which can prove useful in instrumentation (e.g., power measurement, Figure 2) or applications involving passive sums and differences, such as vector subtraction, without additional op amps. Examples of such devices include the AD531 and AD532.

Common-mode error generally manifests itself as an apparent differential input signal δ_{CM}, which is a function of common-mode signal level $V_{CM} = (V_1 + V_2)/2$. It is measured by setting up the multiplier for unity gain (e.g., for an $(X_1 - X_2)Y/10$ multiplier, set $Y = 10V$) and connecting a low-frequency sine-wave generator (20Vp-p @50Hz) in common to both differential inputs (Figure 3). The resulting peak-to-peak output error measurement is the common-mode error. The fractional common-mode error is the ratio of this measured voltage to the common-mode input swing. Common-mode rejection ratio (CMRR) is the reciprocal of the fractional common-mode error, and log common-mode rejection (CMR) = 20 log(CMRR), in "dB."

As a practical matter, it should be noted that the absolute magnitude of the signal level at each of the differential inputs is a composite of the differential signal and the common-mode signal (if they are applied separately). The user should be careful that the combined voltage swing does not result in saturated inputs.

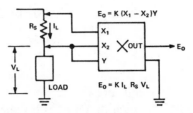

a) High shunt (differential-input multiplier)

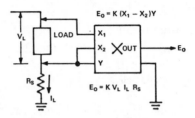

b) High load (differential-input multiplier)

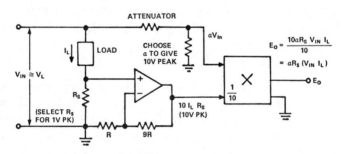

c) High voltage (e.g., ac line)

Figure 2. Multiplier applied to power measurement

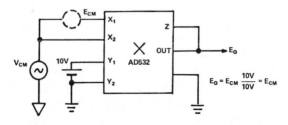

Figure 3. Common-mode-error measurement

3. Balanced Modulator Application (Figure 4)

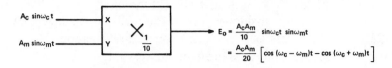

Figure 4. Balanced modulator circuit

The balanced modulator is discussed in Chapter 2-4 (Figure 4a). If the multiplier errors are included, the equation

$$10E_o = A_c \sin\omega_c t \cdot A_m \sin\omega_m t$$

becomes

$$10E_o = (A_c \sin\omega_c t + X_{os})(A_m \sin\omega_m t + Y_{os}) + Z_{os} + f(x,y)$$

$$= \underbrace{A_c \sin\omega_c t \cdot A_m \sin\omega_m t}_{\text{desired performance}} + \underbrace{Y_{os} A_c \sin\omega_c t}_{\substack{\text{carrier} \\ \text{feedthrough}}}$$

$$+ \underbrace{X_{os} \sin\omega_m t + X_{os} Y_{os} + Z_{os}}_{\text{low freq. and dc terms}} + \underbrace{f(x,y)}_{\substack{\text{harmonics} \\ \text{and cross-} \\ \text{products}}}$$

The low-frequency and dc terms are usually inconsequential, because they can be filtered out by a high-pass filter. The carrier feedthrough (linear) term can be adjusted to zero by tweaking Y_{os}; but it will creep back in with changes in temperature. Thus the error will consist of a nonlinear term plus a temperature-sensitive carrier feedthrough.

If an AD530K is considered for this application, its nonlinearity error is 0.5% (50mV) for x and 0.2% (20mV) for y, and its Y_{os} temperature sensitivity is 2mVp-p/°C. For a 20Vp-p signal swing, the nonlinearity error is 0.5% (-46db at room temperature), and the additional carrier feedthrough for a 45°C temperature change is 90mVp-p. If they add linearly (more likely, they will add root-sum-of-squares fashion), the error will be 190mV/20V = 0.95%, or in terms of "dB" (20 log 0.0095), about 40dB of carrier, harmonic, and cross-product suppression.

To make the modulator insensitive to variations in carrier amplitude, some means must be provided either to hard-limit the carrier input or provide some means for controlling its amplitude with an AGC circuit. The AGC approach can be achieved concurrently with the modulation process, if an XY/Z multiplier is used (such as the AD531), as shown in Figure 5. This circuit multiplies the modulating signal (Y input) by the carrier (X input); the gain $(1/V_r)$ is determined by a measure of the carrier amplitude, hence the output amplitude is a function of the modulating-signal amplitude only.

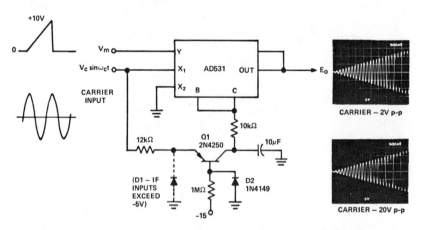

Figure 5. Balanced modulator with carrier level-compensation

The emitter circuit of the transistor acts as a rectifier, to measure the carrier amplitude. Filtering occurs in the collector circuit, and the resulting current, approximately proportional to the carrier input, provides a denominator input current that adjusts the multiplier gain to maintain the output amplitude independent of variations in the carrier amplitude, over a range of 1−10V, and at frequencies down to 20 Hz. The diode D2 maintains a current in the collector circuit to keep the denominator from going to zero for very low values of carrier amplitude (including zero).

4. Squarer Application Notes

When a sine-wave is applied to a multiplier, connected as a squarer, one would expect the time plot of the output to be a double-frequency sine-wave of amplitude $\frac{1}{2}A^2/10$, with a bias of $\frac{1}{2}A^2/10$, from the trigonometric identity.

$$\sin^2 \theta = \frac{1}{2}(1 - \cos2\theta)$$

On an oscilloscope, the time plot and X-Y plot should appear as in Figures 6b and c, at low frequencies.

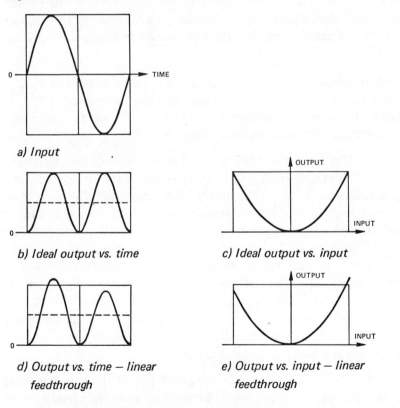

a) Input

b) Ideal output vs. time

c) Ideal output vs. input

d) Output vs. time — linear feedthrough

e) Output vs. input — linear feedthrough

Figure 6. Waveforms of ideal squarer and squarer with feedthrough error

An asymmetry, such as that observed in Figures 6d and e, is due to an input offset, caused either by multiplier feedthrough or a dc component on the sine wave (or both). It adds a linear term to the squared output (plus a negligible offset):

$$(X\sin\omega t + X_{os})(X\sin\omega t + Y_{os})$$

$$= X^2\sin^2\omega t + (X_{os} + Y_{os})X\sin\omega t + X_{os}Y_{os}$$

The offsets of the input (if small) and of the multiplier can be compensated for at room temperature by adjusting either X_{os} or Y_{os} to equalize the alternating peaks.

At high frequencies, the phase shift will be worse for the squarer than for the sine-wave-times-a-constant, because of frequency doubling. Also, as frequency increases, an asymmetry will appear, due to the increase of feedthrough at higher frequencies (Figure 7).

Such response need not be alarming for many applications. It is interesting to note that the average response, as measured by a dc meter, is essentially independent of frequency up to quite high frequencies. What this means to the user is:

A. If you're interested in the average value of the output, as in rms measurements, power measurements, or phase-angle measurements (between the inputs), then the phase shift (and even attenuation) of the double-frequency is of no concern, and ac feedthrough of the fundamental is similarly of little moment. The output is going to be filtered, or integrated; the high-frequency attenuation of the output stage can be viewed as a modicum of pre-filtering. (In fact, if the output-stage summing point is available, a feedback capacitor can be used to provide internal first-order-lag filtering.) What is important is the constancy of the average value, which is quite good over the entire range of frequency for which the unit is rated.

B. If one is interested in the *amplitude* of the ac component of the output, as in frequency doubling or amplitude modulation, then the phase shift is again unimportant, until it is accompanied by attenuation, starting at about 1/10 the rated frequency range.

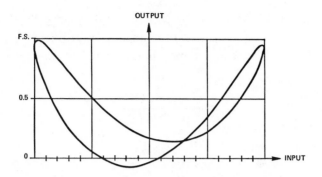

a) Typical plot of output vs. input for squarer at high frequency, showing phase shift and high-frequency feedthrough

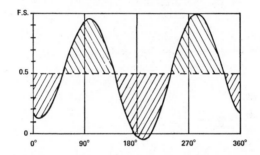

b) Time plot of the above, showing equal areas about 0.5, despite phase distortion

Figure 7. High-frequency response of squarer to sinusoidal input

C. The phase-shift is important if one is interested in the *instantaneous* value of the output, a small but important class of multiplier applications. An example is use of the squarer in linearity correction of high-frequency oscilloscope displays.

In frequency-doubling applications, the output amplitude is proportional to the square of the input amplitude. For most cases, the amplitude of a frequency-doubler output is not of importance. For applications where the amplitude *must* respond linearly (for example, to an input with slowly-varying amplitude modulation), the circuit of Figure 8 may be found useful. The "dc" component of the output of a squarer-connected multiplier (AD531) is filtered and fed back as a denominator signal to control the gain. Thus,

$$E_o = \frac{(1 - \sin 2\omega t)\, E_1{}^2/2}{\overline{E_o}}$$

$$= \frac{E_1{}^2}{2\overline{E_o}} - \frac{E_1{}^2}{2\overline{E_o}} \sin 2\omega t$$

Since

$$\overline{E_o} = \frac{E_1{}^2}{2\overline{E_o}} ,$$

$$\overline{E_o} = \sqrt{E_1{}^2/2} = E_1/\sqrt{2}$$

and

$$E_o = \frac{E_1}{\sqrt{2}} - \frac{E_1}{\sqrt{2}} \sin 2\omega t$$

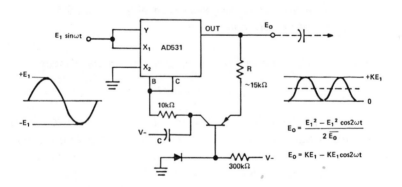

Figure 8. Frequency doubler with linear amplitude response

Therefore, the amplitude of the double-frequency signal is proportional to the input amplitude. The dc component of the output may, of course, be removed by capacitive coupling.

APPENDIX

BIBLIOGRAPHY

The references listed here have appeared in the form of books, manuals, brochures, and articles in technical publications. Individual items have been selected because of their general or specific interest, or because they "fan out" through additional references not included here.

The list is representative rather than comprehensive and is heavily weighted in favor of recently-published material ('70's). In most cases, the practical has been preferred to the theoretical. Within each subject grouping, the titles are listed alphabetically.

The interested reader should, in any event, seek to obtain catalogs and application notes (often voluminous) from manufacturers of the nonlinear devices of interest. Lists of current manufacturers, their products, and their addresses, will be found in such industry guides as *EEM* and the *Electronics Buyer's Guide*. Since the technology is rapidly expanding (and changing), one should also seek to be placed on manufacturers' mailing lists, and to subscribe to at least one of the major semimonthly electronics industry technical periodicals to keep up with new products, new techniques, and new literature.

Readers of this book are invited to subscribe to *Analog Dialogue*, which appears (approximately) quarterly, and is available at no charge from Analog Devices, Inc. Other publications available are the most recent edition of the *Analog Devices Product Guide*, data sheets on individual products, and occasional "Application Notes" on topics of interest.

Copies of certain publications mentioned in this Bibliography, designated by an asterisk (*), are also available upon request (free, except for the *A-D Conversion Handbook*) as of the time this book has gone to press. They will be available as long as the supply lasts. Except for these items, *no other publications mentioned here are available from Analog Devices.*

LINEAR AND NONLINEAR CIRCUITS, GENERAL INFORMATION

Analog-Digital Conversion Handbook, Edited by D. Sheingold, Analog Devices, Inc., Norwood, Mass., 1972, $3.95.

Applications Manual for Operational Amplifiers, Philbrick-Nexus Research, Dedham, Mass., 1965.

"Designer's Guide: Nonlinear Function Modules," T. Cate and others, *EEE*, September 1969.

Linear Applications, National Semiconductor, Santa Clara, California, February, 1973.

Modern Operational Circuit Design, J. I. Smith, Wiley-Interscience, New York, 1971.

Operational Amplifiers, G. Clayton, Butterworth & Co. (Publishers), Ltd., London, U.K., 1971.

Operational Amplifiers, Design and Applications, G. Tobey, J. Graeme, L. Huelsman, McGraw-Hill Book Co., New York, 1971.

A Palimpsest on the Electronic Analog Art, H. Paynter, George A. Philbrick Researches, Inc., 1955, *Out of Print.*

*Product Guide, Analog Devices, Inc., Annual.

"Some Aspects of the Theory of Nonlinear Networks," A. Wilson, *Proceedings of the IEEE,* 61-8, August 1973.

FUNCTION FITTING

"Approximate Analog Functions with a Low-Cost Multiplier-Divider," D. Sheingold, *EDN,* February 5, 1973.

"Design of Nonlinear Networks with a Prescribed Small-Signal Behavior," B. Peikari, *IEEE Transactions on Circuit Theory,* CT-19-4, July 1972.

"Feedback Linearizes Resistance Bridge," R. Guyton, *Electronics,* October 23, 1972.

Fitting Equations to Data, C. Daniel, F. Wood, J. Gorman, Wiley-Interscience, New York, 1971.

"Get Perfect *(sic)* Linearity in CRT's," A. Popodi, *Electronic Design,* 1, January 4, 1973.

"Linearize Analog Signals Continuously," M. Weiner, D. Schneider, *Electronic Design,* 24, November 23, 1972.

"Trig. Function Generators (Applications Brief), National Semiconductor, MOS Brief 10, 1970.

TIME–FUNCTION GENERATION

"Adjustable Sinewave Audio Oscillator Employs Improved AGC for Wide Frequency Range," C. Schwerdt, *Electronic Design,* 4, February 15, 1973.

"Analog-to-Pulse-Width Converter Yields 0.1% Accuracy," N. Robin, *EDN,* November 1, 1970.

"The Basics of Using FET's for Analog Signal Switching," A. Evans, *EDN*, May 20, 1973.

"Current-Controlled Triangular/Square-Wave Generator," S. Franco, *EDN*, September 5, 1973.

*Data Sheets, Models 350, AD351 Comparators, Analog Devices.

*"Digital Sweep Generator," R. Craven, *Analog Dialogue*, 5-4, 1971.

"Function Generators, a Family of Versatile Wave Makers," L. Hunter, *Electronic Products*, November 19, 1972.

"Generate Low-Distortion Sine Waves," J. Vanderkooy, C. Koch, *Electronic Design*, 14, July 5, 1973.

"MOS Analog Function Generator," J. Kubinec, National Semiconductor, MOS Brief 3, 1968.

"Precise Tri-Wave Generator," R. Dobkin, *Electronic Engineering Times*, 1973.

"Triangle-Wave Generator Keeps Slopes Constant as Amplitude Changes," D. Larsen, *Electronic Design*, 20, September 28, 1972.

"Triangular and Square-Wave Generator has Wide Range," R. Burwen, *EDN*, December 1, 1972.

"Using IC's to Generate Waveforms," A. Grebene, *Electronic Products*, March 20, 1972.

"Wide-Range Ramp Generator has Programmable Outputs," C. Wojslaw, W. Buschmann, *EDN*, October 15, 1972.

"Zener Diode" Controls Wien-Bridge Oscillator," W. Crittenden, E. Owings, *EDN*, August 1, 1972.

INSTRUMENTATION AND MEASUREMENT

"The dB-Anything," K. Simons, *Proceedings of the IEEE* (Letters), 61-4, April 1973, pp. 495–496.

"Don't Eyeball Noise," G. Franklin, T. Hatley, *Electronic Design*, 24, November 22, 1973.

"An Electronic Wattmeter for Non-Sinusoidal Low Power Factor Measurements," D. Hamburg, L. Unnewehr, *IEEE Transactions on Magnetics*, September 1971.

"High-Performance Flame Ionization System for Gas Chromatography," D. Smith, *Hewlett-Packard Journal*, March 1973.

"Integrate and Hold Circuit Gives Electrochemical Measurements," R. Warsinski, *Electronic Design*, 1, January 4, 1973.

"JFET Circuit Linearizes Transducer Output," F. Trofimenkoff, R. Smallwood, *IEEE Transactions on Instrumentation and Measurement,* IM-22-2, January 1973.

"Linearizing Thermocouple Amplifiers," L. Garelick, E. Hauptmann, *Advances in Instrumentation* (ISA), Volume 26, Part 4, Paper 852.

"Logarithmic Quantities and Units," C. Page, *Proceedings of the IEEE* (Letters), 61-10, October 1973, pp. 1516–1518, *in re* "dB".

"Logarithmic Readout Attachment for Double-Beam Spectrometers," J. Shepherd, H. Hedgpeth, *Review of Scientific Instruments,* 44-3, March 1973.

"Low-Current Ammeter," R. Dobkin, *Electronic Engineering Times,* 1973.

*"Measuring Airflow Using a Self-Balancing Bridge," J. Miyara, *Analog Dialogue,* 5-1, 1971.

*"Measuring Sine-Wave Amplitudes Without Filtering," D. Jolley, *Analog Dialogue,* 5-2, Spring 1971.

"Narrow Peaks Caught by Better Detector," R. Klatt, *EDN/EEE,* August 15, 1971.

Omega Temperature Measurement Handbook, Omega Engineering, Inc., Stamford, Conn., 1973.

"Referencing Thermocouple Junctions," B. Hollander, *Instruments and Control Systems,* March 1973.

"RMS Voltage Measurements—Which Method Works Best?" R. Chapel, *Electronic Products,* January 15, 1973.

"Some Effects of Waveform on VTVM Readings," B. Oliver, *Hewlett-Packard Journal,* 6-8, April 1955, continued in 6-9, May 1955, and 6-10, June 1955.

*"True-RMS Measurement," L. Counts, *Analog Dialogue,* 7-1, 1973.

"A Wideband Wattmeter for the Measurement and Analysis of Power Dissipation in Semiconductor Switching Devices," F. Schwarz, N. Voulgaris, *IEEE Transactions on Electron Devices,* ED-17-9, September 1970.

SIGNAL PROCESSING

*"Adjustable Low-Pass Filter," R. Burwen, *Analog Dialogue,* 4-2, December, 1970.

"Analog Multipliers Offer Solutions to Video Modulation Problems," G. Shapiro, *EDN,* September 1, 1972.

"Applications of a Monolithic Analog Multiplier," A. Bilotti, *IEEE Journal of Solid State Circuits,* SC-3-4, December 1968.

"A Comparison of Analog and Digital Techniques for Pattern Recognition," K. Preston, *Proceedings of the IEEE,* 60-10, October 1972.

"Demodulate DPSK Signals Coherently Using a Costas Phase-Lock Loop," R. Hennick, *EDN,* July 1, 1972.

"Design of a Noise-Eliminator System," R. Burwen, Audio Engineering Society Preprint 838(B-8), 1971.

"Design Your Own Dynamic Phase Shifter," G. Strauss, *Electronic Design,* 12, June 8, 1972.

"Dual Comparator and R-C Filter Estimate Probability Density Function," J. Sparacio, R. Pierro, *Electronic Design,* 4, February 15, 1973.

Electronic Music Synthesizers: Product Information Sheets and Users' Manuals, ARP Instruments, Inc., Newton Highlands, Mass.

"Fast Amplitude Control of a Harmonic Oscillator," D. Meyer-Ebrecht, *Proceedings of the IEEE,* 60-6, June 1972.

"FET's in R-C Network Tune Active Filter," A. Delagrange, *Electronics,* December 7, 1970.

*"Frequency Modulator," R. Burwen, *Analog Dialogue,* 5-5, 1971.

"A General Analysis of the Phase-Locked Loop," J. Connelly, Harris Semiconductor Application Note 602/A, October 1972.

"Generate Noise-Free Timing Pulses with an IC Peak Sampler," G. Oshiro, *Electronic Design,* 10, May 11, 1972.

"Introduction to the Phase-Locked Loop," D. Jones, Harris Semiconductor Application Note 601, October 1972.

"Making Systems Fail-Operational by Using Multiple Channels with Automatic Voters to Select the Best Signal," P. Rostek, *Electronic Design,* 17, August 16, 1969.

"A Method of Measuring the Peak Value of a Narrow Impulse by the Use of a Voltage-Forced Pulse-Lengthener Circuit," M. Uno, *IEEE Transactions on Instrumentation and Measurement,* IM-22-2, June 1973.

"Modems," J. Davey, *Proceedings of the IEEE,* 60-11, November 1972.

"A New Generation of Integrated Avionic Synthesizers," R. Brubaker, G. Nash, Motorola Semiconductor Application Note, AN553, 1971.

"Op Amps Replace Transformer in Phase-Detector Circuit," A. Gangi, *Electronics,* May 12, 1969.

"Phase-Locked Loop Helps Generate Waveforms with Variable Duty Cycle and Phase Shift," N. Calvin, *Electronic Design,* 19, September 13, 1973.

"Phase-Locked Loop IC's are Ready and Stable," A. Grebene, *Electronic Products,* February 19, 1973.

"Programmable Active Filters," R. Sparkes, A. Sedra, *IEEE Journal of Solid-State Circuits,* Correspondence, February 1973.

"Separate the Signals from the Noise with . . . Voltage Correlator Circuit," T. Cate, *Electronic Design,* 25, December 6, 1970.

"Signal Recovery Using a Phase-Sensitive Detector," P. Danby, *Electronic Engineering* (U.K.), January 1970.

"Simple Linear PLL Demodulator Uses Discrete Components," T. Mollinga, *EDN/EEE,* April 15, 1972.

"60Hz Frequency Discriminator," R. Burwen, *EEE,* February 1971.

"A Synchronized Phase-Lock Loop," R. Bohlken, *EDN,* March 20, 1973.

"Systems Applications for Voltage-Controlled Active Filters," P. Harvey, *EEE,* October 1969.

"Take the Guesswork Out of Phase-Locked Loop Design," D. Kesner, *EDN,* January 5, 1973.

*"30dB Automatic Gain Control," L. Counts, *Analog Dialogue,* 7-1, 1973 (and erratum, 7-2, 1973).

"Try the Monolithic Multiplier as a Versatile AC Design Tool," E. Renschler, D. Weiss, *Electronics,* June 8, 1970.

"Use This Tan-Lock Demodulator," R. Hennick, *Electronic Design,* 25, December 6, 1970.

"Voltage-Controlled Phase Shift of Triangular Waves," K. Kuijk, H. Hagenbeuk, *IEEE Transactions on Instrumentation and Measurement,* IM-22-2, June 1973.

COMPUTING AND CONTROL

"Analog Arithmetic Unit Offers Good Accuracy," C. Wojslaw, *EDN,* July 1, 1972.

"Analog Modules Multiply User's Options," G. Tobey, *Electronic Products,* February 19, 1973.

"Analyzing Low-Frequency Random Phenomena with an Analog Computer," P. Kommineni, G. Smith, *Instruments & Control Systems,* April 1973.

"Analyzing Signals for Information," I. Langenthal, *Instruments and Control Systems,* December 1970, and January 1971

"Applications for Trigonometric Computing Modules," N. Sussman, *EEE*, September 1969.

"Automatic Brightness Control and Linearity Correction Circuits for Large-Screen Color Oscilloscopes," I. Lamoth, *IEEE Transactions on Instrumentation and Measurement*, IM-22-2, June 1973.

"A Computer for the Calibration of Skeletal Muscle Vascular Resistance," R. Zambuto, R. Reder, C. Sanders, W. Powell, *IEEE Transactions on Biomedical Engineering*, BME-19-6, November 1972.

"Detection and Measurement of Three-Phase Power, Reactive Power, and Power Factor, with Minimum Time Delay," I. R. Smith, L. A. Snider, *Proceedings of the IEEE*, Letters, November 1970.

"Distortion Correction in Precision Cathode-Ray Tube Display Systems," Intronics, Inc., 1970.

Electronic Analog and Hybrid Computers, G. Korn, T. Korn, McGraw-Hill Book Co., New York, 1964.

"IC Op Amps Straighten Out CRT Graphic Displays," J. Divilbiss, S. Franco, *Electronics*, January 4, 1971.

"Linearize Your CRT Displays," K. Peterson, *Electronic Design*, *17*, August 16, 1970.

"Low-Cost, Logarithmic Mass Flow Computer," NASA Tech Brief 71-10407, Lewis Research Center, Cleveland, Ohio, November 1971, J. Watson, D. Noga, J. Dolce, J. Gaby, digested in *Instrumentation Technology*, August 1972.

*"Measuring Sine Wave Amplitudes Without Filtering," D. Jolley, *Analog Dialogue*, 5-2, 1971.

"Phase-Locked Loops for Motor Speed Control," A. Moore, *IEEE Spectrum*, April 1973.

"Phase-Locked Loops Provide Accurate Efficient dc Motor Speed Control," L. Milligan, E. Carnicelli, *EDN*, August 1, 1972.

"Real-Time Signal Analysis," R. Rothchild, *Medical Electronics and Data*, March–April 1971.

"Relaxation Time Measurements by an Electronic Method," R. Brousseau, J. Vanier, *IEEE Transactions on Instrumentation and Measurement*, IM-22-1, March 1973.

"Resistance Network Converts Limit Detector to Movable Window Operation," B. Pearl, *Electronic Design*, 6, March 15, 1973.

"Special Report on Signal Averagers," G. Flynn, *Electronic Products*, February 1969.

"Using Analog Function Modules for Measurement and Control," J. Huffman, T. Gibson, S. Rose, *Instruments and Control Systems,* December 1972.

"Variable Sweep-Rate Frequency Response and Vibration Testing," C. Lorenzo, *Instruments and Control Systems,* September 1971.

*"Vector Difference Circuit," L. Counts, *Analog Dialogue,* 7-1, 1973.

LOG, LOG RATIO, AND ANTILOG CIRCUITS

"Bipolar Operation of Paired Log Transistors," D. Sheingold, *Proceedings of the IEEE* (Letters), Vol. 58, pp. 1855–56, November 1970.

"A Circuit with Logarithmic Transfer Response over 9 Decades," J. Gibbons, H. Horn, *IEEE Transactions of the Circuit Theory Group,* CT-11-3 September 1964.

*"Design of Temperature-Compensated Log Circuits Employing Transistors and Operational Amplifiers," W. Borlase, E. David, Analog Devices' Application Note, September 1969.

"Get Wider Dynamic Range in a Log Amp," G. Niu, *Electronic Design,* 4, February 15, 1973.

"How to Specify Parameters of Nonlinear Circuits," G. Osgood, J. Knitter, *EEE,* September 1969.

"Large-Signal Behavior of Junction Transistors," I. Ebers, J. Moll, *Proceedings of the IRE,* December 1954, pp. 1761–1772.

"Logarithmic Analog-to-Digital Converters: A Survey," S. Cantarano, G. Pallotino, *IEEE Transactions on Instrumentation and Measurement,* IM-22-3, September 1973.

"Logarithmic and·Exponential Amplifiers," D. Spicer, R. Mann, Texas Instruments Application Report, Bulletin CA-149, March 1970.

"A Logarithmic Transcoder," D. Degryse, B. Guerin, *IEEE Transactions on Computers,* C-21-11, November 1972.

"Multiplication and Logarithmic Conversion by Operational Amplifier-Transistor Circuits," W. Paterson, *Review of Scientific Instruments,* 34-12, December 1963.

"Transient Response of an Operational Amplifier with Logarithmic Feedback," H. Musal, *Proceedings of the IEEE* (Letters), Volume 57, pp. 206–208, February 1969.

MULTIPLIERS AND DIVIDERS

*"AD555 Monolithic Quad Switches Make 4-Quadrant Multiplying DAC's with 12-Bit Linearity," H. Krabbe, A. Molinari, *Analog Dialogue*, 5-2, 1971.

"Analog Multiplication at High Frequencies with a High Dynamic Range," G. Papadopoulos, *IEEE Journal of Solid-State Circuits*, SC-8-6, December 1973.

"Analog Multiplier Applications," J. Pepper, *Instruments and Control Systems*, June 1972.

"Characteristics and Applications of Modular Analog Multipliers," E. Zuch, *Electronic Instrumentation Digest*, April 1969.

*"A Complete Monolithic Multiplier-Divider on a Single Chip," R. Burwen, *Analog Dialogue*, 5-1, 1971.

"Distortion in Bipolar Transistor Variable-Gain Amplifiers," W. Sansen, R. Meyer, *IEEE Journal of Solid-State Circuits*, SC-8-4, August 1973.

"Don't be Fooled by Multiplier Specs," R. Stata, *Electronic Design*, 6, March 15, 1971.

"Electronic Multiplier-Divider Uses One or 2 IC's," Z. Peled, *Electronic Design*, 21, October 10, 1968.

"An Electronic Multiplier for Accurate Power Measurement," M. Tomota, T. Sugiyama, K. Yamaguchi, *IEEE Transactions on Instrumentation and Measurement*, IM-17-4, December 1968.

"An Error-Stabilized Analog Divider," G. Fitton, *EEE*, February 1971.

"Evaluating and Using Multiplier Circuit Modules for Signal Manipulation and Function Generation," Analog Devices' Applications Brochure, 1970, *Out of Print.*

"Feedback Stabilized 4-Quadrant Analog Multiplier," H. Brüggemann, *IEEE Journal of Solid-State Circuits*, SC-5-4, August 1970.

"FET Conductance Multipliers," F. Crawford, W. Adams, *Instruments and Control Systems*, September 1970.

"In IC Form, Hall-Effect Devices Can Take on Many New Applications," M. Oppenheimer, *Electronics*, August 2, 1971.

"Linearizing Almost Anything With Multipliers," R. Burwen, *Electronic Design*, 8, April 15, 1971.

*"Multiplier Memories and Meanderings," D. Sheingold, *Analog Dialogue*, 5-1, 1971.

"The Multiplying D/A Converter," C. Brown, *Electronic Products,* June 21, 1971.

"New Planar Distributed Devices Based on a Domain Principle," B. Gilbert, *Proceedings of the IEEE Solid State Circuits Conference,* February 19, 1971, pp. 166–7.

"A Precise Four-Quadrant Multiplier with Subnanosecond Response," B. Gilbert, *IEEE Journal of Solid-State Circuits,* December 1968.

"A Precision Current Multiplier/Divider," G. Bredenkamp, *Proceedings of the IEEE* (Letters), November 1972.

"Save Money With Analog Multipliers," R. Burwen, *Electronic Design,* 7, April 1, 1971.

"Top Performance from Analog Multipliers?" T. Cate, *Electronics,* April 13, 1970.

*"Two-Quadrant Multiplier," R. Burwen, *Analog Dialogue,* 4-2, December 1970.

"A Two-Quadrant Multiplier Integrated Circuit," J. Holt, *IEEE Journal of Solid-State Circuits,* SC-8-6, December 1973.

PIECEWISE-LINEAR APPROXIMATIONS, POWERS & ROOTS, MISCELLANEOUS

"AC-to-DC Converters for Low-Level Input Signals," R. Kreeger, *EDN,* April 5, 1973.

"Add FET to Threshold Detector to Improve Hysteresis," G. Oshiro, *Electronic Design,* 12, June 8, 1972.

"Analog Sorting Network," D. Morgan, *Electronic Design,* 2, January 18, 1973, and sequel (Letters), 17, August 16, 1973.

"Analog Switches Replace Reed Relays," D. Fullagar, *Electronic Design,* 13, June 21, 1973.

"Automotive and Industrial Electronic Building Blocks," R. Russell, T. Frederiksen, *IEEE Journal of Solid-State Circuits,* SC-7-6, December 1972.

"Design Features of a Precision ac-dc Converter," L. Marzetta, *Journal of Research of the National Bureau of Standards,* 73C-3, 4, July–December, 1969.

"Extraction of Square Roots . . . a Useful Analog Instrumentation Technique," T. Cate, *Electronic Instrumentation Digest,* January 1971.

"A Hybrid-Circuit RMS Converter," H. Handler, *Proceedings of the IEEE Solid-State Circuits Conference,* 1971, February 19, 1971, pp. 190, 191.

"Ideal Rectifier Uses Equal-Value Resistors," A. Lloyd, *Electronic Design,* June 21, 1967.

"Midvalue Selector Doubles as Precise Voltage Limiter," A. Moses, *The Electronic Engineer,* December 1970.

"Nonselective Frequency Tripler Uses Transistor Saturation Characteristics," R. Lockhart, *Electronic Design,* 17, August 16, 1973.

"Op Amp Art" (Absolute-Value Circuits), S. Rudnick, *EEE* (Letters), June 1968.

"Op Amps Form Self-Buffered Rectifier," J. Graeme, *Electronics,* October 12, 1970.

*SERDEX Serial Data Exchange Modules, Analog Devices, Inc., 1973.

"Simple Control for Sign of Op Amp Gain," S. Franco, *Electronic Design,* 23, November 8, 1970.

"Triodes as *n*th Power Elements," G. Philbrick, *The Lightning Empiricist,* Vol. 1, No. 1, 1952, *Out of Print.*

*"Versatile New Module: $Y(Z/X)^m$ at Low Cost," F. Pouliot, L. Counts, *Analog Dialogue,* 6-2, 1972.

INDEX

Root sum-of-squares, 20, 21, 41, 47, 88, 117-118, 390

RTI, RTO, errors referred to input or output; see errors and specifications for individual devices

S

Sample-hold, 88, 99, 107-108, 114-115, 120, 415-416

Scale-factor specifications, see individual devices

Scale factors, see also Reference, 29-30, 33-37, 45, 57, 63-64, 70-76, 79, 90-97, 110, 114-116, 118-119, 120-122, 129-130, 132-133, 134-135, 138-141, 145-153, 160

SCR, root mean-square vs. average measurements in silicon controlled-rectifier circuits, 392

Segments, see Piecewise-linear

SERDEX^{TM}, SERial Data EXchange systems, 88, 99

Shaping, wave-, see Generators

Signal-
compression, see Compression
conditioning, 29-164
generation, see Generator, Oscillator
shaping, see Generators

Sign-magnitude to bipolar, 27, 363, 365

Simulation, 2

Sine
approximating, 22-23, 43, 49-52, 60-64, 67-68, 158-159
hyperbolic (sinh), 20, 103
inverse hyperbolic (sinh^{-1}), 4, 19-20, 102-104
use of implicit feedback in fitting, 43, 49-52, 60-64
sine-wave generators, 65-68, 78-81, 127-129, 142-144
sine-wave measurements and manipulations, 87-88, 108-117, 124-144, 145-146
testing, uses of sine wave, see individual device tests

Sinh (hyperbolic sine), see Sine

Smooth functions, see Functions

Spectrum analyzers, 141-142

Square-law, see Squaring

Square-rooter, 3, 8, 9, 10, 16-18, 20-21, 40-42, 108-111, 117-119, 149-150, 160-161, 162-163, 369-387, 389-416
inverse vs. implicit, 40
multiplier as, 5, 38
odd-function, 10

Square-wave generator, see Generators

Squaring, 1, 3, 8, 20-21, 32, 40-42, 46-52, 60-61, 63-64, 110-111, 117-119, 126-127, 130-131, 154, 160-161, 162-163, 332, 369-371, 373-380, 485-486, 488
absolute-value in, 8, 21, 24, 118, 485
odd-function, 8

Stationary process, 17, 87, 105, 109, 389-393

Straight-line function, see Piecewise-linear

Strain-gage
linearization, 41-42, 89-92
reference variations, 7, 90-91, 120-121

Summing
linear, 55-57, 263
see also
Root sum-of-squares
Vector

Sweep circuits, 81-82
in device measurement, see Crossplots

Switches, switching
electronic, 2, 3, 4, 13, 14